The MILITARY and POLICE SNIPER

Advanced Precision Shooting For Combat and Law Enforcement

by
Mike R. Lau

with
SFC Kent Gooch, NCOIC US Army National Guard Scout-Sniper School
SGT Sam Chesnut, Special Operations, Oklahoma City Police Department
CPL Sean Little, USMC Scout/Sniper

The MILITARY and POLICE SNIPER
Copyright 1998 by Michael R. Lau

The author, advisors, contributors, and publisher accept no liability whatsoever for any injuries or deaths to persons or damage to property resulting from the application or adoption of any of the procedures, tactics, recommendations, loading data, or considerations presented or implied in this book.

FIRST EDITION: June, 1998
Third printing: September, 2004

Published by:
Precision Shooting Inc.
222 McKee Street
Manchester, CT 06040

CONTENTS

PREFACE

ACKNOWLEDGMENTS

ABOUT THE AUTHOR

CHAPTER 1: THE MARINE CORPS SCOUT/SNIPER and the M40A1 1

CHAPTER 2: THE US ARMY SCOUT-SNIPER and the M24 SNIPER
WEAPON SYSTEM (SWS) 19

CHAPTER 3: TWO CALL OUTS, 10 DAYS. THE POLICE SNIPER 51

CHAPTER 4: THE CUSTOM SNIPER RIFLE and TELESCOPES 71

CHAPTER 5: TERMINAL BALLISTICS 95

CHAPTER 6: INTERNAL BALLISTICS, ACCURACY, PRECISION, and the
DEVELOPMENT of the 7.62x51mm SNIPER CARTRIDGE 105

CHAPTER 7: THE .308 WINCHESTER, 5.56mm/.223, .338 LAPUA,
CALIBER .50 BMG, and MORE 121

CHAPTER 8: EXTERNAL BALLISTICS, INFLUENCES ON ACCURACY,
and BALLISTICS TABLES 149

CHAPTER 9: RANGE ESTIMATION and SIGHT ADJUSTMENT 163

CHAPTER 10: "This Ground Fire Was A Godsend To Us", UPHILL AND
DOWNHILL SHOOTING 179

CHAPTER 11: "Shoot, Damn it, Carlos", WIND READING AND SIGHT
ADJUSTMENT 187

CHAPTER 12: "We'll take 'em", MOVING TARGETS 205

CHAPTER 13: MANAGING THE LOG/DATA BOOK 213

CHAPTER 14: TACTICAL SCENARIOS 243

CHAPTER 15: THE SNIPER MATCH: SUSTAINMENT TRAINING
THROUGH PRACTICAL PRECISION SHOOTING 277

BIBLIOGRAPHY 315

APPENDIX I: Ballistics Tables 323

APPENDIX II: Leads for Moving Targets 333

APPENDIX III: Wind Drift Tables 339

APPENDIX IV: Manufacturers Mentioned in Book 343

INDEX 345

Preface

Military and law enforcement snipers are an intelligent breed with a mind of their own. More often than not they are required to make very quick and difficult decisions using their own judgment based on what they learned and experienced in the past. They are given the power and skill to end a life, or save a life, in a fraction of a second. Because of this, the military and law enforcement sniper strives to gain as much knowledge and experience about his profession as he possibly can.

There are many excellent books on military and law enforcement sniping in print today. They all cover different material and subjects that have anxiously been awaited for years by the serious as well as casual shooter, military and law enforcement personnel, civilian enthusiasts, historians, etc. I did not want to cover the same material as in the other books, but felt that many subjects needed further explanation and emphasis. I know what information many of you lack and don't understand by the questions you ask. I go into much detail about the USMC M40A1 and US Army M24 sniping rifles because this is what interests most. I explain the use of the mil-dot reticle, range finding, and wind reading in great depth, hopefully with very clear explanations along with numerous examples. These are the most important skills of precision long range sniping that many still don't understand fully because they don't have someone to coach them or lack the facilities to practice them. I also cover in much greater detail terminal ballistics, because many books do not emphasize its importance as being fundamental to sniping and for all other types of tactical shooting.

I believe that the young soldiers and police officers of today are smarter and more inquisitive than I was when I was their age. They know that to get ahead they have to increase their skills and know more about the subject than the next guy. They chose to be a sniper because they love firearms and shooting more than the ordinary soldier or police officer. They go to the rifle range on their own time and practice more often. They own more than a few guns. There is nothing wrong with the sniper knowing other methods and more complex details about a skill or a subject. They are not satisfied with just "do it because the manual tells you to." They want to know what a bullet does when it hits glass. They want to know why the M118's bullet shoots harder, has a flatter trajectory, and retains higher velocity at long range than the 168 gr. HPBT, even though the 173 gr FMJ bullet started out slower at the muzzle. They want to know why stainless steel is better for barrels, why geographical altitude changes affects the rifle's zero, and what the bullet actually does to the human body upon impact. They want to know where the magic number came from that corrects their target distance when shooting up a steep slope. With increased knowledge, the military and police sniper can determine if there is some method, equipment, tactic, or skill that is better suited in accomplishing the mission or task at hand. He or she will make better judgments and decisions during times of high stress whether under the complexities of modern high tech combat or old fashioned urban guerrilla warfare. A sniper should know as much about shooting and sniping as he or she can acquire, because in the long run it will make him or her a better shooter, a better sniper, and a better soldier or law officer. Having good shooting equipment helps, but having the right knowledge and experience is worth more.

As you read through this book, there will be times when you may feel overwhelmed by the amount of data and math calculations required to fire a single shot. However, as you gain more practical experience in the field, you will become more cognizant of the importance of having all this data and of making accurate calculations. You will also be criticized by many for shooting with a bipod, using support under the buttstock, and using a telescope sight. They will say experienced shooters don't need all these crutches because they know what dope to put on the rifle to get a hit. The modern military and law enforcement sniper requires help from math, tables, charts, scopes, and bipods to make his shot because he doesn't have the experience or the time to come up to the same level of skill like Master Sniper, Carlos Hathcock. The military and police sniper has to shoot to the same level as an NRA expert or master class rifle shooter in the shortest period of time. He is allowed to "cheat" and use the telescope sight, artificial support, and an observer to help him. When he gets down behind his rifle he is not playing a "game"! His intended target, whether on the practice range or in the field, is another human being. That human being also has the capability to put a bullet through him or her, a buddy, or a bystander. In a real situation there are no sighting shots allowed. Your first shot will be the only one that counts and it is for keeps. There will be no rematch and there are no second place winners.

As you get further into tactical shooting you will find out for yourself what you need in the way of data and also equipment to handle most of the situations. You will find out through experience that keeping things simple was all that was required. You will find out from experience that a simple rifle without a lot of high-tech gadgets was probably better in the first place. Too many accessories hanging off your rifle added too much weight, got in the way when the going got rough, broke or got out of adjustment, served no purpose anyway. You will find out in time what equipment is "classic" and what was only faddish. In time you will find out what works and what doesn't. It is hoped that what you learn from this book will help you get there faster.

Michael R. Lau

ACKNOWLEDGEMENTS

The author wishes to thank the following persons for their advice, contributions, and inspiration that made this book possible. I would also like to thank all others who have contributed and are not mentioned. Appreciation also goes to all my customers who were promised a rifle from Texas Brigade Armory during the period of time when I was writing this book and were told to wait a while longer.

Tony Black
Dave Brennan
Epitacio Carpentier
Norman A. Chandler
Richard Chelvan
Sam Chesnut
Corbon
Dick Davis
Andres Escobar
James Ferguson
Jim Gannon
Brian Gauthier
Kent Gooch
Carlos Hathcock
Jeff Hoffman (Black Hills Ammunition)
LTC Michael Lambrecht (LCAAP)
Sam Lepere
Lake City Army Ammunition
   Plant Employees
Litton Electro-Optical
Sean Little
Guy McCracken Jr.
J. Guy McCracken
Maj. Steve Miles
Bert Mullins
Robert Newton
Dale Pollard
Paul Riggs
John Rogers (Remington Arms Co).
Peter R. Senich
Terry Schleuse
Keith Scullin
Sierra Bullets
Mike Tucker
Chris Thomas
Dick Thomas
Maj. Ron Wigger
Winchester Ammunition

My next book will cover field craft and center mainly on personal camouflage and concealment. I will give explicit examples of all types of ghillie suit construction and detail many individual modifications and the reasons for them as made by dozens of military, law enforcement, and civilian hunters who are using them today. You will be shown exactly how to build or modify the ghillie suit yourself for your own needs, whether it be for military, law enforcement, or sporting purposes. In addition, I will cover basic and advanced field expedient methods of survival, survival equipment, tracking, the sniper's hide, and more.

If you have any unusual or interesting police/military sniper or any tactical shooting incidents, or problems encountered and solved in tactical matches or actual shootings, and you would like to share these with others, please send me the details. I may include them in my next book. Humorous stories will be gladly accepted. Please write me at Texas Brigade Armory, 906 Middle Run, Duncanville, Texas, 75137.

ABOUT THE AUTHOR

I was born and raised in Hawaii. I attended the University of Hawaii in 1968 where I took ROTC. In 1971 I went on temporary duty to attend the Ranger and Airborne schools at Ft. Benning, Georgia. After completing both courses, I returned to Hawaii to finish school and was commissioned as an Infantry officer and assigned to the First Cav Division at Ft. Hood, Texas. After spending only a few months in Vietnam, I returned to Ft. Hood with the First Cav where I continued as a platoon leader in A Company, 2d Battalion, 7th (Airmobile) Cav, 1st Cav Division. In 1973 I returned to Ft. Benning to complete the Officer Basic Course as an honor graduate and once again returned to my unit at Ft. Hood where I continued as a platoon leader and later a company executive officer in 2/7th Cav. In 1975 I volunteered to become the OIC (Officer in Charge) of the Ft. Hood, Texas, Advanced Marksmanship Unit. Besides qualifying as a shooting member of the AMU's rifle and pistol team, I was responsible for all of Fort Hood's competitive marksmanship and small arms training programs, including an informal sniper course for 1st Cav and 2d Armored Division snipers. It was during this period that I developed a real interest in precision firearms and learned how to accurize M-14 rifles and .45 M1911 pistols. In 1977, having married a year earlier and after leaving the service, we moved to Duncanville, Texas, located just outside Dallas. I worked as an Industrial and Manufacturing Engineer with several well known electronics manufacturing companies. Now with two children, I continued to work on building precision firearms, first as a hobby, and in 1990, I started my own gunsmithing business, Texas Brigade Armory. Besides professional gunsmithing, I shoot in registered NRA Bullseye pistol competition and have won categories in State and Regional Matches to place in the Texas State Governor's Twenty. Other interests include collecting and studying ammunition, writing, and spending time with my son to develop his insatiable interest in firearms and shooting.

Mike R. Lau

CHAPTER 1:
THE MARINE CORPS SCOUT/SNIPER AND THE M40A1

Marine Scout/Snipers of 1/23, 4th Marine Division Houston, Texas. From left to right: Lance Cpl. Bennett Thomas, Cpl. James Crawford, Cpl. Donny Dishau, Cpl. Tim Hengst, and Cpl. Sean Little. (Mike R. Lau)

Until a few years ago, many gun enthusiasts outside of the Marine Corps knew very little about the M40A1. Many did not even know it existed. During the past 15 years, Marine Scout/Snipers became involved in numerous conflicts around the world in places such as Beirut, Somalia, Saudi Arabia, and Bosnia. He and his M40A1 rarely appeared in news coverage, and magazines were mostly void of articles on military sniping. Finding info on the rifle and telescope system was not an easy task. It is hoped that this chapter will be the first of its kind to take a critical, indepth look at the Marine Corps Scout/Sniper and his precision M40A1 rifle system.

Wood stocked M40 with Redfield 3x9 "green scope."
(Mike R. Lau)

The Rifle, 7.62mm, M40A1

Now in use for over 20 years, the M40A1 is the result of a product improvement program of the original M40 rifle developed during the Vietnam War. In December 1965, the Marine Corps MTU at Quantico, Virginia, evaluated several different commercial rifles and telescopes as a possible replacement for the M70/Unertl rifles already being used by Marine snipers in Vietnam. Test rifles included 2 each examples, chambered for the 7.62mm NATO cartridge (.308 Win), in the Winchester M70, the Harrington & Richardson Ultra Rifle, the Remington 700 ADL and BDL, the Remington 600, and the custom Remington 700/40X. Final selection resulted in the Remington 700/40X configuration and the Redfield Accu-Range 3x9 scope. Manufacture and assembly of the new rifle system took place at Remington's Custom Shop at Ilion, New York, and 995 rifles, with serial numbers ranging from 168179 to 322769, were delivered to the Corps between 1966 and 1971. Fitted with a 24" medium or "varmint" weight chrome moly barrel, the action was hand inletted into a wooden stock with the barrel free floating. Marine Corps armorers later epoxy bedded the receivers to the stocks before sending them to Southeast Asia. (Note: the commercial Remington 700 was brought out later using this same weight barrel in their Varmint Special and Police Sniper models.) Floorplate and trigger guard was the standard aluminum BDL assembly. Mounted with a one-piece Jr. base, the Redfield 3x9 telescope utilized the Accu-Range range finding system reticle which is similar to Redfield's Accu-Trac found on several models of their currently manufactured scopes. Although the M40 rifle served its purpose well in Vietnam, the rifle had its limitations. The tropical environment rusted the chrome moly steel barrel, and the wooden stock, finished with linseed oil, played hell with the rifle's zero and accuracy when the stock warped and put uneven pressure on the barrel. In addition, the aluminum BDL floorplate assembly and Redfield scope were not tough enough for the rugged "in-country" environment.

As a result, a product improvement program determined that an upgrade of the existing M40 system was necessary to overcome its shortcomings. Utilizing the remaining original Remington 700 receivers, components identified for replacement were the barrel, stock, scope, mount, and the aluminum parts. Replacement of these parts would increase both the ruggedness and the accuracy of the system which was stipulated to shoot 1.5 MOA at both 600 and 1000 yards (approx. 9" and 15" extreme spreads) for three 10-shot groups at each distance. Designated the M40A1, a decision was made to upgrade the new sniper system with specially trained Rifle Team Equipment (RTE) armorers at the Quantico MTU instead of one of the Corps' overhaul depots or by the Remington Arms Co. This decision insured that the MTU had full control of the process and personnel in rebuilding their sniper rifles to match grade quality.

Beginning in late 1976, upgrade of the M40 rifles in inventory was underway and the majority of the surviving M40s returned from Southeast Asia were injected into the program. A heavy stainless steel barrel made by Atkinson Gun Co., Prescott, Arizona, was assembled to the clip slotted receiver. Note: Bill Atkinson sold out to H-S Precision, Phoenix, AZ, and by 1982, M40A1s were only rebarreled with this manufacturer's barrel. When H-S Precision moved to South Dakota, they were unable to meet demand for barrels

CHAPTER 1: The Marine Corps Scout/Sniper and the M40A1

Marine Corps Scout/Sniper Class of Summer 1980, Okinawa, Japan. Note early M40A1s with Redfield 3x9s and early ghillie suits. Former Marine sniper, Kent Gooch, 3rd from left (dark glasses), is now SFC Kent Gooch, NCOIC of the US Army National Guard Scout-Sniper School, Camp Robinson, Arkansas. (Kent Gooch)

so some Hart barrels were used. Today, Hart Rifle Barrels Co. is the main supplier of barrels to the Marine Corps. Contour of the barrel is similar to the Douglas #7. Marine Corps specs call for the 24" barrel to measure 1.18" at the receiver with a taper of about .01159 per inch toward the muzzle which measures around .90 inch. This would indicate almost no straight section forward of the receiver, which would normally be required for bedding 1 or 2 inches forward of the receiver. Twist rate is 1-12" and muzzle is counterbore crowned to a depth of .100 inch and .700 inch wide across the bore. A steep angle at the edge of the counterbore, takes off the sharp corner. The stainless steel barrel is finished black oxide which puts a thin matte black etch on the surface of the metal. While fairly durable, the finish can be scratched off leaving a bright spot, and care must be taken not to deeply etch the inside of the bore with the chemical solutions used during the blackening process.

Because the first two M40 test rifles were built from 40X receivers at Remington's Custom shop, it is assumed that the receivers marked "Remington 700" had 40X tolerances and were precision trued. Besides being fitted with a new barrel, the recoil lug and metal magazine box are TIG welded at spot locations to the receiver. The factory Remington 700 trigger on the M40A1 is adjusted from 3 to 3.5 lb. which is a good safety compromise for a precision rifle used mainly in the field. Original receivers have "U.S." engraved over the receiver serial number.

Experience during the Vietnam War sold the Marine snipers on the sporter style stock because it was more comfortable to carry and easier to use in the field than target type stocks. A tough "Marine proof" stock was desired for replacement and McMillan's fiberglass General Purpose Hunting (HTGP) stock was selected. Many of the M40A1s in inventory today still have the original stocks which

Cpl. Crawford, firing 168 gr. Federal Match in his M40A1, uses his stuffed "butt pack" as support. Ammunition is Federal Match 168 gr. HPBT which shot better than issue M118 Special Ball. (Mike R. Lau)

is a testimony to the durability of the stocks. Dick Davis, of McMillan Fiberglass Stocks, told me last year that only one stock was received back broken that year after falling out of a helicopter. Original M40A1 stocks are in the forest camo pattern made with molded-in-epoxy colors of forest green, pale green, and red-earth brown. This pattern provided the correct colors for most tropical and forest environments the Marines were used to operating in, and the forest green is very dark, appearing almost black. Because the epoxy and fiberglass stock is shaped in a mold, the individual colors on the original stocks appear to be smeared over each other at the edges, with no definite outline on each color's pattern. Also obvious is a very rough seam along the length of the stock caused by the mold halves.

McMillan's newer forest camo HTGP stocks have a more defined color pattern with a less dark forest green and the seam is cleaner looking and less noticeable. Dick said he cannot duplicate the smeared color look of the original M40A1 stocks today because the old epoxy colors are no longer used by the fac-

Right side of Marine Corps M40A1 showing old style McMillan stock in forest camo with molded-in-epoxy colors appearing "smeared". Camouflage scope/receiver cover under rifle has elastic band sewn around edge and is attached to scope with black Velcro straps seen between scope rings and scope turret body. (Sam Lepere)

CHAPTER 1: The Marine Corps Scout/Sniper and the M40A1

Cpl. Sean Little catches fired case from M40A1. Using his "buttpack" for support. The camouflage scope cover, next to the web gear, has an elastic band and Velcro for attaching to the scope. M40A1 rifle and Unertl scope were painted tan for field exercise at Schofield Barracks, Hawaii, because the tall dead grass on the open plains were light brown to tan in color. (Sean Little)

tory. In addition to using the fiberglass/epoxy and ground-up Kevlar, the butt of the sniper stock is filled with "sniper-fill" which is a denser epoxy mixture that adds both strength and weight to the stock. The 1-1/4" sling swivels made by Wichita Engineering Company are very sturdy and made with heavy gauge wire loops. Held into the stock with large imbedded nuts, the threaded swivel studs are designed to not pull out even when you are running with the rifle slung over your back. Buttpad is the .60" brown Pachmayer Presentation with the black spacer, and although not as recoil absorbent as the 1" pads, the heavy 14 1/2 lb. weight of the rifle reduces recoil considerably. Macho Marine snipers don't complain of the weight or recoil of the rifle and the extra heavy weight helps in steadying the rifle. Currently being evaluated by the Corps is McMillan's new A3 tactical stock with a straight pistol type grip, squarer forend, and a hand stop on the bottom side of the buttstock where the rear sling swivel is located. It is unclear if the Corps will accept the McMillan A-3 stock at this time. More on the A-3 stock will be found in chapter 4.

Very often the Scout/Sniper is authorized to do a specific custom job to his M40A1 rifle: paint it to match the combat environment. Although the rifle metal is matte black and the stock color is forest camo, many of the recent world events, requiring Marine involvement, called for the rifles to be painted desert camo. This included such areas as Beirut, Somalia, Bosnia, Saudi Arabia, and even the hillsides of Camp Pendleton, Calif., with its tall yellow grass that sometimes looks almost reddish gold. After masking off the scope knobs and receiver area, the rifle is spray painted. CPL Little recalls that when 1/23 went to Schofield Barracks, Hawaii, for training, they painted their rifles desert camo because the tall grass on the plains were also tan in color. All this may sound like fun, but it

Modified Winchester Model 70 steel floorplate and trigger guard on the M40A1 replaced the aluminum Remington BDL assembly which was prone to breakage on the Vietnam-era M40 rifle. (Mike R. Lau)

doesn't come off as easily as it is put on. CPL Little says removing the paint is done with acetone and Q-tips and makes for very tedious work.

Floorplate and trigger guard are modified steel Winchester Model 70 parts. This is a good low cost replacement for the weaker aluminum Remington factory assembly found on the M40s and BDL 700s. Modifications to the M70 parts are not complex and are easily modified with a milling machine and file.

Devcon 2-part catalyst epoxy, impregnated with steel particles, is used to bed the barreled action to the stock. 1/4-28 socket head screws at the front and rear of the receiver are used to assemble the rifle which are torqued to 65 inch lb., while the small center screw in front of the trigger guard is finger tight. Bedding of the barreled action extends 1-11/16" forward of the recoil lug to support the heavy barrel with the remainder of the barrel left floating. A few commercial gunsmiths believe that accuracy is better without this additional bedding, but without this barrel support, the heavy barrel stresses the receiver and can cause accuracy problems through uneven vibrations throughout the barreled action. An additional ninety-nine clip slotted Remington 700 receivers were purchased from Remington

Using the "butt pack" for support. (Mike R. Lau)

during fiscal year 1992 for rebuild into M40A1s. Serial numbers ranged between C6711636 and C6711764.

The M40A1 Rifle Today

So how does the upgraded rifle rate now that it has been in use for over 20 years and seen combat in many different areas of the world? The M40A1 has been admired by many throughout the world and is the standard by which others have developed or chosen for their own sniping systems. Even the US Army changed to the bolt action M24 with its Remington 700 receiver, and some in the Army had even desired that the USMC M40A1, built by the Marine Corps, would have been a better choice. M40A1s do have minor faults attributed to the fact that it is built from a commercial rifle originally designed for sporter use. However, I consider these faults as minor and rarely experienced to be detrimental in combat if proper precautions are taken before going to the field.

Possible problems noted with the M40A1 system were gathered from interviews and examination of rifles used by Marine Scout/Snipers of 1/23 (1st Battalion, 23rd Marine Regiment, 4th Marine Division), a reserve unit headquartered at the Naval Marine Corps Reserve Center in Houston, Texas. A weapon record book is kept with each M40A1 system and round count and problems encountered are noted in the book. Rifle/scope systems are sent to a maintenance depot in Waco, Texas, every six months for bore inspection, disassembly for cleaning, and repair. If the facility cannot repair a system, it is returned to Quantico and a different M40A1 system is issued to 1/23. This scheduled maintenance and replacement procedure works well since the Scout/Sniper is not allowed to tighten the stock screws or remove the barreled action from the stock for cleaning.

A couple of potential problems are associated with the use of the modified Model 70 Winchester floor plate. CPL Sean Little, 1/23, told me that a few floorplates have been found to be poorly fitted and a few may rattle even when latched with the push-button floorplate lock in the M70 trigger guard. Marine Corps stocks are not pillar bedded so there is the possibility that the trigger guard and floorplate could change depth in the stock after the screws are torqued. The floorplate could also be machined a little too short to correctly reach the trigger guard causing the floorplate to not engage the lock button securely like it should. Corps M40A1s are not pillar bedded which, if done, would increase the strength of the bedding and solve a lot of the fitting problems resulting from years of stress on stock by the receiver and trigger guard/floorplate components. Another potential problem is that the push-button floorplate lock is outside the trigger guard and faces forward. It could accidentally be depressed by a tree limb or other object when going through heavy brush or low crawling. To overcome potential problems with the floorplate, CPL Little told me that the floorplate is taped with OD duct tape when taken to the field, just in case. An all steel trigger guard similar to the Remington BDL assembly with the floorplate release latch on the inside of the trigger guard would be a good fix, but is an expensive item.

Also noted, but not a problem is that the magazine capacity of the M40A1 may vary from 3 to 5 rounds. The two guide "legs" on the bottom of the follower are ground off to allow the follower to depress further, but sometimes 5 rounds still cannot be loaded into the magazine box. As a commercial builder of reproduction M40A1's, I have solved this problem by cutting the front extension on the trigger guard right up to the center screw countersink and then beveling the cut. With this modification and correct assembly of the follower spring with the longer leg of the spring under the follower, the bends at the rear of the

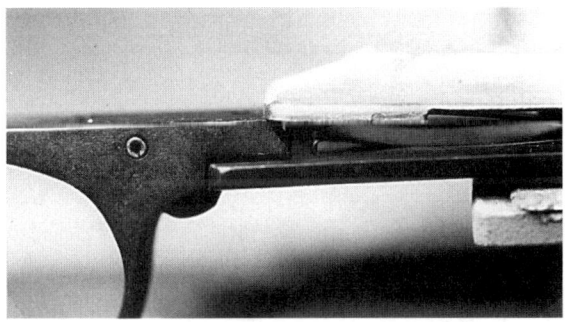

Author's modification of M40A1's trigger guard allows 5 round magazine capacity by keeping follower spring to completely clear the guard extension. Note that the legs on the bottom of the follower are also ground off. (Mike R. Lau)

compressed spring will sit on the beveled edge and forward of the trigger guard extension. This method always get 5 rounds into the magazine with a little room to spare.

Often perceived to be a weakness of

the Remington 700, is the small extractor located in the bolt face. I have seen many 700's extract stuck cartridges without damage to the extractor, while I have seen Sako type extractors pull out of the bolts in the same situation. The Remington extractor is very strong and is held in place by half of the rim of the bolt face. Those extractors that do pull out of the bolt face or break, happen because they were incorrectly installed, or were reused after being removed once before, and have been bent open to restore spring tension which weakens them. Extractor breakage was not a problem with the M40s in Southeast Asia and it was not a major issue addressed in the upgrade to the M40A1. Many shooters praise the claw extractor of the Winchester M70 and like the idea of controlled cartridge feed. This would be OK except it is not a problem to not have one. Many modern military small arms today don't have controlled feeding and the claw extractor would actually weaken the breech end of the rifle by not allowing full enclosure of the cartridge head.

Another feature that could be a possible problem is the bolt stop and release. If one of these pieces binds due to rust, foreign matter, or rubbing against the inside of the stock, the bolt stop may not return into the bolt raceway and cause the bolt to be pulled out of the receiver the next time it is opened. This could become a problem since the Marine S/S is not allowed to remove the barreled action from the stock for cleaning.

A weakness of the M40A1 is the safety lever which has been known to break occasionally. M40 receivers used in the 1960's have trigger safeties that lock the bolt when the safety is engaged in the safe position. Trying to open the bolt with the safety "on" can break or bend the safety. A safety lever was observed to break on an M40A1 used by 1/23 at Ft. Wolters when the safety was engaged with a live round in the chamber. The safety had to be disengaged with a pocket knife blade stuck down into the stock through the safety opening before the bolt could be opened. Later Remington 700 trigger assemblies have safeties with the bolt locking lever removed and CPL Little noted many of the older M40A1s now have the newer safety lever.

The most serious fault I find with the M40A1 is its 24" barrel length. Although handy in the field, the shorter barrel cannot always keep the 173 gr. M118 bullet, or most .308 Win loadings, out of the transonic region

Left side of USMC marked Unertl 10X. Scope mount is Marine Corps/Unertl with lugs on underside. Rings are 1" Redfield with bottom screw assembly. (Mike R. Lau)

beyond 800 or 900 yards. This is especially true in colder climates where the bullet may leave the muzzle at less than 2550 fps. The fix for this is a longer 26" barrel to start the bullet at 2700 fps from the muzzle under most temperatures. More on ammunition later.

The Unertl 10X Scope

When you pick up the Unertl 10X MIL-Dot scope, the first thing you notice is the coldness and the very heavy steel scope pressing into your hand. Weighing 2 lb. 3 oz., the 12 1/2" long scope has a 1" tube and approximately 40 mm objective. Finished with a thin light gray parkerized coating, one can see the lathe turned tool marks on its entire length. Its unusually large elevation and windage knobs have markings that can be read from a few feet away even by someone like me wearing contacts who forgets to bring along his reading glasses. Elevation is marked in even 100 yard increments with the numerals 1 to 10, corresponding to yards out to 1000 with a single turn of the dial. A cam inside the scope moves the internal reticle up or down to compensate for the M118's bullet trajectory at the corresponding yardage marked on the elevation knob. There are no intermediate click stops on the elevation knob between the 100 yard increments. However, there is a second, smaller elevation knob, just below the main elevation knob for fine tuning that will do + or -3 MOA in 1/2 MOA increments for minor adjustments due to impact changes in ammunition lots, temperature, geographical elevation, etc., that cannot be corrected by rezeroing at the time. Scout/Snipers of 1/23 sometimes turn the main elevation knob to positions between the even hundred yard detents when estimating odd distances, but there are no click stops so the knob position could be lost easily. On older scopes, the windage knob allowed for 4 1/2 MOA in either L or R directions, while later scopes provide a total range of 16 minutes. Zeroing is normally done on a 300 yards target as follows. With the elevation knob set on "3" and windage at "0", fire a three shot group. Unlock the lock set screws on the side of the main elevation and windage knobs. Turn the large allen screw in the top center of each knob until the reticle is aligned with the point of impact of the shot group. Retighten the locking screws.

Lenses are coated with a hard film of HELR (High Efficiency Low Reflect) which increase the light gathering capability of the lenses to approximately 91% of the available light. If the lenses were not coated, much of this light would be reflected away from the image in the scope and that image would appear dimmer. The approximately 40mm objective lens would produce an exit pupil of 4 mm which is not very large, but the optics appear to be excellent and Marines say they can make out targets under bright moonlight out to 600 yards. There are no provisions on the mount or scope for attaching a night vision device. Eye relief is 3 inches and the field of view of the 10X scope is 11 feet at 100 yards.

Focus adjustment on the Unertl is the same as commercial scopes while parallax adjustment is not. CPL Little recalled a time when he was having trouble grouping shots at 300 yards. At first he was not aware that the reticle was out of focus while forcing himself to focus on the target image. Another Marine brought over a white sheet of paper and held it in front of the scope. After focusing at a distant target for several seconds, CPL Little quickly looked into the scope and saw that the crosshair was not in focus so he adjusted this by loosening the lock ring turning the eyepiece focus ring until the crosshair was in focus and then retightened the lock ring. The cross hairs were then placed on the 300 yard target with the rifle in a stable position on a sandbag.

Left side view of a law-enforcement Unertl 10X mounted on M40A1 built by author. Only a few Unertl 10X scopes were made for law enforcement and are marked "10X SNIPER" instead of "USMC SNIPER". Marine Corps Unertl's have not been surplused and have limited commercial availability. Only a few have been sold to law enforcement, the FBI, and the Canadian Armed Forces. (Mike R. Lau)

Close up of author's M40A1 with law-enforcement Unertl scope. Unertl mount on author's rifle has 30mm rings with 1" insert adapters. Red arrows on elevation and windage dial indicate direction of impact. (Mike R. Lau)

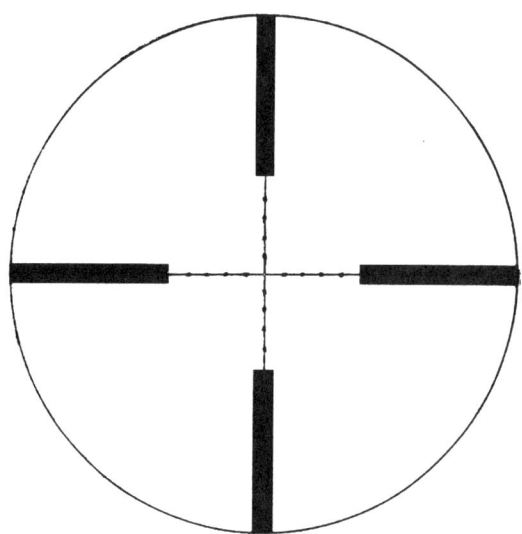

Fig.1.1. The USMC mil-dot reticle has wide outer posts and very thin cross wires with round or football shaped dots. The space between two dots represent one Milradian or approximately 3.6" per 100 yards and 3.9" per 100 meters. The dots or footballs are very small and cover approximately 3/4 moa or 3/4" at 100 yards.

Looking through the scope and moving his head from side to side, he noticed the crosshair move away from the target each time his head went to one side of the scope. Parallax was adjusted by turning the adjustment ring located on the very front of the objective bell with the issue aluminum spanner key. This corrected the problem and Sean's rifle grouped noticeably better.

The USMC MIL-Dot Reticle

Developed by the Marines in the late 1970's, the MIL-Dot reticle in the Unertl 10X scope aids the scout/sniper in estimating distances. A series of dots on the reticle are spaced equally apart and each spacing represent a specific height at a specific distance. The reticle allows the sniper to accurately estimate distance to target at long ranges. For a detailed explanation on the theory and use of the Mil-dot reticle see the chapter on range estimating.

The Unertl Scope Mount

The scope mount, produced by the Unertl Optical Company, is a heck of a piece of secure scope mounting equipment. Machined from a solid piece of steel, the base has two lugs on the underside which fit into the bridge clip slot and the radius cut at the rear of the receiver ring for the bullet nose clearance. Scope rings are Redfield 1" bottom screw type that are brazed and pinned to the base. The rings are well centered to allow the windage adjustment on the scope to be fairly centered and also for the 100 yard zero to align with the mark on the elevation knob. All four factory screw holes on the top of the receiver are used to attach the base to the receiver. This mount allows for very secure attachment even if one or two of the base screws come loose. The mount is serial numbered to the receiver to insure the mount stays with rifle. Finish of the mount is matte black with all sharp corners and edges neatly rounded.

The USMC M40A1 is the standard by which many other sniper systems are measured. The Remington 700 action, McMillan stock, Winchester M70 parts, and Unertl scope have proven itself in combat and to be "Marine Proof," simple, yet very rugged, has low maintenance cost, and capable of excellent accuracy. There are no adjustable buttplates or

CHAPTER 1: The Marine Corps Scout/Sniper and the M40A1

Lance Cpl. Thomas engages target at 600 yards while Cpl. Hengst notes wind conditions and calls windage adjustment. Chocolate brown Eagle drag bags were special contracted by the Marine Corps and are not available commercially in this color. (Mike R. Lau)

The 600 yard firing line at Ft. Wolters, Texas. Cpl. Hengst is shooter with Cpl. Crawford as spotter. Barely visible in photo, farthest wind flag on left shows 15 mph left wind, while two flags on right show zero wind. Mirage seen through spotting scope is better indicator of wind between shooter and target. (Mike Lau)

cheekpieces which can break or get out of adjustment in rough combat environments. Still simply the best and still in use after 20 years.

Part II: Shooting the USMC M40A1

With 8 M40A1 rifles assigned, the Scout/Sniper Platoon of 1/23 has a normal strength of 16 enlisted and one non-commissioned officer. Attached to the Headquarters and Service Company (H&S Co) of the Infantry Battalion, the S/S platoon supports the battalion's rifle companies or can be deployed at the discretion of the Battalion Commander for recon or other special operations. For a while, Marine Corps sniper platoons were labeled the "STA" Platoon, acronym for Surveillance and Target Acquisition. While the terminology may sound politically correct to some, 1/23 Marines disliked the use of that name and renamed the unit "Scout/Sniper Platoon."

Marine snipers operate in the field and on the range as a two man team. The observer, or spotter, using an M49 20X telescope, is located close to the shooter's right side (for a right handed shooter) and behind the rifle where he can observe the "vapor trail" of the bullet going down range. Besides assisting in locating the target, the spotter is responsible for verifying the target distance, wind conditions and mirage for determining windage corrections. He asks the shooter for the MIL height or length of the target and calculates the distance to the target using the distance formula or uses the chart found in the log book. The Marine Scout/Snipers learn to estimate target sizes within 1/10 of a MIL in order to get very precise target distances.

For moving targets, the spotter will also verify the lead in MILs for the shooter after noting the windage and target distance. While the sniper is sighting on the target, the spotter continues to observe the wind condi-

Two Marine Scout/Snipers during exercise at Quantico, Virginia. (Kent Gooch)

tions and will correct the shooter's windage as conditions change. When the sniper has the windage and distance dialed in on the Unertl he will tell the spotter he is "ready." The spotter continues to observe wind conditions and makes corrections as necessary. When no changes in wind conditions are noted the spotter tells the shooter to "hold center of mass" and "ready" or "fire." If a change in field conditions are noted, the spotter tells the shooter to "hold" and makes further corrections to the shooter. If the spotter notes a sudden change in wind conditions and the target must be engaged immediately, the spotter can correct the shooter by telling him to hold to the right or left edge of center of mass or give a MIL correction for "Kentucky windage."

Marine Scout/Snipers (MOS 8541) are taught to shoot the M40A1 in both supported and unsupported positions. In a prone position the sniper can use his butt pack, a sandbag, or construct a tripod from tree branches to use as support. The firing hand grasps the rifle firmly. Individual preference will determine if any rearward pressure is exerted by the firing hand. Excessive firing hand pressure can transmit trembling and also cause trigger control to be difficult. Sideways pressure should not be exerted on the grip of the stock or the rifle will move to the side as the rifle is fired. Some shooters put hardly any pressure

Jim Gannon, Army rifle shooter, observes Cpl. Crawford shoot while Cpl. Hengst observes. Note how low the supported position is with non-firing hand under toe of stock. (Sam Lepere)

CHAPTER 1: The Marine Corps Scout/Sniper and the M40A1

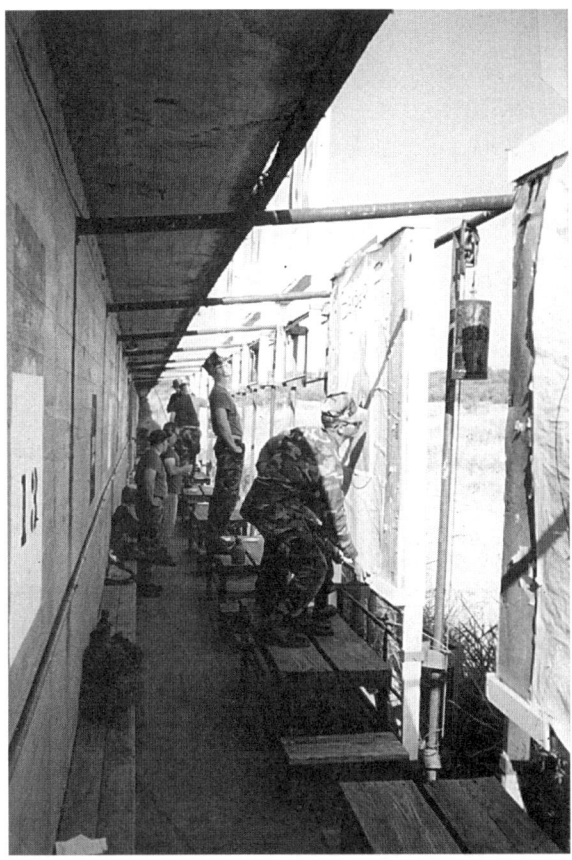

600 yard line, Range 2, Scout/Sniper School, Quantico, Virginia. Instructor is SSGT Dan "Ghengis" Geltmacher now Marine Chief Warrant Officer. (Kent Gooch)

Marine Scout/Snipers of 1/23 do their fair share of work pulling targets in the "butts" at Ft. Wolters. Target board in foreground has "Semper Fi" written above target. (Mike R. Lau)

on the grip with their firing hand to give them maximum trigger leverage. This will cause the rifle to recoil more freely with more muzzle whip, and it needs to be consistent from shot to shot. The non-shooting hand grasps the rear sling swivel and sling forming a fist or laid flat on the ground and the rear of the stock can then be lowered or raised with this hand to change the elevation. This hand is also used to help pull the rifle rearward into the shoulder.

Correct body placement to obtain the natural point of aim and to absorb recoil properly is very important to insure that the rifle does not jump sideways off the support during recoil. Scout/Snipers of 1/23 study a film showing the slow motion vibration of the recoil moving along the full length of the body from shoulder to foot and how improper body positioning causes changes in point of impact on the target. The 14-1/2 LB weight of the M40A1 helps in steadying the aim and also to reduce recoil Shooting from unsupported positions are also taught using the hasty sling. Keep in mind that the sniper will eventually develop his own technique of firing that provides him the greatest accuracy with the easiest and most natural position for his body.

Formal courses for Marine Scout/Snipers are taught at Quantico, Virginia, at Camp Pendleton, California, at Camp Lejeune, at the Special Operations Training Groups (SOTG), and at the Marine Corps Security Force Battalion (MCFSBn). A typical qualification course of fire on the rifle range for a S/S team is fired at stationary and moving targets from 300 to 1000 yards using a full size silhouette target which has a 24" wide upper body. For

Early days of the Quantico's Scout/Sniper school. Building was extended to accommodate a classroom. Cpt. Tim Hunter, OIC, points to real estate sign modified by Gy "Snake" Latimer the NCOIC. Classes were held in the barracks across the road in the Weapons Training Battalion (WTBN) Lounge or "Smudge Pot". (Kent Gooch)

Kent Gooch's unit, 3rd Bn/5th Marines, STA Platoon, then at Camp Margarita, during an M49 spotting scope exercise. Tripods were not available because of defense budget cuts. (Kent Gooch)

a moving target, the "E modified" has its shoulders and sides cut off, creating a 12" wide target across the upper body to present a side view. On the known distance range, three moving and two stationary targets are engaged with one round each at 300, 500, 600, and 700 yards. At 800 yards, two stationary targets and three short exposure time targets are engaged with one round each. Five stationary targets are fired at each range of 900 and 1000 yards. Moving to the unknown distance firing range, targets are engaged out to 600 yards. A total of 60 rounds are fired by the team and a qualifying score is 80% hits for all targets.

Qualification for the sniper candidate at the First Marine Division Scout/Sniper School at Camp Pendleton, California, consists of firing 25 rounds at the 20x40 "E" silhouette at varying distances out to 840 yards. This course is repeated over 3 days and to qualify, the sniper must hit the target 20 out of 25 times on two of the three days.

1/23 stresses the importance of engaging targets in the combat environment at 600 yards because this distance puts the sniper team out of the effective ranges of most enemy small arms and also allows the team a higher probability of hitting the engaged target than at longer ranges. Many consider a law enforcement sniper's shot to be more critical because of the possibility of hostages or bystanders being endangered if a target is missed. Marine Scout/Snipers, however, make every effort to make their first shot count also, because they don't want to make follow-up shots. Not only may they lose an opportunity for a second shot, but they will come under fire by enemy personnel immediately if detected and this retaliation may include heavy machine gun and artillery fire. If the Marine sniper team has to make another effort on another day to take out a target that was missed, the situation would be made more dangerous by the enemy's increased alertness and security. For these reasons, Marines of 1/23 are also taught to fire only one round at a selected target from a concealed position during a combat mission. The M40A1's magazine is full when the sniper is on the move in case the team makes enemy contact and gets in a firefight, while the spotter is armed with an M16A2.

Marine Scout/Snipers of 1/23 are taught to keep the Unertl scopes and rifle receivers clean in the field at all times. This was evidenced on the firing range at Ft. Wolters by 1/23's constant use of their camouflage scope covers when the rifle was not being fired. They also kept their bolts fully

CHAPTER 1: The Marine Corps Scout/Sniper and the M40A1

The Physical Training at Quantico included knife fighting. "Ghengis" Geltmacher watches as one student impales another. (Kent Gooch)

While the new classroom was being built, the Scout/Sniper school was moved into the indoor pistol range. Students bring their M40A1s and ghillies to the classroom. (Kent Gooch)

forward with bolt handle raised and the receiver covered to keep dust out whenever a cease fire was called. Range safety personnel were told that the M40A1 chambers were clear and the Scout/Snipers preferred to continue to keep their bolts closed and receiver covered. For cleaning the bores, 1/23 uses bore guides and sectioned brass cleaning rods because that is what the unit has inventory. Field manuals authorize LSA and CLP, but Scout/Snipers are authorized by the commander to use Shooters Choice. 1/23 Scout/Snipers understand the need for one piece nylon coated rods and these have been ordered through normal supply channels, but no telling how long it will take to get them. Proper methods and equipment for cleaning stainless steel barrels appears to not have caught up with the Marine Corps yet and is not a high priority. Lack of funds to purchase proper cleaning equipment is a major problem, not only for 1/23, but Marine Snipers in active duty units have told me that just getting authorized cleaning equipment can be difficult. Many Scout/Snipers, who understand the value of proper maintenance of stainless steel barrels, buy coated rods, solvents, and bore guides with personal funds. Operator's manuals recommend that the bore of the M40A1 be cleaned after every 20 rounds because bullet impact may change from 1/2 to 1 MOA elevation if this is not done. Camp Pendleton's school recommends cleaning the

Going stalking. Ghillies and M40A1s. (Kent Gooch)

Early drag bag. Cut off one of your buddy's camo utility trouser leg. (Kent Gooch)

bore every 15 rounds if time is permitted. Sometimes the M40A1 barrels are cleaned at the end of the day of firing which may result in 80 to 100 rounds being fired. This is a carry-over of the type of cleaning normally done with the standard M-14s and M16A2s and unit commanders should discourage this and emphasize the importance of proper caring for precision stainless steel barrels. Manuals also stress that the bore and chamber be cleaned of all solvents and oil prior to firing to prevent a high first shot, smoke signature on first shot, or excessive chamber pressure. Rifles are cleaned again on each of the three consecutive days after firing. As mentioned before, Scout/Snipers and Battalion armorers are not allowed to remove the barreled action from the stock, and if a stock screw is found loose, the rifle must be sent back to the RTE shop at Quantico for inspection and reassembly.

A historic moment. General Day addresses the Scout/Sniper class whose top graduate would receive, for the first time, the Dale E. Bertoli (who was killed in the South Pacific during WWII) trophy. Seated next to flags is Col. D.J. Willis. Kent Gooch was an instructor at the Quantico Scout/Sniper School at the time, 1984. (Kent Gooch)

One of the graduating classes when Kent Gooch was instructor. Pete Dordal, behind rifle, was top graduate. (Kent Gooch)

SNIPER AMMUNITION and ACCURACY of the M40A1

Any rifle is only as accurate as the ammunition that is fired in it. This is unfortunately true with the M40A1 as the full potential accuracy of the precision rifle is held back by the use of the not-so-accurate M118 Special Ball (SB) ammunition, the only recommended ball ammunition for this rifle. Fortunately, the Marines are attempting to correct this problem with the current development of a new round known as the M118 Long Range (LR). The cartridge is similar to the SB except it is designed for greater accuracy at longer ranges. As manufactured by Lake City Army Ammunition Plant (LCAAP), the components of the LR are of match quality. The bullet is the new Sierra 175 gr. hollow point boattail (HPBT) which gives considerably greater accuracy at long range than both the M118 SB and the M852 Match cartridge. The Marines and LCAAP are hoping to get consistent 1 MOA accuracy at 1000 yards with this new ammunition when development is completed and LR is standardized. This would increase the probability of the Marine sniper to hit targets at 1000 yards and a little beyond. For right now the Marines have to be content to use the M118 Special Ball. For complete information on the development and accuracy of the new M118 LR see chapter on ammunition. Marine Corps Scout/Snipers are definitely one of the world's elite fighting forces. With additional training being received today in special operations tactics, the Marine sniper can also engage in hostage rescue and anti-terrorist threat situations anywhere in the world. His mastery of the rifle and field craft makes him the predator on the battlefield.

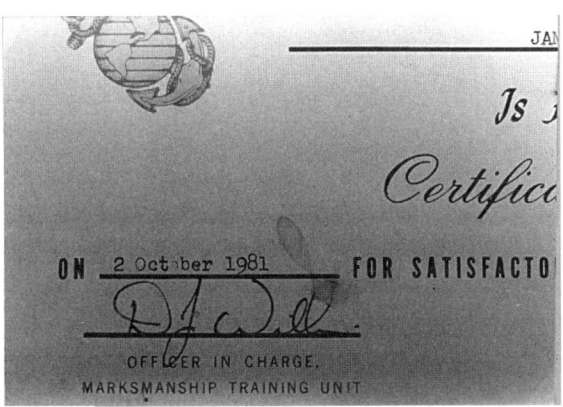

Marine Scout/Sniper James E. Furgeson, graduated from the Quantico, Virginia, Scout/Sniper School on 2 October 1981. His graduation diploma was handed to him by Carlos Hathcock, who was retired at the time, but came to visit the graduating class. Col. D.J. Willis, the C.O. and a Marine's Marine, spattered beechnut tobacco juice on his signature and then wiped it off, smearing the signature, before handing it to Carlos. (James E. Furgeson)

Author with Carlos Hathcock at the 1996 Soldier of Fortune show in Las Vegas. (Mike R. Lau)

New Marine Corps sniper ammunition is labeled M118 Special Ball, but is loaded with the new 175 gr. Sierra HPBT. Label has lightly printed Marine Corps emblem in background with "USMC" overprint. Box does not have "NOT FOR COMBAT USE" warning on it. (Mike R. Lau)

CHAPTER 2: THE US ARMY SCOUT-SNIPER AND THE M24 SNIPER WEAPON SYSTEM (SWS)

Author flanked by SFC Kent Gooch (left), NCOIC, Scout-Sniper School and Major Ron Wigger (right), MTU Staff and OIC 1997 Wilson Sniper Matches, Camp Robinson, Arkansas. (Mike R. Lau)

The M24 SWS Development and Background

When the United States pulled out of Vietnam in the mid-1970's, the M21 (accurized 7.62mm M14 rifle with telescope) continued to be the Army's main sniper weapon. Efforts were begun to develop a new Sniper Weapons System (SWS) soon thereafter, but lack of funds soon led to lack of interest. SWS projects to develop a replacement for the M21 sniper system went away for a few years until major world events made the Army search for a new sniper rifle system again.

During the 1980's, acts of international terrorism and instability of many foreign countries threatened the security of the United States and the world. The US military had to reorganize as they saw an increasing need for special operations units such as Special Forces, Rangers, Light Infantry units, Navy SEALs, Air Force and Army special operations aviation, and Marine Expeditionary Unit Special Operations Capable (MEUSOC). Even today, units in Marine Infantry Battalions are now training for special operations missions to deal with anti-terrorism and hostage rescue.

In 1982, the El Salvadorian Army formed sniper units with H&K G3 rifles and ART-MPC scopes in its counter-guerrilla strategy. So effective was this addition, that M21s were issued to special operations El Salvadorian recon units with formal sniper schooling initiated. By the late 1970's, US Marines had their M40A1's with Unertl scope as they had long seen the value of the sniper in both conventional and unconventional warfare. To come up to speed, US Army Special Operations units began to study the success of the sniper's role in El Salvador. By 1985, the Army began to look hard at the SWS program again. There was a shortage of M21s and to manufacture or rebuild and accurize the M14 for sniper duty was expensive and time consuming. Requirements for a new SWS was developed by the Special Operations Target Interdiction Committee (SOTIC) at Ft. Bragg, NC. The new rifle was to meet requirements for special operations units, airborne, and airmobile units, and have the capability of engaging targets beyond 1000 meters. To meet this last requirement, the rifle would have to shoot a cartridge much more powerful than current issue 7.62mm M118 Match or Special Ball. The SWS would have to easily convert to .300 Winchester Magnum without changing many of the major components such as the telescope, scope mount, stock and receiver.

Two M24 rifles with M3A day-optics were provided to author for close examination by SFC Gooch. He broke them down for author to examine parts. Tan painted rifle is SFC Gooch's favorite. (Mike R. Lau)

CHAPTER 2: The US Army Scout-Sniper and the M24 Sniper Weapon System (SWS)

It was also desired to have an "off the shelf rifle." Selection for the new rifle was approached by means of the Non Development Item (NDI) route because the Army needed the rifle right away and procuring the new SWS by regular R&D was expensive and took too long. Desired specifications were sent out to firearms manufacturers who would then submit samples to the Army for testing. Numerous commercial manufacturers, including several foreign firms, participated in the initial meetings. However, because of the many stringent and limiting Army requirements, including the requirement of being able to convert to .300 Win Mag, only Remington and Steyr submitted sample rifles when bidding requests went out. During the initial testing, the Steyr SSG (Scharfschutzengewehr) was eliminated due to the superior accuracy of the Remington prototype.

It is not real clear why this was so because the short action SSG has a good reputation for accuracy. There was speculation that the test rifle was built in haste on a longer action and the test rifle may not have been up to Steyr's normal quality. Accuracy tests for the XM24 trials called for the rifle to shoot an average mean radius of 1.3 inches or less at 200 yards and 1.9 inches at 300 yards with M118 Special Ball. The Army's acceptance standards were based on five targets with 10 shot each when fired from machine rests. Remington got the contract and the M24 Sniper Weapon System was on the way. Cost for the first contract for 500 systems including development and first article testing was $4995 each. According to John Rogers of the Remington Arms Company, later production, in option increments of 500 systems, cost around $3500 per unit. Remington delivered the first M24 SWS in October 1988 and 2,535 rifle have been purchased by the Army to date. This number includes prototype and test sample rifles. The US Army recently approved an Israeli Government contract to purchase 890 M24SWSs from Remington.

When reports of the Army's M24 sniper rifle first began to appear in popular military and firearms magazines in the late 1980s, the Army and the new rifle were both

M24 Sniper Weapon System is currently being used by U.S. Army Snipers. M144 spotting scope is Bausch and Lomb 15 to 45X with 60mm objective and is Army issue. Lower power is excellent for low light observation and higher power allows for observation of mirage at cooler temperatures. (Mike R. Lau)

severely criticized. Everything from development, design, cost, procurement, accuracy, and maintainability was negatively criticized. Much of this bashing created a lot of bad vibes and downright hostility toward the M24 and the Army. Even I got stuck on the faults of the M24 without considering the numerous good things associated with it. This same feeling increased the love and affection by many for the Marine Corps M40A1 sniper rifle and made it seem like the best sniper rifle that ever set foot on earth. But the Marine Corps M40A1 has its shortcomings too. The Army has now had the M24 SWS for 10 years. So how is the Army and the M24 SWS getting along today? I observed Army National Guard snipers and the M24 rifle go through some very tough shooting situations and field exercises at the 1997 Wilson Sniper Matches at Camp Robinson, Arkansas. SFC Kent Gooch, NCOIC of the Army Sniper School at Camp Robinson, provided the bulk of the material to the author. Besides a graduate of the Army's Sniper School at Ft. Benning, GA, SFC Gooch is a graduate of the 3rd Marine Division Scout/Sniper School in California and the Sniper Instructor School at Quantico, VA. In addition, he was a sniper in 3/5 Marines and an instructor at the Scout/Sniper School. Here is a man that is familiar with both the Army and Marine Corps sniping programs and has experience with the M40A1 and M24 sniper weapons. He will give you straight answers on the subjects without speculation. After studying the use of the M24 and interviewing many competitors, I can honestly say that the SWS and the Army's approach to maintainability has proven to be a success. The weapon system meets a variety of combat requirements, is tough, accurate, field maintainable, and well liked by the Army snipers. The Army's M24 Sniper Weapon System will be around for a long time.

M49 spotting scopes are still in inventory also. Sgt. James T. Kringle behind rifle with SFC Tim Weber spotting. M24 rifle is painted camouflage tan over the black finish. (Mike R. Lau)

CHAPTER 2: The US Army Scout-Sniper and the M24 Sniper Weapon System (SWS)

US Army's M24 Sniper Weapon System

SWS consists of seven major components:
1. M24 rifle in 7.62 NATO caliber with M3A day-optic (telescope)
2. Soft case (drag bag)
3. M1907 military sling
4. Iron sight and day optic case
5. Deployment kit with tools and cleaning equipment
6. Protective shipping case
7. Operator's manual, TM 9-1005-306-10

The M24 is a 5 shot bolt action rifle in caliber 7.62mm capable of engaging targets to 800 meters. Muzzle velocity is 2600 fps. Average length of rifle is 43 inches and weighs 14.95 lb. with the Leupold M3A day optic, bipod, full magazine, and sling. Total encased weight is 64 lb.

Fitted to the action is a 416R stainless steel barrel of approximately the same contour dimensions as the Marine Corps M40A1. The 24" free floating barrel averaged between .90 to .92 inches at the muzzle which has a pronouncely rounded front edge and counter bore 90 degree crown. On the left side of the barrel, approximately 7 inches forward of the receiver, is the "7.62 NATO" caliber stamping in 1/8" high single spaced numerals and letters. According to John Rogers of Remington Arms, the original prototype M24s had barrels made by Mike Rock. The Army wanted a prototype rifle in 45 days and Remington did not have mandrels for their own hammer forge machines to produce the 5R groove barrels. The Rock Barrel Company was able to produce the prototype barrels which had 5 grooves, angled rifling, and 1 turn in 11.25" twist. This type of rifling was copied from the Soviet AK-47 rifle and gives superior accuracy. Angled rifling allows for cleaner bores by not having the normal buildup of fouling in the bottom corners associated with regular cut rifling. This rifling type also allows for less wear on the bore and pressure according to John. However, Mike Rock did not have the capacity to produce large quantities of barrels for Remington. John mentioned that Mike had hired additional persons to produce the barrels, but the quality of these were not to Mike's standards. Nevertheless, Remington got the 5R mandrels soon after the prototypes were made and they now produce the M24 barrels with their own hammer forge machines. Remington has twelve machines and this process of barrel making is done differently than the cut or button rifled methods. A .308 caliber mandrel with the reverse impression of the rifling is inserted into the hole of a stainless steel blank. The machines hammer the outside of the blank with tremendous amount of force to shape and elongate the short, fat, steel chunk until the barrel is formed inside and out. It only takes a few minutes. John mentioned that the 5R barrels had improved accuracy by 20% over the standard 6 groove rifling. A few Police Sniper (PSS) rifles were made with Remington's 5R barrels and their accuracy was impressive. John also said, that possibly due to manufacturing cost, the plant decided not to switch all of their barrel making to the 5R rifling. All of the

Stainless steel free floating barrel is 24" long and has 90 degree recessed crown of .500 inches diameter and .040 inches deep. Muzzle's outer edge is noticeably beveled with corners rounded. Muzzle diameters measured ranged from .90 to .92". (Mike R. Lau)

Army's rifles have Remington made barrels on them now since the prototypes were also rebarreled. A second First Article test was required because of the barrel change from Mike Rock's to Remington's and this is also reflected in the initial high cost of the first 500 rifles.

The only major fault with the barrel is the same as that with the Marine M40A1: the barrel needs to be 26 inches long instead of 24 inches. With that 2 extra inches, improvements in 7.62 ammunition, and additional training, the sniper can increase the range of the M24 to almost what SOCOM wants by converting to .300 Win Mag.

To accommodate the possible conversion to .300 Magnum caliber at a future date (SOCOM requirement), the M24 was designed around the long action Remington 700. John Rogers stated that several M24s were converted to .300 Win Magnum for Army Special Operations Command, but no requirements for regular infantry units have surfaced. Original bolts supposedly had uncheckered bolt handle knobs, but I only saw checkered bolt knobs on the M24s I examined.

The barrel, receiver, and bolt are coated with a black epoxy finish like powder coating. The finish is electrostatically applied and thermally cured. The Remington developed finish is commercially marketed as Rem-Tuf (r) and is used on some sporter rifles. The bolt is also coated with a second finish that acts like a lubricant so that the Rem-Tuf (r) does not have to be removed from the camming and sliding areas. The coatings give the M24 a durable matte finish. Several of the rifles examined had been repainted with flat epoxy paints. Every shade of color from brown to dark greens to chartreuse can be seen when attending a sniper match with M24s coming from nearly every state in the country.

A new flash suppresser has been developed for the M24, but none were seen or used. The suppresser attaches to the barrel and has long slits cut into it sort of like the M14 flash suppresser according to one sniper who has seen one.

Factory trigger is modified with an externally accessed trigger weight adjustment screw. Sniper is allowed to adjust his trigger pull anywhere between 2 to 8 lb. with the 1/16" hex wrench provided in the tool kit. (Mike R. Lau)

The standard factory trigger assembly is modified with a second allen adjustable screw with spring accessed from the outside for changing the trigger pull weight. To provide a bearing point for this second trigger weight spring, the housing of the trigger assembly has an extension block that extends from the rear bottom of the trigger housing. Trigger pull weight can be adjusted from 2 lb. to 8 lb. by the sniper himself. The normal trigger weight adjusting screw, located at the front of the housing, is adjusted by the factory to break at 2 lb. minimum so that the sniper cannot adjust the trigger weight to below 2 lb. Sniper is not allowed to adjust the sear engagement, overtravel, or pull weight screws that are on the front and rear of trigger housing. He is also not allowed to disassemble trigger housing or remove from receiver.

The trigger guard is a machined steel casting made by Dakota. Located on the right side of the TG is the capital letter "H" in a circle which is hidden when the unit is assembled to the rifle. This signifies a magnaflux inspection which is an Army specification. Magazine floorplate is released by the latch

CHAPTER 2: The US Army Scout-Sniper and the M24 Sniper Weapon System (SWS)

Right side barrel markings. Circle with M designates second Magnaflux inspection. REP in oval is factory proof mark and stands for Remington English Proof, a carry over from the old days. The "55" is the gallery operator who test fired rifle. The very first mark to the left of the Circle M is a triangle mark which indicate 1st magnaflux inspection performed. (Mike R. Lau)

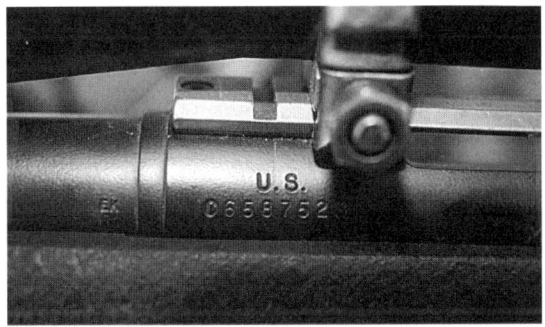

Left side markings include "U.S." over serial number. Letter "EK" would be the final assemblers code. No date code shown, but is usually on this side along with the assembler's code. (Mike R. Lau)

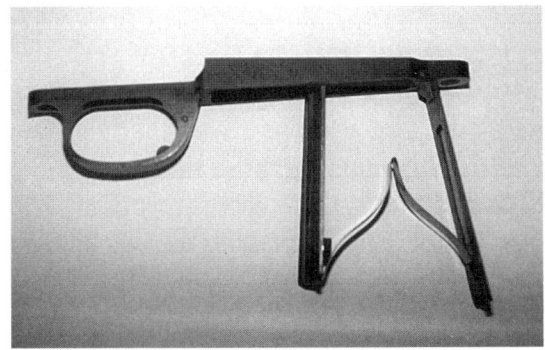

Floorplate is steel alloy casting made by Dakota. Part is beefier than aluminum part found on standard. (Mike R. Lau)

Stock screws are standard Remington factory oval head allen. (Mike R. Lau)

"M24" is added to the "MODEL 700" below Remington logo. (Mike R. Lau)

Issue torque wrench is precalibrated for 65 in-lbs. 1/2 inch socket is for scope ring hex nuts and 5/32 allen wrench is used on stock screws. (Mike R. Lau)

located inside the front of the trigger guard. This is a plus when dragging the rifle on the ground during a low crawl. Follower is machined steel and blackened. Follower spring is standard Remington. Box magazine is Remington BDL long action type except it is secured at the top rear to the receiver with a small slotted head screw in the manner of the blind magazine ADL type. Magazine capacity is 5 rounds and because of the long action, 7.62mm ammunition must be loaded to the rear of the magazine to prevent cartridges from tipping downward during feeding. There is

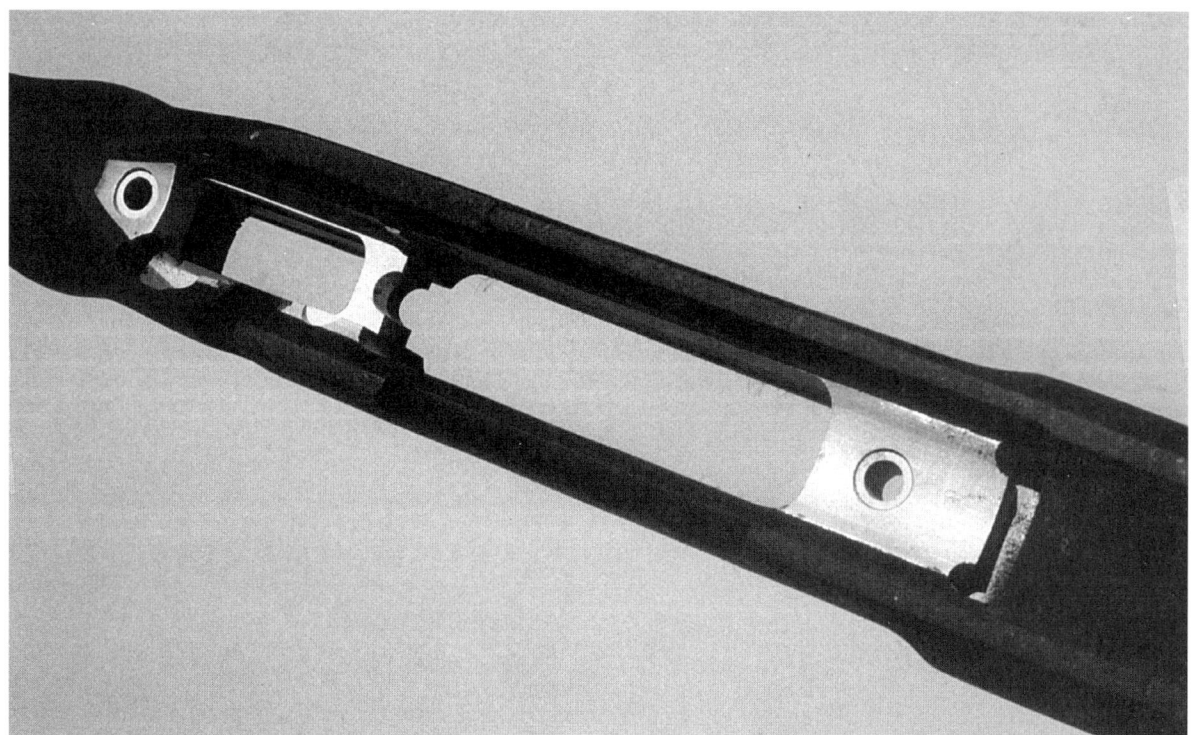

Aluminum bedding block in stock allows for complete interchangeability of receivers and stocks without the use of normal pillar bedding and epoxy bedding materials. (Mike R. Lau)

no filler block in the magazine to take up the extra space. This procedure is clearly explained in the operator's TM. (Note: An advantage to having a long action for the short 7.62/.308 cartridge would be that the handloader would be able to load long Berger or JLK VLD match bullets to touch the lands and still feed through the magazine.)

Stock on the M24 was designed by Remington and H-S Precision, Inc. and is commercially sold by H-S Precision under the trade name "Pro-Series Sniper" stock. The stock is made from a combination of Kevlar, graphite, and fiberglass, all epoxied around a polyurethane foam inner core. Imbedded into the receiver area of the stock is an aluminum "bedding block" that is contoured to allow the barreled action to be pulled into the bottom center of the block when secured with the allen stock screws. The bedding block allows the interchangeability of stocks and is the factory's way to "pillar bed" without the use of metal pillars and epoxy bedding materials like

Stock has an adjustable stock length assembly made entirely of aluminum including the adjusting wheel and screws. Although easily turned by hand, the main adjusting wheel has holes for using tool to turn it. A thinner wheel behind the adjusting wheel is used for locking the adjustment wheel. Because of aluminum construction, care must be taken not to put too much stress on assembly when fully extended. Buttpad is 1" with rounded edges and may have H-S Precision or Pachmayer name. (Mike R. Lau)

CHAPTER 2: The US Army Scout-Sniper and the M24 Sniper Weapon System (SWS)

Height of comb can be increased in height by any means. Here we have foam padding from a sleeping pad taped to stock with OD duct tape. (Mike R. Lau)

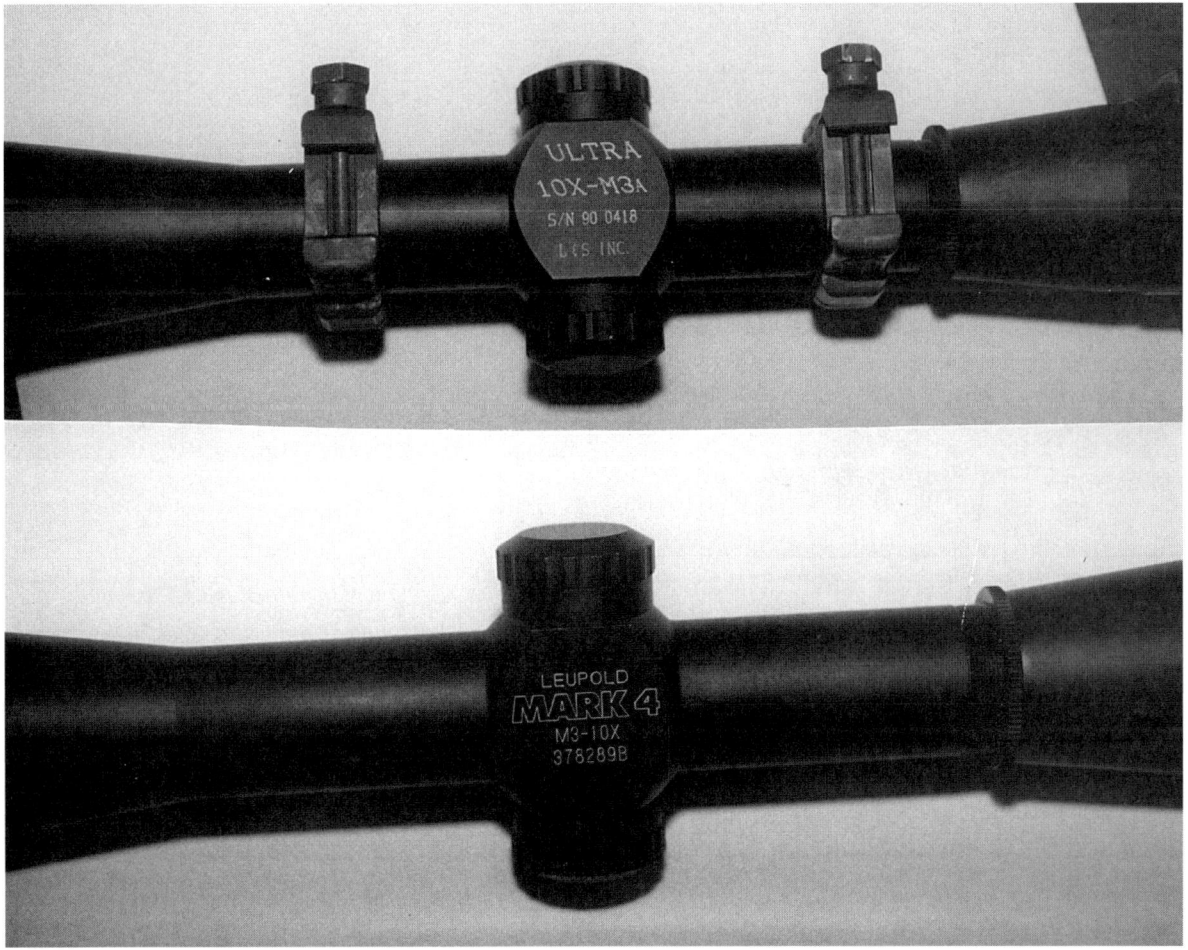

Markings on Army M3A Ultra and current commercial Leupold Mark 4 M3. Both 10X scopes are identical except that M3A has etched reticle on flat glass, while Mark 4 M3 has wire reticle. (Mike R. Lau)

Devcon or Bisonite. There is a slight gap noticed between the sides of the receiver and the stock. One of the M24s was noticed to have one side of the barrel touching the stock in the forend. This is easily remedied by taking a piece of sandpaper to the stock barrel channel and making the clearance. Both John Rogers and SFC Gooch told me that loose bedding blocks are not a common problem. A few of the snipers objected to the large grip, but most didn't mind it. Small hands would have trouble getting a firm grasp around it. The large grip could also make it difficult to keep the trigger finger from touching the stock when trying to get a good grip. The forearm is fairly wide and the bottom is rounded so it is not uncomfortable when carrying the rifle. A square bottomed forend will slow blood circulation into the fingers when carrying or when shooting with hand under forend. Quick release sling swivels studs are imbedded into the stock.

Action and stock assembly screws are the allen oval head type which are identical to that currently used on Remington 700 sporter rifles. Operators are allowed to remove receiver from stock for maintenance and cleaning, but should recheck zero. Torque on the screws are set at 65 inch-lb. by the pre-set torque wrench provided with the rifle.

Butt of the stock is fitted with an aluminum alloy adjustable buttplate with an attached black rubber recoil pad. The large aluminum adjusting wheel allows the butt pad to extend the length of pull from 12" to 14". The adjusting wheel is locked by the narrower wheel on the screw. This feature is a plus as it helps the sniper to get proper head and shoulder positioning during firing and his neck won't be cramped up or stretched when trying to maintain correct eye relief. A drawback of the adjustable buttstock is that it may be subject to combat rigors such as using the butt of the rifle to break a fall when getting to a prone position. It can also get out of adjustment or can collect mud and foliage. Several of the M24s observed had OD duct tape wrapped around the open part of the butt assembly. Outside finish on the stock is a rough texture and it is painted matte black.

The M3A Day-Optic

The term "day-optic" is the Army's nomenclature for "scope." Unlike the steel constructed Marine Corps Unertl, the main 30mm tube of the Leupold Ultra M3A is machined from a solid piece of 6081-T aircraft aluminum and has a thickness of 1/8". The inner (or erector) tube is the same size as a normal 1" scope erector tube. This allows for greater elevation and windage adjustments and

Adjustment knobs on Ultra M3A are the same as that on the Mark 4 M3. BDC on M3A is graduated in 100 meter increments and calibrated for M118 Special Ball. Elevation dial has 1 MOA click adjustments so sniper can adjust between 100 meter increments. Windage knobs have 1/2 MOA adjustments. (Mike R. Lau)

Commercial Mark 4 M3 scopes comes with several BDC dials. Note .308 Federal Match dial is in yards instead of meters and it is calibrated for 168 gr. bullet at 2600 fps, .300 Win Mag dial is for 220 gr. at 2650 fps, .30-06 dial is for 180 gr. bullet at 2700 fps, and .223 dial is for 55 gr. at 3200 fps. (Mike R. Lau)

CHAPTER 2: The US Army Scout-Sniper and the M24 Sniper Weapon System (SWS)

a thicker outer tube for the M3A. Leupold's Multicoat 4(r) lens coating results in superior edge to edge sharpness, precise resolution, minimal distortion, and optimum low light visibility. A side by side comparison of Leupold's 3.5x10 1" scope with the M3A, reveals that the M3A has a much brighter image. As explained by Chris Thomas of Premier Reticles, the fixed 10 power of the M3A has less lenses than a variable 3.5x10 so there is less light loss through reflection in the 30mm scope. Although the M3A does not have a steel tube it is very rugged and the Army put it through some very tough tests. Dual leaf springs oppose each adjustment in the M3A versus single bent 45 degree springs in 1" Leupolds. Internal adjustments on the M3A are more positive than the 1" tube because of this increased resistance. Turret caps get lost easily and are left behind when moving off the firing line on a range or from a shooting position in the field. At 21 oz., the M3A is much lighter than the 35 ounces of the Unertl 10X.

The bullet drop compensator (BDC) elevation dial is graduated in 100 meter increments up to 1000 meters. It takes only a 3/4 revolution of the dial to go from 100 to 1000 meters and is designed specifically for the trajectory of 7.62mm M118 ammunition. Full adjustment travel on elevation dial is 75 MOA. Many military and police known distance ranges are in yards so you will need to make a conversion to meters or use your mil-dot reticle and estimate the range anyway even if distance is known and then use your regular come-up for meters. (This will be covered in the chapter on range estimation.) 1 MOA click stops are provided in the elevating mechanism so that the sniper can fine tune the distance. It is not as precise as the 1/2 minute adjustment on the Unertl 10X secondary elevation dial. The windage knob has 1/2 minute clicks with a full adjustment travel of 70 minutes. Besides the elevation and windage knobs, there is a third knob on the left side of the turret housing which adjusts for the parallax by focusing the image. This is an extremely important and useful feature as the mil-dot reticle must be in focus to make accurate range estimations. For greatest accuracy, as much parallax as possible must be removed at each distance. Most fixed scopes have the parallax adjusted for only a specific distance usually 100 yards or, as in

The rings are normally attached to the M3A with the nuts on the left side of the receiver, but SFC Gooch says that the torque wrench for the ring nuts scuffs and burrs the Cloward base where the Palma sight attaches. So he turns scope rings around with nuts on other side. Note plug screws in upper holes of rear sight base. These are kept in place to keep mud and grit from filling threaded holes. Simrad base on front ring is for AN/PVS-9 night sight attachment. (Mike R. Lau)

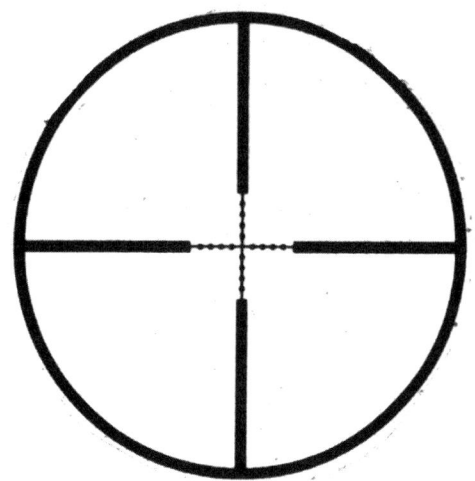

Army mil-dot reticle has narrower posts than the Marine Corps reticle and works the same way as the Marine Corps reticle. (Mike R. Lau)

the case of the Unertl, 300 yards only. Many shooters are unaware of the impact of parallax on bullet group size at different ranges because many scopes don't have this adjustment feature. Experienced shooters know that if the focus and parallax are not adjusted properly, you get larger size groups. The focus knob is easily accessed and will focus from 50 yards to infinity. Although the dials are not as large or easily readable as the Unertl 10X, the M3A adjustments can be read and accessed without getting off the rifle. As you read throughout the book, you will see that the sniper rifle needs to have easily seen and reached adjustments on the optics.

The day-optic is sealed and tested at the factory to determine if it will keep moisture from entering. This is done at the factory by submerging the scope in 120 degree water in a vacuum environment to cause the internal pressure of the scope to become higher. If the scope is not sealed properly the escaping air in the scope would be detected.

Mounting of scope to the rifle is done with Leupold's Mark 4 one piece long action base which uses all 4 screw holes on top of receiver. The long action base used to be made by the Brookfield Precision Co. that went out of business which is why you cannot get the long Rem 700 scope base and also their fine M1A/M14 mount commercially. The massive Mark 4 rings are attached to the base, ala Weaver style. Ring clamp hex nuts are torqued to the mount base with 65 inch-lb. Upper half of rings are secured by 4 each 8-40 screws per ring. (Note: Newer rings have screws with Leupold's new Torx(r) Head Drive System.)

Army's Range Estimating Mil-Dot Reticle

Many of the day-optics on the M24's are the old Ultra M3A models. The original Ultra scope is not like the current commercial Mark 4 M3 scope in that it has the mil-dot reticle etched onto flat glass and the dots are round instead of football shaped.

The Army's version of the range estimating mil-dot reticle is slightly different when compared to the Marine Corps reticle in that the M3A reticle has thinner posts on the outer portions of the wires. According to Chris, the newer Mark 4 M3 10X scopes are now being purchased by the Army. The football dots are supposedly easier to use when measuring mil-heights because of the points at each end of the "footballs" are more defining.

Chris Thomas stated that the drawback to the flat etched glass is that it can reflect some light whereas a curved lens will reflect or scatter the light. The US and some foreign countries possess surveillance equipment that will

External Mount Assembly (EMA) laser filter is provided with the M24. (Mike R. Lau)

CHAPTER 2: The US Army Scout-Sniper and the M24 Sniper Weapon System (SWS)

Killflash Anti-Reflection Shield is the honey comb filter that will reduce reflection of standard lens or EMA filter. Item is not standard issue, but noted on some M24s. Holes in filter are long so a laser surveillance device must be looking almost directly into the front of the telescope to get a reflection. Device does reduce light gathering into scope and image appears slightly darker. (Mike R. Lau)

scan terrain with a laser light that can detect other optic devices. The flat glass with the etched reticle in the M3A will retro-reflect that surveillance laser and give away the presence of the sniper.

EMA, Sunshade, and Dust Caps

An External Mount Assembly (EMA) laser filter is provided with the M24 and mounts on the objective bell of the day-optic. Looking like the Leupold 40mm sunshade, this device uses an amber colored lens to reduce the effect of Directed Energy Weapons (DEW) and to protect the sniper's eye against unintentional laser damage due to increased number of laser devices used on the battlefield. (Note: The flat etched reticle glass of the M3A is also more easily detected with specific surveillance equipment.) Disadvantages of using EMA is that it has a highly reflective shine and also makes the true colors of the image blend together which could make target ID difficult. When attaching the EMA to the scope, the indicating tick mark on the assembly should be located to the 12 o'clock position before tightening the lock ring. This will insure that the laser filter is pointed slightly downward directing most sunlight reflections toward the ground when you are aiming and viewing targets in the general direction of sun. Use of the EMA device also causes point of impact to change by as much as 1 MOA and a different zero must be used. By making sure the indicator mark is always in the same position when removing and reattaching the EMA, you should not have to rezero but just make the corrected elevation and windage adjustments.

Many of the M3As examined did not have the EMA laser filter attached, but instead were fitted with sunshades and/or the Butler Creek dust caps.

Iron Sights

Although seldom used, a set of aperture iron sights are issued with the M24 and can become a life saver if the day-optic becomes disabled in the field. Stored in the op-

Redfield Palma iron sight showing vernier scale. (Mike R. Lau)

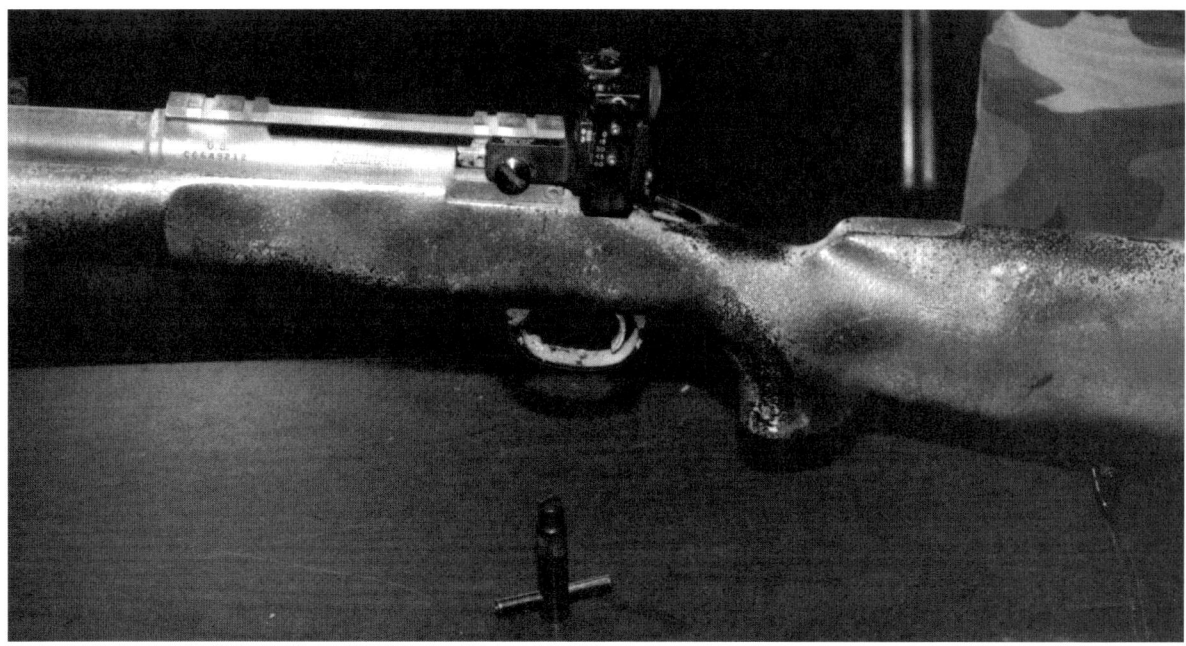

Redfield Palma mounted on SFC Gooch's rifle. (Mike R. Lau)

tic case, the rear sight is the Redfield Palma and the front is the Redfield International "Big Bore" globe with interchangeable posts and aperture inserts of varying widths. Both are provided to the Army via Redfield. Base for attaching the Palma sight is similar to the commercial Cloward except slightly taller in height. It is attached to mounting holes drilled and tapped in the left side of the receiver bridge and requires a clearance cut in the stock. Front sight base is a standard sight width male dovetail and is .175" in height and .910" in length. The sight bases are kept on the rifle even when the iron sights are not used.

Both windage and elevation knobs on the Palma sight have 1/4 MOA adjustments and one full turn equals 3 MOA. On the left side of the sight body is a 3 MOA vernier scale and on the aperture slide is an elevation scale plate showing graduation marks from 0–60 MOA in 3 MOA increments. The windage scale on the top of the slide has a similar vernier arrangement except the scale plate indicates a total of 36 MOA for windage in 3 MOA increments. The vernier scale is a little difficult to read because of the small tick marks which are very close together. The way a vernier scale reading is taken you must locate two opposite tick marks that line up with each other to get a value between each 0 and 3 MOA increment. Because the click stops in the knobs are in 1/4 MOA increments you can lose count of the adjustments made if you have to really crank on the knobs to get way out there. Now you have to look at the vernier scale and see how many MOA you cranked up. The sniper keeps his iron sights zeroed and he also has a come up table from 0–1000 meters (or yards) in his data book. Sight should return to zero when taken off and put back on base, but shooter should check to see if any change actually occurs.

The M24's iron sights are not as grunt proof as the sights on the M14 or M16A2, but it is better than nothing. Sight is rugged and has all steel construction so is good off-the-shelf choice for Army's rifle. With proper care and maintenance it will serve the intended purpose. If the sniper desires to use his M24 in rifle competition his iron sights are adequate.

According to Redfield the International "Big Bore" front globe has not been produced as normal stock item for a few years

CHAPTER 2: The US Army Scout-Sniper and the M24 Sniper Weapon System (SWS)

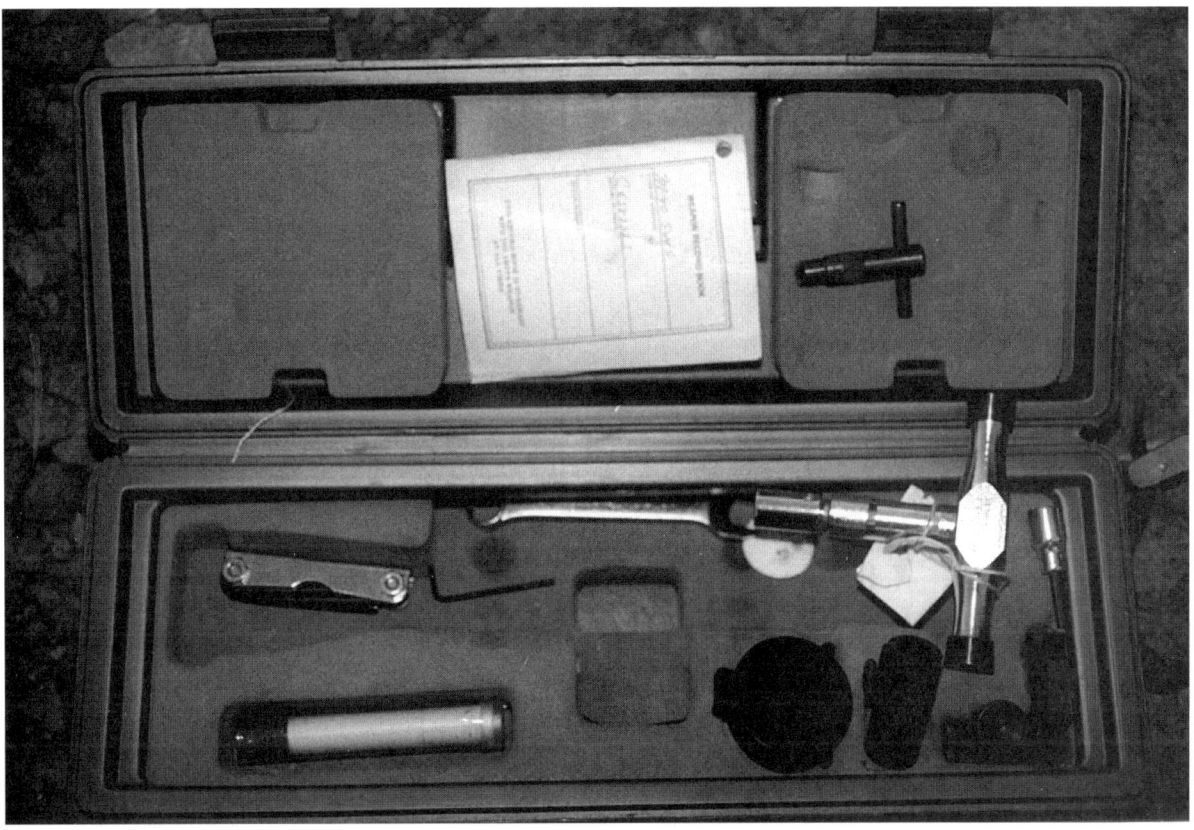

Day-optic case is used to store M3A scope, EMA, iron sight set, and optic cleaning kit. Sunshade is on rifle when EMA not used. (Mike R. Lau)

Day-optic case. (Mike R. Lau)

AN/PVS-9 mounted on SFC Gooch's M24. Unit is attached to the Simrad base and assembled into objective bell of M3A. (Mike R. Lau)

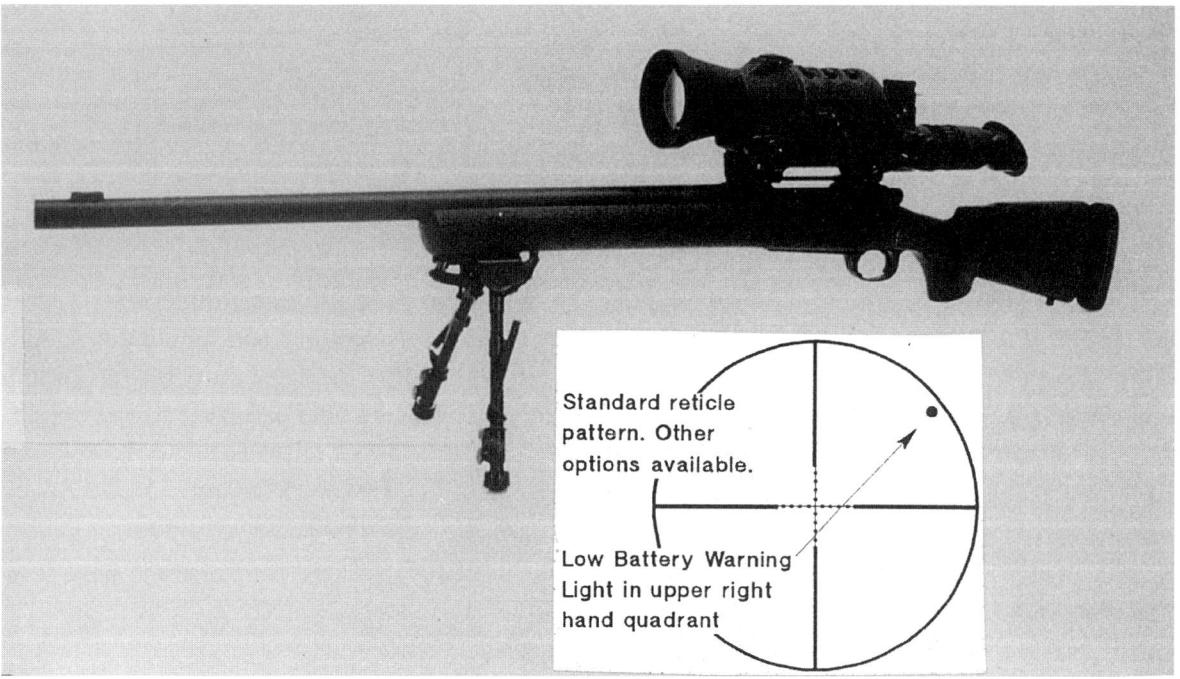

AN/PVS-10 day-night-optic on M24. (Litton Electro-Optical Systems)

now but has been built on special contract. Redfield intends to bring back the "Big Bore" commercially and is planned for their 1998 catalog.

Night Vision Optics

There are two types of night vision devices, active and passive. Passive devices do not require the device to transmit any light energy to the object or viewing area for it to work. An example of the passive device is the image intensifier known more commonly as the starlight scope. This night-optic uses the available light given off by stars, the moon, or any other external light sources including other laser or infra-red sources, and intensifies the viewed image seen in the unit. Another passive device is the thermal imager which senses the differences in temperature of various objects being viewed. This device can see through fog, light vegetation, and camouflage. The second type of night vision optic is the active device such as the infra-red, (IR). IR devices emit a spectrum of light that cannot be seen by the human eye unless this light is magnified by the metascope component of the same or with another separate device.

More often used by Special Operations Units than by regular infantry, the AN/PVS-9 is basically the Model KN250 made by Simrad Optronics A/S of Oslo, Norway. The device is assembled to the M24 by mounting to an add-on Simrad base that replaces the upper half of the front scope ring. On the under side of the unit is a projection that is inserted and locked into the objective bell of the M3A scope. By being an add-on unit to the M3A day optic, the image as seen through the M3A, is intensified. This allows the sniper to use the M3A in the normal manner with mil-dot reticle and the rifle's zero. Although the KN200 has a maximum detection range of 1400 meters, practical target engagement is 400 meters with starlight and 600 meters with moonlight. The AN/PVS-9 can be fitted with either the Generation II image tube or upgraded with the higher resolution Generation III image intensifier. SFC Gooch has engaged targets effectively as far as 800 meters with this device under night conditions. The battery powered unit weighs 2.2 lb. and has a 1X magnification.

Currently under evaluation by both the Marines and Army is the AN/PVS-10 Sniper Sight manufactured by Litton Electro-Optical Systems of Garland, Texas, and Tempe, Arizona. Contracted by the US Army, Litton has developed this long range optic to replace the M3A day scope on the M24 that would be used for both day and night. The AN/PVS-10 is designed to mount directly on the M24's existing scope mount rail and will allow the sniper to maintain the same shooting techniques such as eye relief, cheek position, and eye level above comb. Besides having the standard Army mil-dot reticle and bullet drop compensator, the scope also has controlled reticle illumination and an image brightness control which can maximize the image contrast and clarity. These two functions are controlled by two push buttons located on the top of the unit which can be accessed by either the right or left hand. Weighing 4.7 lb., the optic has magnification of 8.5X and a 125mm objective lens. Elevation has 1 MOA adjustments and windage has 1/2 MOA adjustments.

Care and Maintenance of the M24 SWS

Unlike the Marine Corps Sniper, the Army Sniper can disassemble his rifle from the stock for cleaning and maintenance. He can also remove the day-optic from the mounting base for cleaning and can also tighten the mount base screws. He is allowed to remove the firing pin and bolt sleeve assembly from the bolt body for cleaning and oiling. A spare firing pin assembly is included with the deployment kit should the firing pin break. Other allen wrenches are provided for tightening the iron sight bases and for removing or adjusting the BDC dial on the day-optic.

A very complete set of tools and cleaning equipment are included with the deployment kit.

36 *The MILITARY and POLICE SNIPER*

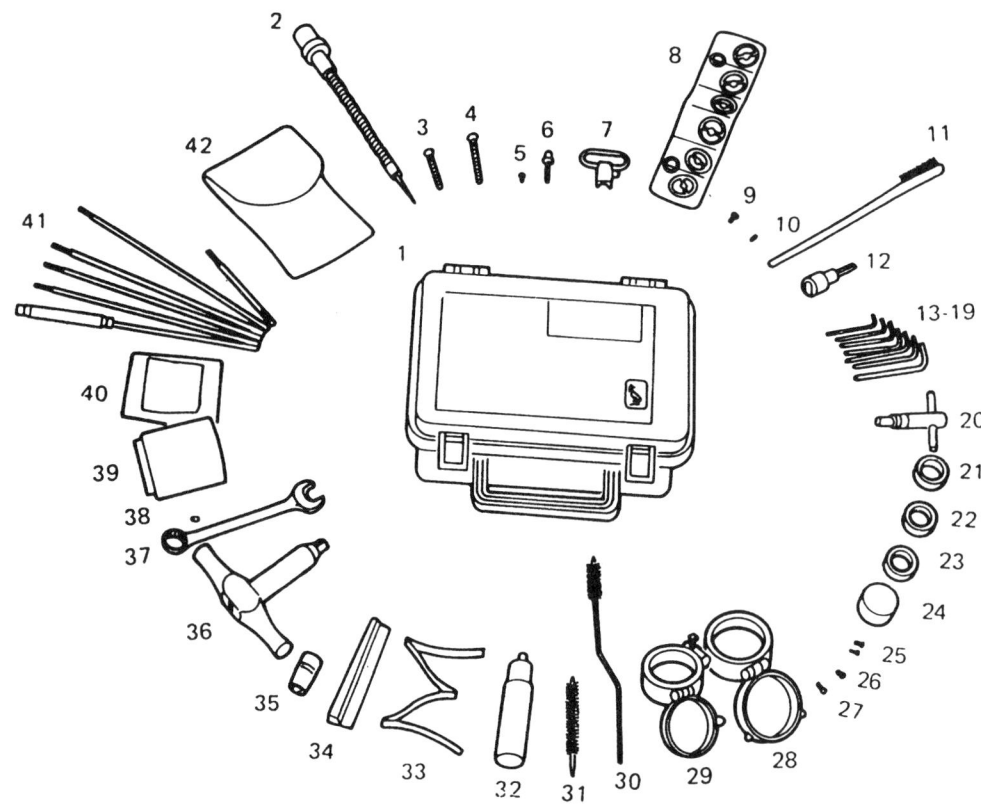

DEPLOYMENT KIT

1	Deployment Case	24	Day Optic Sight Adj. Dial Dust Cover
2	Firing Pin Assembly	25	Day Optic Sight Ring Screws
3	Front Guard Screw	26	Day Optic Sight Base Screw Front
4	Rear Guard Screw	27	Day Optic Sight Base Screw Rear
5	Front Sight Base Screw	28	Day Optic Sight Dust Cover, Front
6	Swivel Screw	29	Day Optic Sight Dust Cover Rear
7	Swivel, Sling	30	Brush, Chamber
8	Front Sight Insert Kit	31	Brush, Bore
9	Rear Sight Base Screw	32	Oil Bottle
10	Trigger Pull Adj. Screw	33	Magazine Spring
11	Brush, Cleaning Small	34	Magazine Follower
12	Socket Wrench Attachment 3/8" Drive Hex Bit 5/32"	35	Socket, Socket Wrench 1/2"
13	.050" Key, Socket Head Screw	36	T-Handle Torque Wrench
14	1/16" Key, Socket Head Screw	37	Wrench, Box and Open 1/2"
15	5/64" Key, Socket Head Screw	38	Rear Sight Base Plug Screw
16	3/32" Key, Socket Head Screw	39	Day Optic Sight Sunshade
17	7/64" Key, Socket Head Screw	40	Swabs, Cleaning, Small Arms
18	1/8" Key, Socket Head Screw	41	Cleaning Rod Kit
19	5/32" Key, Socket Head Screw	42	Lens Cleaning Kit
20	T-handle Combo Wrench		
21	Day Optic Sight Windage Dial w/Screws		
22	Day Optic Sight Elevation Dial w/Screws		
23	Day Optic Sight Focus Dial w/Screws		

(Photo 2.29)
Deployment Kit. (U.S. Army)

CHAPTER 2: The US Army Scout-Sniper and the M24 Sniper Weapon System (SWS)

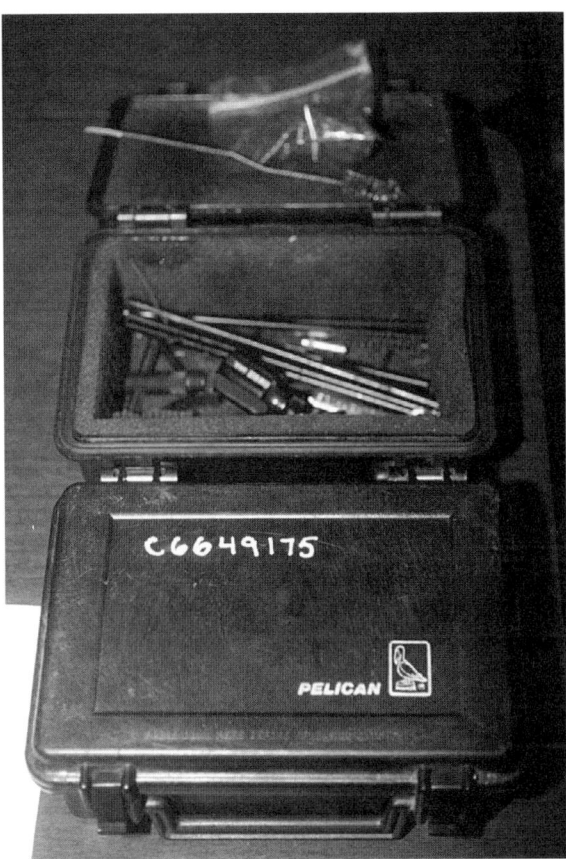

Deployment Kit case holds the tools and rifle cleaning equipment. Although sectioned cleaning rods are not recommended for cleaning rifles, the military continues to use them. Case is small Pelican Travel Vault, Model 3715, and measures 8 1/2 x 6 1/2 x 6 1/2. (Mike R. Lau)

The Army sniper is allowed to do much more disassembly and adjustment on the M24 than the Marine sniper can do on his M40A1. Looking through the Army Operator's Manual, TM-9-1005-306-10, the sniper can fix many items on the rifle until he reaches a point where he cannot fix an item such as when the dayoptic is damaged, trigger pull weight needs adjustment or the extractor, ejector, or safety breaks. SFC Gooch told author that the drawback to the Army maintenance policy is that the entire M24 SWS must be returned to Remington Arms for the repairs. On the other hand, the Marine sniper is not allowed to do simple adjustments such as tightening the stock screws or leveling the scope crosshairs. SFC Gooch believes the Marines should be allowed to do things like this. If "This is my rifle ..." then I should be able to fix it. The Marines, however, have their own in-house maintenance with their Rifle Team Equipment (RTE) armorers.

Need to replace broken Rem 700 extractors? Gunsmith's buying a lot of expensive Rem 700 extractor replacing tools ? I can remember replacing only one non-functioning extractor over a course of several years while working exclusively with 700s. The reason for it being replaced was because after truing the bolt face I tried to put the original extractor back in and it broke. I put a brand new second replacement extractor in, but did not realize that a small piece of the first replacement extractor was left under the new second extractor. It shot well for a while until the customer sent me back the broken second extractor and the piece from the first. Remington's snap-in extractor is easily replaced if broken and requires only a pin punch or flat tip screwdriver to "snap" one into the bolt face. The old style riveted extractor in standard caliber bolt heads were phased out by Remington in 1983 so the Army M24s all have the new snap-in type. The old style riveted extractor requires a special small anvil for flattening the small rivet on the outside of the rim on the counterbored bolt face. If the M24 is ever converted to .300 Win Mag, the bolts would have the rivet type extractor. In magnum bolts, the rim of the counterbore in the bolt face is thin and cannot be undercut deeply to hold the snap-in type extractor. I have removed and installed new extractors in nearly every Remington 700 that comes through my shop because I lathe true the bolt face and not because of broken extractors. As for the possibility of a cartridge getting stuck in a rusted chamber, the new 416R stainless barrels reduces the possibility of this problem.

When I asked John Rogers what the greatest maintenance problem with the M24 was, he stated that it was to replace shot out barrels. Remington guarantees the accuracy

SNIPER WEAPON SYSTEM PARTS LIST (Cont.)

CLEANING ROD KIT

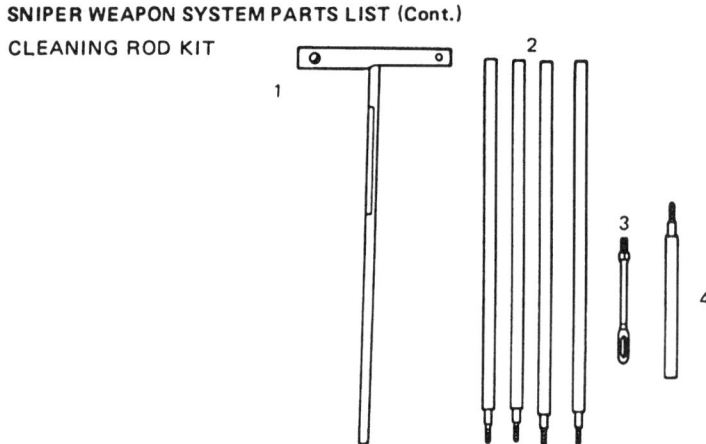

NO.	FSCM/CAGE	PART NO.	NOMENCLATURE	NSN
1	3A703	96092	T-handle Section	1005-01-271-3856
2	3A703	96093	Cleaning Rod Section	1005-01-271-3861
3	19204	11686237	Swab Holder	1005-00-937-2250
4	3A703	96095	Adapter	1005-01-271-3857

LENS CLEANING KIT

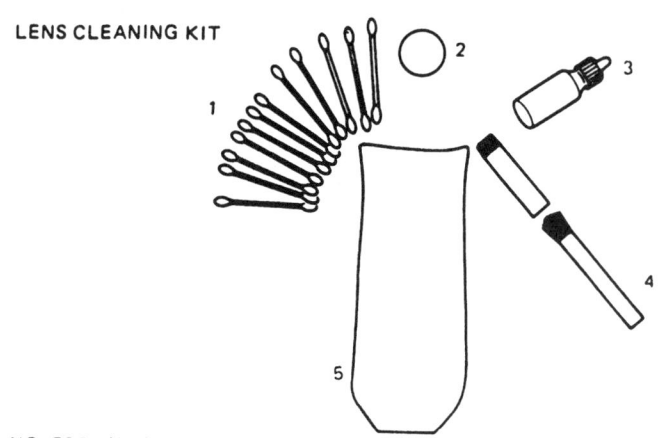

NO.	FSCM/CAGE	PART NO.	NOMENCLATURE	NSN
1	81348	GG-A-616	Applicator (Q-Tips)	6515-00-234-6838
2	81348	NNN-P-40	Paper, Lens (Cleaning Tissue)	6640-00-663-0832
3	81348	L-B-56A	Bottle (For Containing Isopropyl Alcohol)	8125-00-824-9058
4	81348	H-B-118	Brush, Artists (Cleaning Brush)	8020-00-224-8010
5	3A703	96041	Case, Lens Cleaning	1005-01-260-2661

Cleaning Rod Kit and Lens Cleaning Kit is part of Deployment Kit. (U.S. Army)

CHAPTER 2: The US Army Scout-Sniper and the M24 Sniper Weapon System (SWS)

Army National Guard Scout-Snipers prepare for unknown distance firing at Camp Robinson, Arkansas. (Mike R. Lau)

of the M24 for 10,000 rounds. Most of the rifles turned in for barrel replacement have had somewhere near 20,000 rounds shot through them according to John. Many of these rifles come from Ft. Benning. Round counts are not required to be kept by the Army sniper, but many do so. John said the rifle is shot until the firer determines the rifle will not group like it used to and bore wear is suspected.

SFC Kent Gooch stated that proper care and maintenance of the M24 will usually prevent most parts breakage and unreliability. The use of the drag bag is recommended when in the field. John Rogers told author that M24s leaving the Remington factory are guaranteed for 18 months. H-S Precision guarantees the stock for a longer period. A rifle needing repair is usually shipped directly by the using Army unit to Remington who will ship it directly back after repair. A time and material contract, for repair and spare parts for a period of 5 years, was negotiated between the Army and Remington during the initial purchase. This was to allow the Army to build up it's own in-house maintenance program during the contract. However, it seems that the Army will continue to let Remington do maintenance/repair on the M24s because a third 5 year contract was just negotiated.

Ammunition and Accuracy of the M24 Rifle

As mentioned earlier, the acceptance standard for the M24 rifle is an average mean radius (MR) of 1.3 inches or less at 200 yards or 1.9 inches at 300 yards. This equates to a group size of approximately 4.22 inches at 200 yards and 6.17 inches at 300 yards or 2.11 MOA at 200 yards and 2.06 MOA at 300 yards. Accuracy normally expected from M118 Special Ball (brown box) is usually 1 to 1.5 MOA @ 100 yards. With M118 Match (white box), accuracy is improved and has been observed to be around 1/2 to 1 MOA. With M852 Match

Torque stock screws to 65 in.-lb. Torque scope ring clamps to 65 in-lb. (Mike R. Lau)

and M118 LR, accuracy is improved to between 1/2 to 3/4 MOA @ 100 yds. This, of course when fired from accurized M1A/M14s and heavy barreled sniper bolt rifles. The M24, by the nature of its components and maker, is a 1/2 to 3/4 MOA rifle with good ammo. In comparison, the factory acceptance standard for the standard caliber Sendero, .300 Win Mag Sendero, 700 Police Sniper, and the 40X is 1.5 MOA, 2.0 MOA, 1.5 MOA, and .75 MOA respectively @ 100 yds with required factory hunting ammunition. In actuality, at 100 yards right out of the box with good commercial match ammunition, we have seen sporter Remington 700s shoot under 1 MOA and PSSs and 40Xs shoot 1/2" groups or better at 100 yards with factory match ammunition or good handloads. The expected accuracy of the M24 should be at least the same as the 40X or PSS.

The Army is currently evaluating the Marine Corps M118 LR ammunition made by Lake City Army Ammunition Plant with the Sierra 175 gr. HPBT match bullet. See chapter on Ammunition and Ballistics.

At the Army National Guard Sniper Matches held during the 1997 Winston P. Wilson Matches at Camp Robinson, the author witnessed the accuracy of the M24 by experienced Army National Guardsmen at long range shooting. During the Cold Bore- 2 shot matches, unknown distance targets were engaged at approximately 567 meters and 725 meters. Almost all of the snipers and observers made first round hits on silhouette size metal targets at these ranges. A four inch "V" ring was also painted on the center of the silhouettes and a good number of competitors made "V" hits.

The Army Scout-Sniper

A well trained Army Scout-Sniper team is one of the deadliest weapons on the battlefield today. Six Scout-Snipers are assigned to each recon/scout platoon of Light Infantry units. These soldiers usually have

CHAPTER 2: The US Army Scout-Sniper and the M24 Sniper Weapon System (SWS)

System Case is made by Hardigg Industries, South Deerfield, Maryland. (Mike R. Lau)

Army National Guard Scout-Snipers compete at Camp Robinson, Arkansas, with their M24s.

other primary skills so are considered "dual-hatted." Two designated snipers are usually assigned to headquarters of the Mechanized Infantry Company. The sniper supports the commander by delivering long range precision fire on key targets and reporting battlefield information. A sniper team properly employed can disrupt enemy movement, observation, and infiltration, plus influence the enemy's decisions and actions. The sniper team can instill fear and confusion, thus lowering the enemy's morale.

Army snipers are normally deployed as a two man team because they usually operate independently for long periods of time. They must maintain continuous observation, assist in range and wind estimations, observe and adjust rounds, and provide security for each other.

During offensive operations, the scout-sniper team is usually employed one or two days prior to the main unit moving out. Information about enemy positions, routes used for attack, and security for the main force are provided by the team. During the attack, the Army snipers provide counter-sniper support, take out key enemy personnel, eliminate crew served weapons including anti-tank weapons, provide flank security, eliminate small isolated pockets of resistance, or prevent enemy reinforcements from joining the main enemy force. In addition, the Army sniper team can provide security or directly engage the enemy during raids and ambush missions. When operating independently, team should avoid situations that would involve an all out fire fight with a larger enemy force. In defensive roles, Army snipers provide observation, security, and counter-sniper missions.

Army snipers are taught how to assess and make their own decisions on engaging targets. When selecting key targets for engagement by precision fire there are six factors in the decision process and a method for determining priority of target. The following outline reveals how much is involved in order to fire just one round. This subject is taught in the Army's sniper schools and is very important because it can apply to almost any type of military sniping mission and can be modified for law enforcement purposes also.

CHAPTER 2: The US Army Scout-Sniper and the M24 Sniper Weapon System (SWS)

Scout-Snipers engage targets at 600 yards with the M24.

Selecting Key Targets

A. Decision Factors

1. Threat that the target presents to the sniper. The danger to you, the sniper, can be an immediate threat, such as an enemy unit walking toward your position or a future threat such as an enemy sniper or a dog tracking team that could find you.

2. Probability of first round hit.
   a. Distance to target determines chance of first round hit.
      90% @ 600m
      70% @ 800m
      60% @ 900m
   b. Direction and velocity of wind.
   c. Visibility in the target area.
   d. Amount of target that is exposed.
   e. Amount of time that target is exposed.
   f. Speed and direction of target movement.

3. Target's impact on enemy. The sniper must consider what impact the elimination of the target will have on the enemy emotionally, physically, and his fighting ability. The sniper must determine that the target is the one available target that will cause the greatest harm to the enemy. The acronym "CARVER" is used to assess this impact. A point value of 10 to each element if the factor is in the sniper's favor, 0 points if in favor of the enemy, and 5 points if could favor either way.

   Criticality. 10 points if essential to enemy that target functions.

   Accessibility. 10 points if target is easily accessible.

   Recuperability. 10 points if the target is difficult to replace.

   Vulnerability. 10 points if team has capability to reduce target.

   Effects. 10 points if target will have negative effect on enemy.

   Risk. 10 points if risk is in sniper team's favor.

4. Enemy's reaction to fire. The team must consider what the enemy will do once shot is fired. The team must be prepared for such actions as immediate suppression by indirect fires and enemy sweeps of the area. Use

SGT James Kringlie spots for SFC Tim Weber. Team represents the Minnesota National Guard in sniper matches held at Camp Robinson.

the acronym "CIAS" to analyze the enemy's possible reactions.

    Counter-sniper
    Indirect fire
    Assault Teams
    Search Teams

5. How will the engagement affect the overall mission. The mission may be one of intelligence gathering for a certain period of time and firing will not only alert the enemy to sniper's presence, but may also terminate the mission if the team has to move from the location because of engagement.

6. Certainty of the target's identity. The sniper must be certain that the target he engages is the right one.

B. Key Target Identification

Identified by actions or mannerisms, positions within formations, rank or insignia and/or equipment worn or carried. Targets can also include weapons and equipment. Examples of key targets in order of priority:

1. Snipers and scouts
2. Dog tracking teams
3. Officers (military or political)
4. NCO's
5. Vehicle commanders and drivers
6. Communications personnel
7. Weapons crews
8. Optics on vehicles and aircraft
9. Communications and radar equipment
10. Weapons systems

Army Sniper Marksmanship Fundamentals

Snipers are taught to use the best available support which is usually the prone supported position. Support must be artificial by way of bipod, sand bag, back pack, etc. SFC Gooch describes to author the basic prone firing position for the sniper:

"At the ARNG school we teach a good,

CHAPTER 2: The US Army Scout-Sniper and the M24 Sniper Weapon System (SWS)

The Army Scout-Snipers operate as a two man team on the range and in combat.

firm grip on the pistol grip with the whole hand grasping the stock (thumb included) with slight pressure to the rear. Non-firing hand provides support for the butt of the stock by a) grasping the rear sling swivel, b) forming a fist for the stock to rest on, c) holding a sand sock for the stock to rest on, or d) placed on the shoulder, forming a "V" for the stock to rest in (primary for positions other than the supported prone). The weapon is placed firmly in the pocket of the shoulder and the cheek is placed firmly on the stock (usually with a cheek pad duct taped on the stock) forming a solid stockweld. Elbows are firmly planted with shoulder muscles relaxed. While incorporating bone support, muscular relaxation and body alignment, the shooter lays in-line with the rifle in either the traditional prone with the belly flat on the deck, feet together, ankles flat or in a modified cocked leg position. The cocked leg is modified in that the strong side leg is not pulled up as far as it is normally for a competition shooter but is up enough to get the shooter's diaphragm and sternum off of the deck reducing pulse beat and movement from breathing when shooting under stress. This is a compromise dictated by the fact that a) the observer needs to be close to the shooter and b) it reduces the size of the target that the sniper presents to the enemy. Overall, the position should be firm, solid, and comfortable enough so the sniper can lay in it for long periods of time."

Sometimes getting a good position may be impossible during a stalk and you may be in a very unnatural position when you must take your first shot. Try anyway to get a good shooting position if it doesn't compromise your concealment to make the extra movement. In a practice stalk with blank ammunition this is not a problem. In the real thing, you could miss your target even at 150 yards. Time your aiming and the shot with your natural breathing pause. This pause will last from 2 to 3 seconds but you can extend this to 8 to 10 seconds. Getting all of the above correct won't get you a good shot if you jerk the trigger. The tendency is that when you see a lot

U.S. Army Scout-Sniper teams and the M24 SWS have proven their capability and accuracy on the battlefield and on the range. Having engaged the enemy in places such as Panama, Saudi Arabia, and Somalia, the Scout-Sniper Team and the M24 SWS is still the mainstay of the Army's precision individual weapons. The US Army had recently approved a contract for the Israeli government to purchase 890 M24 SWSs directly from Remington Arms. (Mike R. Lau)

of target around the center of the reticle you think you can get by with jerking the trigger and sometimes you do. This is especially so if you don't have a lot of time. You will develop the habit of jerking the trigger and continue to do this even on far targets which is why you miss them even with no wind and all other fundamentals are correct. With practice and experience the sniper develops his own methods of shooting, but the fundamentals remain basically the same.

Qualification Course at Camp Robinson, Arkansas

SFC Kent Gooch teaches a sniper qualification course that is modified from that which is taught at the Army Sniper School at Ft. Benning, GA. The following is summary of that course.

Actual instruction includes a day on the operation of the M24 and M3A, and the fundamentals of marksmanship and ballistics (internal, external, and terminal). The student spend half a day zeroing at the 100 yard line then confirm zeros at 300 yards. For two days the students will fire on the NRA 200 yard repair center at the 300, 400, 500, and 600 yard lines. We then move to the 600 yard NRA center and fire the 700, 800, 900, and 1000 yard lines. SFC Gooch explains why he uses the NRA bullseye target. "I started using the bullseye centers back in 1983 or 1984 when I was the Primary Marksmanship Instructor at the Marine school in Quantico. At the time we were using the FBI-TRC silhouette for all known distance (KD) shooting. I thought this allowed the students to become sloppy with their zeros. The other reason for going to the

NRA bullseye centers was that the student should learn precision shooting while on the KD ranges for training and then move to silhouettes on Unknown Distance (UKD) ranges for qualification. Scores went up on the KD range when the students went back to fire on silhouettes because their were now a lot tighter.

Now students qualify the same as at FT. Benning using the silhouette pistol target where you either hit or miss that target. SFC Gooch intends to change the course where qualification will now require either the bullseye target or the British Figure 11, 12, and 14 targets. The last are camouflaged figure with bullseye scoring rings. This gives the shooter an indistinct aiming area, but has scoring rings, demanding precision aiming for higher scores.

The version of US Army KD qualification course Camp Robinson consists of 3 stationary and 2 movers at the 300, 500, 600, and the 700 yard lines. Each first round hit is worth 10 points, while a second round hit scores as 5 points. A score of 160 points is required to move to the next qualification phase which is UKD.

UKD training firing consists of firing on the "F" (20"x20") and "E" (40"x20") silhouettes located between 250 to 800 meters. On a typical day, the student will fire on 2 different lanes consisting of two F targets between 250 - 400 meters and eight E targets at 400 - 800 meters. Eight of the ten targets will be between 250m - 600m, the "meat and potatoes distance". Students are allowed to fire on each target until he obtains good center shots and he will record that data in his score book for zeros at those distances.

For qualification, the lanes are changed, the two F and eight E targets are renumbered, and a time limit of 40 minutes is allowed to engage all 10 targets twice. The student begins with the closest target, works his way to the farthest, then his way back to the first target. This gives the student a total of 20 targets for qualification and two chances to hit each target. A first round hit is worth 10 points, while a second round hit is worth 5 points. 160 points is required to pass this phase of qualification.

What makes it interesting is that a student cannot graduate with the minimum of 320 points acquired on the KD and UKD ranges. The student needs 350 points. He either fires above the minimums on these ranges or he can acquire more points in the "Final Shot". All students participate in this exercise, but not all may need the extra points. A Ft. Benning animal, the Final Shot requires the sniper to work by himself. He will be observed by two instructors. The student will be given a single target, usually an F at around 500m or an E at around 700m. The student is given 10 minutes to obtain the range and engage target. He gets 100 points for 1st round hit and 50 points for 2nd round hit. The instructors will spot his 1st shot and if it's a miss, the student will be told where he missed. Ft. Benning used to require all the students to attend this exercise and hit this target in order to graduate. One of SFC Gooch's instructors had attended the Ft. Benning course and had only 2 students graduate because the target was 740m and there was a stiff wind. It no longer is a requirement for graduation to hit this target as long as the student gets 350 points.

Each student fires around 300 rounds during KD and 300 rounds at UKD. He will also fire a night fire using the M16A2 with AN/PVS-4 and M24 with M3A during low light conditions and also under M203 illumination.

The US Army Sniper Team

Delivering precision fire involves many tasks requiring numerous skills. Army snipers work as two man teams with overlapping tasks. Both team members are trained as firer and observer and who gets to do what will depend on their own skills and how they operate best. The observer positions himself to the side and slightly behind the shooter so that the trace and impact of bullet can be observed. Normally the observer will ID target, estimate the range, and assist the shooter in locating target. He then determines the wind speed and direction, mirage, temperature, and other field conditions that will affect first round hit. Elevation and windage calculations are made and conveyed to the shooter. After locating target, the firer gets into good firing position. He confirms info relayed by observer and adjusts the sights. He then tells observer he is "ready" or "up". Observer will tell sniper to "fire" if conditions have not changed or "stand by" if corrections are needed. The observer can change target on shooter if higher priority target noticed or adjust windage, elevation, or lead. After firing, the observer will announce "target hit" or observe impact of miss. If target was missed the team decides if another shot should be made and the observer will determine what adjustments are to be made from the noted impact. Army sniper teams are taught not to fire more than 3 rounds from the same position when they are operating independently.

Army snipers are taught the competitive shooting mindset of "mental management". This technique provides the sniper with the ability to focus on the task of firing highly accurate shots under conditions of stress commonly found in combat. The basic principle behind this technique is that the human mind is similar to a computer and that it can be programmed to perform certain tasks automatically (subconsciously) upon a certain que. The sniper uses visualization techniques to program his mind. When in training, the sniper establishes a cue thought or word to start a firing sequence. For example, after receiving the sight adjustments from the observer, the sniper will repeat the adjustments to the observer who says "roger". This is the cue for the shooter to start automatic sequence. He will tell himself to "relax" and begins to relax while carefully making his sight adjustments. He then gets his head down on to the rifle while checking his firing position. He adjusts parallax and then closes his eyes imagining the sight picture and how he is going to slowly press the trigger. He then opens his eyes rechecks his sight adjustments and tells the observer "up". The observer tells shooter to "fire". The shooter goes through the sequence again: checks for safety off, checks grip of firing hand, places cross hair on target, inhales, exhales, places finger on trigger, pauses, presses trigger. Repeating this procedure time after time will produce the conditioned reflex that will help the sniper to maintain his cool and perform under the stress of combat.

Carlos Hathcock once put it this way. "It never rains on the range. It never gets cold or hot. You aren't hungry or thirsty. You are never uncomfortable or lose you attention for any reason when you are in position. When you can mentally remove outside influence from you concentration, you are in your 'bubble'. When I was in Vietnam, the temperature would soar to 120 degrees. It would pour down rain or the wind would blow the stink of rotten vegetation into your face. But I'd get into my 'bubble' and wouldn't notice these things. My only thought was directed towards that one well aimed shot."

The Army sniper with the M24 is an elite fighting soldier. His personal and military skills, intelligence, and attitude are a cut above normal. The Army sniper is chosen for his maturity and judgement, and is a well trained soldier. He will be called upon to operate independently and he will have to make his own decisions using his own judgement of the situation. He must be able to communicate his ideas and thoughts to his leaders and maintain rapport with all in the unit. Snipers represent themselves in a positive manner and maintain their physical fitness and increase their skills because they want to. He is chosen because he *doesn't* have the "snipers eat their dead" or "kill them all and let God sort them out" mind set. His personal discipline tells him not to shoot even when a target presents itself because the commander's mission

CHAPTER 2: The US Army Scout-Sniper and the M24 Sniper Weapon System (SWS)

SFC Kent Gooch (left) with his favorite M24 SWS, "Baghdad Betty", was the top military sniper competitor in the Canadian Forces Small Arms Competition, Connaught, Ottawa, in 1996. Next to him is his friend, Andy Weber, owner of Armament Technology of Nova Scotia, and the overall sniper champion. Kent's partner was SFC Dwight Peck of the 29th ID in Virginia, another Quantico trained Army Guard sniper. (Kent Gooch)

would be jeopardized. The wearing of blood and guts tattoos, loud macho-ego actions, and the like, are not necessary because the Army sniper has a high self-esteem and knows he is good. He doesn't have to show off to impress someone else. Because of these qualities he earns the respect and admiration from all ranks who will seek him out for advice and ask him to help with providing leadership. The unit commander can rely on his well trained and mature Army sniper to operate independently and get the job done.

CHAPTER 3: TWO CALL OUTS, 10 DAYS. THE POLICE SNIPER

SGT Sam Chesnut, Oklahoma City Police Special Operations, with his Ruger Mini-14. (Sam Chesnut)

Sam Chesnut was a two year veteran with Special Operations of the Oklahoma City Police Department, when he received a call on his pager at 6:30 a.m. Responding to the page, the dispatcher told him that a report came in about an armed man taking a convenience store clerk hostage in southern Oklahoma City. As other officers blocked the streets and took up positions, Sam placed himself behind a dumpster, 35 yards from the store, and off to one side where he could still cover the front. He was armed with his .223 Ruger Mini-14 topped with a 4X Burris scope.

Armed with a shotgun, the drunk gunman began throwing beer cans out the front door. Showing himself several times, he kept the hostage at arms reach, but not close. Sam suspected the man wanted the police to shoot him. Suddenly, the suspect came out by himself with the shotgun and pointed it toward one of the other police units. Picking him up in his scope, Sam aimed for the suspect's head, but at that close range it was difficult to align the crosshairs of the 4X Burris on the bobbing head. Sam lowered his aim to the chest area of the man's white T-shirt and fired one round. The suspect fell. Two days later, he died of the gunshot wound to the right lung.

Sam was suspended from duty and his Mini-14 was taken to storage for ballistics lab tests. The 55 gr soft point bullet was never recovered since it passed completely through the suspect. Homicide detectives conducted an investigation on Sam, just like they would on any other person involved in a killing. A complete packet of findings was made up and sent to the District Attorney to be reviewed for any criminal action on Sam's part. On the 8th day after the shooting, Sam received a letter from the DA. The shooting was reviewed and found to fall within State statutes and Oklahoma City policy and procedure. On Saturday, Sam was back on active duty and his rifle was returned to him.

At 6:49 a.m., Sunday morning, Sam received another page. The dispatcher told him to report to the scene of a barricaded suspect in an apartment building. In the ground floor apartment, was a suspect who had mental problems and was angry because of domestic problems.

Having just come off the previous shooting, only days before, Sam was to let other officers handle the situation. However, due to a mix up at dispatch, only half of the Special Operations team responded. Sam positioned himself about 15 yards from the suspect's front door at the end of the breezeway between the two apartment buildings. Taking cover in a flower bed, he could see right down between the two apartment buildings and the suspect's door which opened into it. Another SpecOps officer had placed himself on the far end of the area between the buildings where he could observe the suspect's door from a 45 deg angle. An assault team had positioned themselves in the area around the apartment hallways in the building directly across the suspect's door.

The suspect broke windows and pulled down curtains. He came outside with a knife, and with his shirt off, began to slash his arms and chest. As he drew a circle on his chest with his finger, he shouted, "Shoot me here, but don't make it hurt." Sam tried to calm the suspect down, but he only cursed at the officers saying, "F*#* it, I'll make it easier for you." With that he went back inside and slammed the door shut. Sam attempted to call for negotiators but the situation could not wait. No sooner than he'd gone back inside, the suspect came back outside with a semi-auto .22 caliber rifle.

Because of the problems he had last week while trying to get the crosshairs on the previous suspect's head, Sam had removed the scope off his Ruger Mini-14. The current suspect now moved away from Sam and his partner, toward the officer on the far side of the causeway between the buildings. Sam took aim at the gunman's head watching to see what happens next. When the suspect raised his rifle at the officer, Sam fired three rounds quickly. All three of the bullets hit the suspect in the area around the ear. Initially there was doubt if all the bullets hit the suspect. Autopsy later proved that one of Sam's bullets went right through the suspect's ear canal.

Sam remembered to fire until the suspect goes down. The other officer that the suspect pointed the rifle at, also fired three times with an AR-15. Only one of his bullets found its mark, a chest hit. Oklahoma City police are taught to continue to fire until the suspect can no longer harm anyone and to aim for the instant incapacitating areas.

Sam was suspended from duty for 10 days this time. Homicide filled out the "blue sheets" once again, and the DA cleared him. When he visited the police psychologist, the

CHAPTER 3: Two Call Outs, 10 Days. The Police Sniper

LCPL Thomas, USMC, spotting for SGT Sam Chesnut. Sam's rifle is M40A1 built by Texas Brigade Armory. Scope is Leupold 3.5x10 tactical with USMC mil-dot reticle. (Mike R. Lau)

doctor asked him how he felt about killing a man. Sam's reply was: "I'm not proud of killing somebody, but I'm not ashamed of it either. I trained for it, so I was prepared for it. I hope it doesn't happen again."

In the Beginning

On that fateful day in 1963, President John F. Kennedy was assassinated by Lee Harvey Oswald with a rifle fired from a building window in Dallas, Texas. Neither the Dallas Police nor the Secret Service deployed counter snipers. In 1966, crazed gunman, Charles Whitman, climbed the tower of the University of Texas campus in Austin and began shooting at the people below. The Austin Police sent no counter snipers to the scene.

In 1973, black militant, Mark James Robert Essex, held police at bay for ten hours from inside a Howard Johnson's Motor Lodge in downtown New Orleans. Setting fires in rooms and using fire crackers for diversion, he caused numerous casualties to police, firemen, and motel residents, with a Ruger .44 Magnum Carbine. He was finally killed by Marines firing an M60 machine gun from an overhead helicopter. The following day, police continued to search and fire shots into rooms, believing that there were more persons involved, but Essex was by himself. Police snipers were not called out.

In these incidents, no trained police counter snipers were deployed because there were none to deploy. Throughout the turbulent years of war protest and racial violence in the 60's and 70's, many law enforcement agencies in the U.S. faced similar situations and had similar problems. They were not prepared. These agencies not only dealt with crazed gunmen, but also terrorism by political and religious fanatics. The number of terrorist incidents increased dramatically going into the

70's and early 80's, and were becoming more violent, resulting in increasing numbers of casualties among bystanders and hostages. Law enforcement agencies were forced to arm and train special units to meet these threats and in their infancy, these special units were sometimes not effective.

During the 1972 Olympic Games in Munich, Germany, the Palestinian terrorist organization, known as Black September, took over the Israeli athlete's living quarters in Olympic Village. Three Israeli's were killed outright and nine more, who could not escape, were held hostage. The Palestinians demanded the release of fellow Arabs who were being held as prisoners in Israel. Negotiating with the terrorists, the German government flew the gunmen and the hostages by helicopter to a NATO airbase outside Munich. The terrorists were told they would be flown to Cairo by passenger jet when they got there. At the air base, waiting West German government snipers were told to kill the terrorists. When the two helicopters arrived, the snipers opened fire. Several snipers aimed at the same target so only three of the eight terrorists were hit when the first rounds were fired. A strange battle had erupted. One of the terrorists exploded a hand grenade in one of the helicopters, killing all of the persons aboard. The other helicopter was riddled with bullet holes. When it was over, the nine Israelis were dead, along with five terrorists, and one German policeman.

Many other incidents of terrorism with greater loss of lives happened in the past, but the Munich incident was a turning point in a new era of modern terrorism. In the US, the Munich disaster was brought into the home of many Americans by the media that were already in Germany, covering the Olympics. It literally shook up the sense of security of the US citizens and that made the attack an even greater success for Black September. As terrorists acts increased in number, both in the US and to American citizens abroad, the Government began to look harder at different ways

*"Terrorists!" From the video, **USMC Scout Sniper**, by Video Free America. Author's good friend, Don Parish, is on right with AK-47. (Mike R. Lau)*

SSG Barry Owens (left) and SSG Leslie Dolan (right), Arkansas National Guard, participate in the 1997 Wilson Sniper Matches at Camp Robinson, Arkansas. The 188th Tactical Fighter Wing out of Ft. Smith, Arkansas, has 20 M24 SWSs for air base security and for training other air base sniper teams. Dolan is a former USMC sniper who served with the 1/3 Marines in Desert Storm. (Mike R. Lau)

to counter the threat. By the mid 80's people were actually afraid to travel overseas and were encouraged not to by the Government.

The War on Terrorism Today

One of the actions by our government to counter terrorists attacks, was the creation of Delta Force (Special Forces Operational De-

CHAPTER 3: Two Call Outs, 10 Days. The Police Sniper

tachment - D or SFOD-D). Based at Ft. Bragg, NC, Delta Force originally included units from Army and Air Force special operations units, Navy SEALs, Marine recon units, units from the 101st Airborne, and was supported by the 23rd Air Force. Standard Operating Procedures (SOP) for the hard core, no-holds-barred Delta Force is Top Secret. Prior to the 1996 Olympic Games in Atlanta, Georgia, the old baseball stadium in Arlington, Texas, became a training ground for Delta Force. Since the stadium was actually going to be leveled for new construction, Delta Force was allowed to practice explosives entry and counter sniper tactics with live munitions inside the stadium.

Working closely with Delta Force is the FBI HRT or Hostage Rescue Team. FBI snipers have .308 Winchester sniper rifles with Marine Corps Unertl scopes. FBI HRT also has access to Delta Force's state-of-the art, live fire training facility, known as Range 19, at Ft. Bragg, NC. About the time of the incident with the Branch Davidians in Waco, Texas, the HRT consisted of 50 personnel. After the incident, the FBI increased the number of agents to create additional HRTs of the same size. The actual number of personnel is classified, but it is generally known that six designated geographical regions in the US, known as "war zones," were each to have its own full strength HRT. In 1995, an airstrip was constructed at Quantico for the FBI Academy. Its purpose was for greater rapid response to situations using Army aircraft from Special Operations Aviation Regiment 160, or aka Task Force 160, a covert operations unit. Persons recruited for the FBI HRT are usually young military types. They are sent directly to the HRT from the Academy without regular street agent work according to an unnamed Academy source. For a law enforcement agency, criminals are only suspects, and every attempt is made to negotiate and arrest them. It is up to the court to decide if the suspects are criminals and what punishment they should receive.

The Secret Service, and other government agencies were also gearing up to combat terrorism attacks in the U.S. from both international and domestic organizations. A recent example is a contract for fifty L.O.D. sniper rifles by the Department of Energy. The Air Force has also recently added the M24 SWS and sniper training to all air base security forces in the continental U.S. Many police departments, even small town agencies that only have a few full time officers, are also equipping and training for both terrorist tactics and heavily armed barricaded suspects. However, many police departments don't want to have the appearance of being heavily armed and training for war in the eye of the public.

Police Special Operations Today

During the 1997 shoot-out involving two gunmen in the North Hollywood area, the Los Angeles police officers found themselves without rifles or shotguns. The gunmen were heavily armed with full auto AK-47s. The officers found themselves outgunned because none of them had rifles in their car trunks and it was department policy for patrol officers to not carry shotguns because they looked too intimidating. It was good fortune and bravery of the officers that finally brought the gunmen down. One of the officers who took cover behind his car fired at the feet of one the gun-

Highland Park, Dallas, Texas, SWAT snipers practice long range shooting at Ft. Wolters. (Sam Lepere)

Billy Martin explains the fine points of his L.O.D. rifle to Brig. General Rainville, Adjutant General of the State of Vermont. (Mike R. Lau)

men from under the cars with an M16 rifle obtained later. There are still some police forces throughout the US that suffer this syndrome and only time, and unfortunately casualties, will cause them to change.

Today, many major police departments, including the Los Angeles PD, have special units to handle situations calling for highly skilled and heavily armed police officers. These units are known as SWAT (Special Weapons and Tactics), TAC Team (Tactical), SpecOps (Special Operations), HRT (Hostage Rescue Team), SRT (Special Response Team), ERT (Emergency Response Team), and a host of other politically correct names.

Special Operations units are usually divided into 4 sections with each having specific duties and skills requiring specialized training. Overall control and responsibility of the entire unit rests with the command element. As few as one individual may make up this section who supervises deployment, timing, and final actions taken by all of the team members. The second element is the negotiations section. This is the group that we all hope can resolve the situation without violence. Members are trained negotiators that are sometimes assisted by psychologists and psychiatrists to assist in the negotiations.

A third element comprises the sniper/observers. This unit has several functions besides providing precision shooting. Special training requires the police sniper to be the SWAT team's real-time intelligence. Officers in this unit are trained to use "field craft" or stealth, cover, and concealment, in selecting a position to both observe the situation and support the other teams with precision fire when needed. Movement may be nothing more than entering the back of a building and finding a location inside. On the other hand, it may require low crawling several hundred yards in a ghillie suite to traverse open ground to get to a good observation position.

CHAPTER 3: Two Call Outs, 10 Days. The Police Sniper 57

Federal Government Joint Task Force Training allows for military and law enforcement to combine training for drug enforcement and interdiction purposes. Marine Corps Scout/Snipers of 1/23 and the Dallas Police Swat team leaders share long range shooting skills at Ft. Wolters, Texas. Front row, kneeling, from left to right: Sr. Cpl. Tony Black (DPD), author, Sr. Cpl. Sam Lepere (DPD), Sgt. Bob Newton (DPD) Standing from left to right : LCPL Bennett Thomas, CPL James Crawford, James Gannon (range armorer and Palma shooter), CPL Donny Dishau, CPL Tim Hengst, and CPL Sean Little. (Mike R. Lau)

The fourth element is the assault section. When the sniper/observers and negotiators cannot resolve the situation or quick action is required for some reason, the assault section goes to work. With specialized weapons, equipment, and training, this unit becomes the "final option" when all else fails. Sometimes SWAT team members are cross-trained in other section's skills and become very valuable assets to their agencies.

When a critical situation occurs, the first step police officers attempts to do is to contain the situation so that others don't become involved and negotiations can happen.

A police sniper/observer team can sometimes handle this first contact with the suspect or suspects in the "inner perimeter," but that team is usually no longer hidden from the suspects. However, sometimes just the presence of a sniper team made known to a criminal may cause him to surrender.

As patrol officers arrive, they will form an "outer perimeter" to help contain the situation by keeping bystanders from entering the area or assist them out of the area. As the commander and negotiators work with the felons, they also buy time for the assault team to plan an assault if needed. The entire team must be very careful to assess the felons as they may be desperate, heavily armed, and well trained themselves. Any assault action by SWAT could end up disastrous if not planned carefully.

Small agencies that do not have SWAT or specialized tactical units, usually still have snipers to contain and observe hostile situations until special units can respond. Sometimes, precision fire from the police sniper is enough to end a deadly situation. In the past, old tactics of assaulting a barricaded position was costly in lives of hostages, officers, and criminals. As today's criminal becomes more deadly and violent, even the courts are getting more lenient toward law enforcement snipers as they see it as a quick and less destructive way of ending a situation.

A Mid-Western City Special Operations Sniper Unit

Today, Sgt. Sam Chesnut has been an Oklahoma City Special Operations sniper for six years with a total of 15 years in Special Operations. Besides being a sniper, Sam is also a firearms instructor for the City and the Oklahoma County Sheriff's Department. In 1996, Sam was one of the Top Gun (non-military) graduates of the National Guard Scout-Sniper School at Camp Robinson, Little Rock, Arkansas. He assists in the training of law enforcement snipers with SFC Kent Gooch at the Camp Robinson Scout-Sniper school.

Sam assists in the training of 6 police snipers that range in age from 25 to 32. Two are new members and going through training. The others have at least 3 years experience as snipers. To become a tactical sniper with the

Some of Sam Chesnut's equipment is the night observation device (NOD), 550 parachute cord, M22 Steiner binoculars, and Sam's data/log book. Bottom is Federal 168 gr. Gold Medal Match ammunition, two way radio, and spotting scope. Many police officers purchase their own special equipment so don't expect to find high dollar, state-of-the art fancy equipment. The NOD is a Russian import that Sam bought from a Sam's Wholesale store. All of this equipment is more than adequate for the task at hand. (Sam Chesnut)

CHAPTER 3: Two Call Outs, 10 Days. The Police Sniper

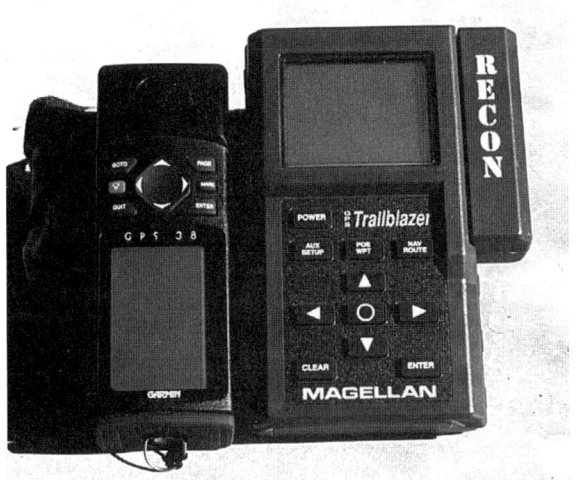

*Grid Positioning System (GPS) devices are becoming more popular as prices come down and units become more available. The one on the left is the Garmin that Sam Chesnut bought at a Walmart for $130. The other is the Magellan Recon which has been discontinued. Sam bought this one out of a **U.S. Cavalry** catalog and paid around $350. According to Sam, the Garmin has led OK City PD to the seizure and destruction of millions of dollars worth of illegal narcotics. GPS devices are fairly accurate and will give you a universal map coordinate based on your position as sensed by at least three orbiting satellites. Sensors in the devices will actually indicate to you how strong a signal you are receiving so you can move about until you get a stronger signal or read more than three satellites. The government uses grid positioning in cruise missiles and say that the positioning is slightly off and is supposed to be corrected in 1998. However, Sam says that the positioning coordinates are very accurate to within plus or minus a couple of meters. Sam also says, however, that there is no substitute for basic land navigation skills and common sense when traversing terrain. (Sam Chesnut)*

Oklahoma City PD, the officer must have at least 3 years duty with the Department and must not be on disciplinary probation. To qualify as one of the team's snipers, the officer must pass a physical and firearms test which leads him or her to go before a board. If the candidate is selected, he or she will attend a 1 week basic course. It may sound easy, but competition for these slots are really tough.

City owned rifles are the Remington 700 Police in .308 Win with Leupold 3.5x10 Tactical scopes. Ammunition used is the Federal 168gr HPBT Gold Medal Match. Like many police departments across the country, the OK City Special Ops sniper can use his own personal rifle. Use of personal weapons usually have to be approved by the Police Chief and the Firearms Instructor. One of the officers uses a custom M70 Winchester with McMillan fiberglass stock and Leupold 4.5x14x50mm scope. Sam Chesnut still has his old Ruger Mini-14 for those special situations calling for it. His choice for a main sniper weapon now is a Texas Brigade Armory M40A1 in .308 Win caliber. His preferred telescope is the Leupold 3.5x10x40mm. Sam also has two other TBA rifles which he uses for duty and competition.

Oklahoma City covers about 650 square miles and its population is only about 1 million including the surrounding enclaves. This population density makes it about 50% rural. In addition to the city limits, the OK City's SWAT team of 20 members and 4 supervisors assists all of the other agencies in Central Oklahoma including the Oklahoma Highway Patrol. Oklahoma is divided down the middle by Interstate 35. One of two OHP TAC teams is responsible for each half of the state. When a town or county outside of OK City requests SWAT support, the OHP is the first response. Due to the distant locality and consequent delay in response by the OHP Tac Teams, the call for help may be turned over to the OK City SpecOps. For this reason, Sam's sniper teams are prepared for both city and rural call outs. Sniper deployment in rural areas can increase target ranges to over 100 yards and Sam has been in situations that actually required the use of ghillie suits. Prior to one drug raid in the inner city, Sam and his partner actually put on their ghillie suits and staked out a drug house that was surrounded by wooded areas. One of the druggies got within 3 feet of the two law officers and never saw them. In another incident, Sam and his partner responded to a complaint about a man walking down the road in a rural area firing his rifle. The nearest house to the suspect's residence was 300 yards away so there was plenty of open ground and some cover and concealment. As they arrive separately, each puts on their ghillie suits. The two officers deployed separately, about 45 degrees and about 125 yards from the front of the house, to observe the suspect until the negotiators and assault team arrived. The suspect was a mental patient and had been drinking too much. Sam had been out here before, when the suspect was drinking too much and had caused a disturbance. This time he was shooting his chickens behind the house. While Sam and the other police sniper kept tabs on the suspect, the assault team arrived and deployed around the chicken coop, a modular building, and a garage behind the main house. The suspect pulled down curtains and broke windows before he set fire to the house which forced him outside. With an armful of firearms and ammunition, he moved to some nearby railroad timber. The two police snipers watched as the man loaded his rifles. About 30 minutes later, he fell asleep behind the railroad tie. Sam confirmed this to the assault team that finally moved in on the suspect's position. This time, luckily, there was no shooting by any of the police officers.

The Need for Long Range Sniper Capability

Oklahoma City has about 11 barricaded armed suspect incidents a year. The city's police snipers normally back up assault and negotiating teams in these situations. The majority of the incidents involving Special Operations are drug raids. An assault or entry team usually does the forced entry and arrests, while the police snipers usually provide cover for the entry team. Some departments have entry team members wear infra-red marking devices at night which allows the sniper with an IR device to identify the good guys. Oklahoma City PD, like many other cities, have police snipers on regular patrol so that one is available to go to any site soon after the call out.

Andres Escobar is founder of International Body Guard Caribbean, Puerto Rico. IBA is headquartered in England and is one of the largest private bodyguard agencies in the world. IBA was originally organized by Major Lucien Ott in 1956, who formed an elite personal protection unit made up of ex-French special forces veterans, called "Les Gorilles" (the gorillas), to provide personal protection for General De Gaulle. Today, unlike many other private protection agencies whose members have minimal training, IBA personnel receive from 100 to 500 hours of specialized training in bodyguard and anti-terrorist tactics. IBA has modified and improved many of the VIP protection methods used by the U.S. Secret Service. Among their clients, IBA has trained KGB in the former USSR and the bodyguards for the leaders of the Afghanistan Islamic government under the Mujihadeen. Many IBA members have been assigned to personally protect well known royalty and diplomatic VIPs. Andres Escobar uses his Texas Brigade Armory .300 Win Mag M40A2 rifle and Leupold Mark 4 M1 in 16X for training and counter-sniper security work. (Andres Escobar)

OK City PD also handles airport security so there is the possibility of a long, 700–800 yard shot. There are numerous drainage ditches and tunnels surrounding the airport which would allow cover and concealment for a gunman. For this reason, OK City snipers practice long range shooting and have a 600 yard facility. Large airports like the Dallas-Fort Worth airport, has its own separate police force. They are backed up by the Arlington City PD which is located south of the DFW airport area and both police forces combine rifle training together. DFW police have a 300 yard range and can also practice at Fort Wolters, Texas, on the 1000 yd KD range.

Dignitary Protection

When President Clinton visited the Oklahoma City Bombing Memorial in April 1996, the Secret Service requested Oklahoma City to provide four counter-sniper teams to assist their own four sniper teams. Sam said this was an unusual request, because the Secret Service did not normally ask a city for counter-sniper support and the Secret Service prefers to handle this particular responsibility themselves during a Presidential tour.

Several weeks prior to the visit, Secret Service agents came in and surveyed the area to plan security. They are very thorough in their planning and don't miss a thing. Sam had not seen their rifles during the briefings, but had seen them several years ago when he had trained with them. Rifles were 7mm Remington Magnums with heavy barrels and thumbhole stocks with adjustable buttplates. Federal Cartridge Company ammunition was used at the time. Secret Service rifles and ammunition used for a security mission are thoroughly tested before coming out to the location. Cartridges are randomly selected from 500 round case lots and are tested in a lab for ballistics and accuracy by actual firing through the selected rifles. Average local weather conditions and expected weather are obtained from the city where the Presidential tour is to take place. A computer printout is generated for each rifle with these conditions and the resultant effects on each rifle's trajectory and other ballistics. A copy of the printout is given to each Secret Service sniper team using the particular rifle and specific case lot ammunition.

All buildings along the tour route are inspected and users and owners are advised to keep all windows closed. Secret Service and police sniper teams were employed along the roof tops and are not to show their weapons during the tour unless it has to be used. Teams are assigned specific areas of responsibility and prepare range cards. Sam and his partner, Lt. Gonshor, were assigned a position from on top of the Federal Courthouse building which is located behind the memorial site where the Murrah Building once stood. The crowd attending the President's ceremonial speech was located across 5th Street behind the old abandoned Journal Record Building. Sam's team was assigned observation and engagement responsibilities which included the old abandoned Record Journal Building at a distance of 237 yards, the YMCA building at 196 yards, an apartment building at 480 yards, and a church at 200 yards. To obtain the yardages, Sam used his Bushnell Laser Range Finder. It took only several minutes to sight on the buildings, get the readings, and give the distances to Lt. Gonshor who finished the range card.

Sniper teams were given instructions that if a real threat was observed, they are to "just put a round though him." All buildings at the site were secured by Secret Service agents at the door a day prior. During the tour, if a window on a building is opened, the observing sniper team will report the location of the building and window to a Secret Service Counter Assault Team (CAT) or to a police Cover Assault Team (CAT) on the ground. A prearranged system labeling each building and window is used to prevent screw-ups in finding the opened window by the snipers and CAT. During the President's speech at the memorial, one onlooker sat on the edge of a rooftop with his legs dangling to get a better view of the President. The Secret Service quickly brought the man back inside the building. Providing "high cover" protection for dignitaries by local law enforcement has become a proven strategy. The President's tour went off well and there were no other incidents.

Police Counter-Sniping

Sam said the old Journal Record Building would have been a very likely position for a criminal sniper to use. Being abandoned, it

62 *The MILITARY and POLICE SNIPER*

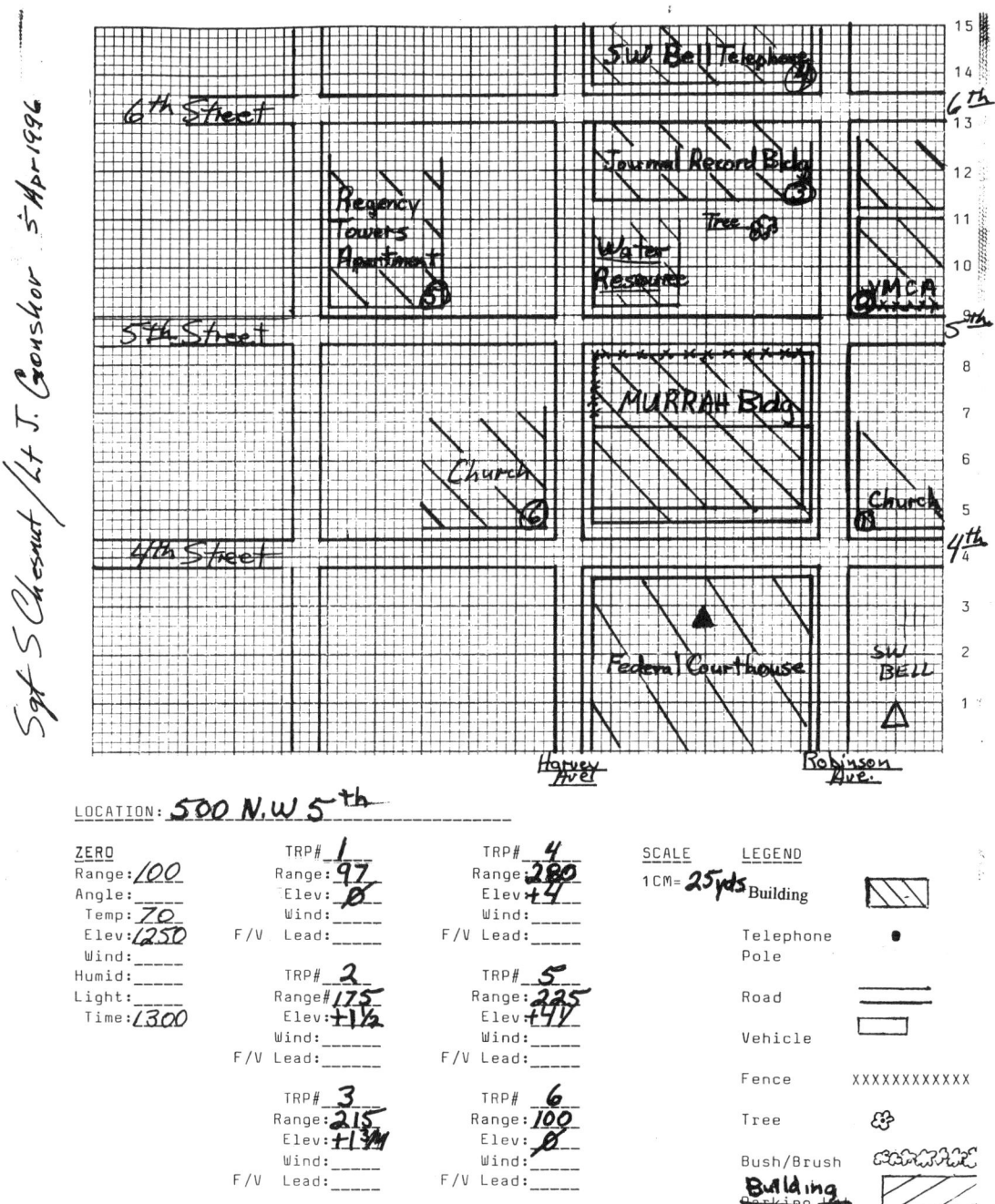

(Fig 3.1) This is the actual range card created by Sam Chesnut and his partner, Lt. Gonshor, during President Clinton's visit to Oklahoma City in April 1996. Introduced to Sam by the Marine Corps, the centimeter grid graph paper allows for a scale drawing to be made of the target/building locations and distances. It also allowed the sniper team to move to a new position and use the same range card by measuring the new distances to the buildings with a centimeter ruler. Horizontal ground distances are given so correction for high angle firing from on top of a tall building to the ground is corrected with the range card. (See chapter on angle firing). The Water Resource Building was also destroyed during the blast, killing several persons in the area across the street from the Murrah Building. The serial numbered axle of the Ryder truck, used to transport the explosives by Tim McVeigh, was found in front of the Regency Tower apartments. The force of the blast blew out the windows of the churches on both sides. Bricks were pulled from the side of the Journal Record Building as the energy wave from the explosion was channeled between the buildings along Robinson Avenue. (Sam Chesnut)

CHAPTER 3: Two Call Outs, 10 Days. The Police Sniper

Overlooking the memorial site is the Southwestern Bell Telephone building where Sam and Lt. Gonshor's counter-sniper team set up observation for Vice President Al Gore's visit to the site on April 19th. The building is located in the lower right hand corner of the range card and the counter-sniper's position noted by the open triangle is 225 yards to the center of the memorial site fence. (Sam Chesnut)

President Clinton visited the memorial on April 5, 1996. Sam and Lt. Gonshor occupied a position on top of the Federal Courthouse located in the block behind the destroyed Murrah Building. Sam asked that the American flag, seen above the "Angels Among Us", be lowered to half mast because it actually blocked his view to the memorial area. Lt. Gonshor determined that the flag was supposed to have been flown at half mast during that time anyway so he lowered the flag. (Sam Chesnut)

had numerous open and broken windows and was unsecured until a day prior to the President's tour. Someone could have easily slipped in a few days prior to any dignitary visit and hid in a ceiling with a rifle and gone undetected by any inspection. The assailant could sight from his perch in the ceiling down through a window opening overlooking the memorial sight without exposing himself in the window. This would have made it impossible for any counter-sniper team on the outside to have detected the assailant's firing position.

Criminal snipers and terrorists are very ingenious in their methods and the military and police sniper/observer has to be very wary of these techniques. A window in a building that appears closed to an observer from a distance, can be completely open and not detected by making it look closed. One way is to completely remove the glass so that the pane is still in its closed position. The appearance of a curtain opened and a darkened room with no light on will make it look closed. A black cloth draped from the top of the window to the criminal sniper's shooting table in the back of the room is also very deceptive. The criminal sniper will try to match the appearance of the surrounding windows so that his window's appearance looks like the others.

A clever sniping technique was used by the IRA on British infantry in the streets of Northern Ireland. The sniper would position himself in the attic of a house. A hole, large enough to sight and shoot through, would be cut in the plywood sheets or slats that were supporting the composition or wood shingles. A shingle would be loosened so that it could be moved up or slid up to allow the sniper to sight and shoot. So as not to appear obvious, the shingle was left down to cover the hole

The counter-sniper team's nightmare! Eerie and ghostly looking is the long abandoned Journal Record Building located across the street from the Murrah Building. Where you are standing to get this view is "ground zero" where the Ryder truck filled with explosives was parked. At 9.02 AM. on April 19, 1995, several persons were killed in this parking area while sitting in their cars behind the Journal Record Building and where the Water Resource Building once stood to the left (see range card). President Clinton addressed the crowd in this area behind the JR Building during his visit. There are more than 80 openings in the building that could have provided cover to an assailant to engage a dignitary visiting the site. The tree in the parking lot is known as the "Survivor Tree". (Sam Chesnut)

until the British soldier was to be fired upon. This usually required another person in an outside position who could signal to the sniper when the exact moment was to occur.

With today's efficient long range weapons, it would be easy for the criminal sniper to position himself much further than what we normally would see as a likely sniper's position. Lee Harvey Oswald used an Italian 6.5 Mannlicher-Carcano with a telescope to kill President Kennedy. Not efficient nor accurate when compared to today's standard sniper weapon. Oswald fired from a window of the School Book Depository building overlooking the street used for the tour. Because the Dallas Police and Secret Service at the time had no counter-sniper teams, Oswald went undetected from his position only a short distance away. Modern sniper rifles with synthetic stocks, heavy barrels, and precision telescopes, can accurately shoot beyond 600 yards.

Sr. Cpl. Sam Lepere, Dallas Police Dept., and Lance Cpl. Thomas discuss long range shooting during USMC/Law Enforcement Joint Task Force training at Ft. Wolters, Texas. (Mike R. Lau)

A criminal sniper can position himself two or three blocks away and shoot between buildings. The "crack" of the bullet passing a nearby building would sound like the shot was fired from that building. If the criminal sniper chose the correct position, he would not have to have an accomplice notify him when the target is nearing.

No more "Green Light"

In the past, police snipers involved with barricaded suspects, with or without hostages, used to have to wait for their commander to give the go ahead or "Green Light" to fire on the suspect. Today, almost all law enforcement agencies have done away with this practice or use modified versions. Once the sniper is in position and has placed the suspect in his sights, he can fire at will if he sees it necessary. The reason for the discontinuance of the "Green Light" practice was that it did not give the police sharpshooter the same right as a patrol officer in the use of deadly force. Patrol officers have always had the right to use their firearm to shoot at a suspect anytime the officer felt his life or that of another was in danger of receiving serious bodily injury or of being killed.

Training and Selection of the Special Operations Sniper

Training for the law enforcement sniper usually begins with the 40 hour basic course followed up with advanced courses to hone skills. Training goals for departments vary, but there are realistic standards which most schools incorporate.

(1) Place a single cold bore rifle shot into a half inch aiming area on a face target at 100 yards in 15 seconds. Begin in the prone position with the chamber empty, magazine loaded, and bipod folded up.

(2) Place a single rifle shot into a one inch target area from 200 yards within 20 seconds from the prone position using the same beginning weapon condition.

(3) Fire five shots into the correct target area of a face target (more on this in the chapter on terminal ballistics) from 100 yards with larger time limit. Five shot group should not exceed 1 inch and be within the correct target area on the face.

(4) Fire five shots at 200 yards at the face or body target within the correct target area. Group size should not exceed 2 inches extreme spread during a reasonable time limit.

(5) Locate and setup an observation and firing location in both an urban and field area without being seen by the target at anytime. Distance to the target will depend on the specific environment. Use of the ghillie suit is recommended for one of the situations. Maintain observation from that position for 2 hours and report information back to the commander.

(6) Advanced courses should stress longer range shooting out to 400 to 500 yards and include range estimation, wind reading, and sight adjustments.

Most SWAT teams conduct tactical exercises with their personal equipment. They know what equipment works and what doesn't when in a shooting position and how to adjust for it. Besides body armor, headset, and equipment vest, they also practice shooting with the gas mask worn. Besides using pepper gas on suspects and terrorists, LE officers are vigilant to terrorists and criminal suspects using chemical and bio agents against themselves and civilians as well. Military personnel in Saudi Arabia during Desert Storm wore biological/gas masks for long periods of time while performing normal duties. Like military personnel, LE snipers should occasionally be required to conduct range shooting with his or her protective mask worn.

The police sniper usually has an affinity for precision shooting and weapons of greater precision. He or she is usually selected because of excellent marksmanship abilities. Snipers are in excellent physical condition and many refrain from smoking cigarettes or drinking alcohol. The sniper is selected because he or she displays control of temper and has the intelligence and ability to make good judgments during highly stressful situations. They must have a strong desire toward perfection and display utmost patience. The LE sniper may be asked to maintain a field position for hours and may not be relieved during an entire day's incident.

A department that has shooter/weapons systems capable of higher accuracy can increase the accuracy requirements of training to ensure both are maintained at peak performance. Maintenance and increasing of

Here is some other basic equipment Sam takes into an observation position. From top left to right: Poncho, canteen, food, rope sling and snap links. Next row: Plastic bags, flashlight, pencils, paper, compass and map. On bottom row: calculator, first aid kit, Multi-plier/Leatherman tool, and Bushnell laser range finder. Not shown is a small pruning shears which comes in handy if you have to cut vines and thin gauge wire. (Sam Chesnut)

skills and knowledge is very important for the LE sniper. Departments are encouraged to send their snipers to outside schools and sniper matches frequently. This enables the sniper to experience new situations and learn other tactics and shooting skills.

The USMC Scout/Sniper school out of Quantico has put together a very extensive training course for law enforcement that is continuously updated and revised with new information and ideas as they are learned or evolve. One of the activities taught at the school is the use of the M40A1 in *.22 Long Rifle* as a 100 yard training tool. The rifle is described in the lesson plan as a Remington 40X with scope to simulate the firing of the M40A1 except without the power and recoil. There is a complete firing exercise with the 100 yard target details and rules of engage-

Until a few years ago, the Dallas PD SWAT snipers used wood stocked Remington 700 BDL Varmint Specials. The Department now has 12 synthetic stocked Varmint Specials in .308 caliber. Texas Brigade Armory installed 3.5 x 10 Leupold Tactical scopes and adjusted scope mounts on these rifles for 1000 yard shooting. (Mike R. Lau)

CHAPTER 3: Two Call Outs, 10 Days. The Police Sniper

The AR-15/M16A1 and Ruger Mini-14. Long time police standard rifles for close-in sniper/counter sniper support of assault teams. 5.56mm cartridge from these weapons are more effective than the 9mm cartridge fired from MP-5 sub-machine guns for both precision shots and effective terminal ballistics. The AR-15 has a 3.5x35mm Trijicon ACOG Tritium red dot sight. The Ruger has a Weaver 2.5X telescope. Both are excellent close range telescopes. (Mike R. Lau)

ment. Law enforcement SWAT agencies that do not seek training from the Marine Corps are missing out on some very detailed and valuable training as the Marine Scout/Sniper school goes to great lengths to learn, investigate, create, and distribute all of the latest information on LE sniping from many different sources. The Marines make it a point to try to be more informed about LE sniping than any single LE agency. The Dallas PD Tactical Unit is one of those agencies that takes advantage of the knowledge provided by the Marines.

Tactical Rifle Considerations for the Police Sniper

The Close Distance Sniper Rifle

Today's sophisticated criminal gunman and organized drug dealer, can pack some pretty awesome firepower. They carry not only high capacity semi-auto pistols, but shotguns, and full auto assault rifles and/or sub-machine guns. For most patrol officers that just happen into such situations, such as in the North Hollywood, California shoot-out, all the officer can do is take cover, try to contain the situation, and call SWAT. A .308 Win sniper rifle with a fixed 10 power scope would put you at a disadvantage if the suspect is barricaded in his home and the only good cover position is behind a dumpster 20 yards across the street. At the same time, if you had to shoot a terrorist highjacker in a plane's doorway from 650 yards out, your probability of hitting him with a submachine gun or short barreled AR-15 will be pretty slim. This is why most SWAT members carry an array of weapons to the scene.

For close in shooting, a very common choice is the semi-auto Colt AR-15 or full auto M16A1/A2. These rifles shoot the common variety 5.56/.223 cartridge and can also give the officer firepower for suppressive fire if needed. The Ruger Mini-14 in .223 is popular if it can shoot inside 1" at 50 yards. If a scope is used, it should be of low power, like 2.5X or 3X and have wide posts or cross hairs for use in dim light or at night. **Because of the high sight height on the AR-15, officers should practice shooting on the face target**

For all around LE sniping/countersniping the police standard is the Remington 700 Police (shown) and the similar Remington Varmint Special (VS). In .308 Win caliber, both rifles have a medium weight barrel known as the 5.5 contour with a muzzle diameter at around .850". The police model comes with the synthetic H-S Precision stock and the Varmint Specials can be purchased with either the synthetic or laminated wood stocks. Remington has just introduced the 700P in the .300 Winchester Magnum caliber in late 1997. Several years ago, both the Police and Varmint rifle could only be had in 24" barrels. Remington realized the demand for better ballistics and the synthetic VS rifles, 700Ps, and the new Senderos have 26" barrels as standard. The police rifles are very accurate and many owners report their rifles will shoot inside 1/2 moa at 100 yards right out of the box. However, trigger pull on the Remington 700P is a hefty 7-8 lb. and needs adjusting. Luckily, the Remington 700's factory trigger is one of the best factory designs and is designed to be easily adjustable. 700Ps also come in a detachable clip model. The scope is a Leupold 6.5x20x50mm, with target knobs. (Mike R. Lau)

between 15 to 50 yards. There are other good close quarter rifles that can be used, but they all need to be accurate, reliable, and have quality adjustable iron sights in addition to the scope.

The Long Range Arm of the Law

For longer distances and for all around use, the police sniper has a really wonderful array of weapons to choose from. Considerations for police agencies purchasing sniper rifles should be (not necessarily in this order):

(1) The rifle must be very accurate to make precision shots even under 100 yards. Preferably 1/2 MOA or better. For police work it doesn't matter if the rifle is a bolt action or semi-auto, as long as it gets you the desired accuracy. The need for this kind of precision will be explained fully in the chapters on terminal and external ballistics.

(2) The rifle and scope must be able to maintain zero during weather changes so that the only correcting factor is temperature (more on this in later chapters). It must also be able to withstand the even higher temperatures of the car's trunk during the summer without losing its zero. Synthetic stocks are preferred for this reason. Well sealed, epoxy finished wood stocks have already been proven not to work as well.

(3) The rifle and sighting system must be fully reliable and made of quality components. Misfires, jamming, or parts breakage, are not acceptable in a tactical shoot-out. Police rifles see a lot of use because many departments practice with it a lot. It does not have to be "Marine proof" but does have to withstand everyday handling because rifles are taken in and out of the car trunk frequently, if not daily. The system will also have to withstand a lot of vibration and jostling when carried in the

police cruiser every day and should be stored in a quality padded gun case.

(4) The rifle should be chambered for a standard cartridge such as the .308 Winchester. Ammunition should effectively handle most police type targets with consideration toward reducing risk of ricochet, penetration through walls, and bullet fragmentation to bystanders or hostages. Other ammunition requirements should be carefully considered based on accuracy, cost, and desired terminal ballistics results. LE ammunition is discussed in following chapters.

(5) The rifle should not be excessively heavy and should be easy to carry and comfortable to shoot in unsupported positions. It should be heavy enough to provide a stable firing platform by reducing the transfer of some of your own body movements, that caused by the wind, and other external vibrational sources. Barrels should be at least medium weight to reduce heating affects and inconsistent barrel vibrations which result in less precision.

(6) The rifle and telescope sight must cost within the agencies budget, but systems that meet all the requirements are usually expensive and could run over $2000. The agency or department should demand to get the best weapons system possible. Luckily, there have been great improvements in over-the-shelf tactical sniper weapons available today so that a good system can be had for around $1000.

(7) Due to the necessity for high precision/accuracy, and maintenance of correct zero, police sniper rifles should not be shared. It is a requirement that the sniper rifle be zeroed for the assigned user at all times.

(8) The rifle and scope should be within all other department standards.

In the next chapter we will look at custom rifles available to the LE sniper.

"We are in the business of hitting what we aim at....It is not a game."

Each year in the early spring, Carol Shelton holds an NRA High Power Rifle Course at the Fort Worth Rifle and Pistol Club. In 1997, this particular class had the Arlington, Texas, SWAT snipers join in with the regular civilian students. The 7 police snipers use their duty rifles and shoot in the Open "F" class which is for any rifle, any sight. One of the matches was won by Officer Mark Garber with Officer Keith Scullin taking second place. After that match, one of the civilians who shoots regular high power rifle with sling and aperture sights, came over to talk to Keith and Mark. The civilian made a comment that the officers cheated during the match because they use scopes and bipods and should shoot with only the sling for support. Mark Garber replied, "We are in the business of hitting what we aim at. We use every means possible to insure we hit what we shoot at." "It is not a game," Keith told author. "It would be good to train with the sling, but to get police snipers up to the level of an experienced competitive shooter is time consuming and expensive. Scopes, bipods, and support under the toe of the buttstock gets the police sniper to a high level of accuracy quickly."

Arlington, Texas, Police Tactical Snipers, Keith Scullin (left) and Paul OConnor with Texas Brigade Armory M40A1s. Mark Garber, another Arlington PD Tac Team sniper, also uses a personally owned TBA M40A1 for duty use. (Mike R. Lau)

Chapter 4:
THE CUSTOM SNIPER RIFLE and TELESCOPES

What Tactical Rifle Design is Best?

Of all the chapters in this book, this one was the most difficult to write. I rewrote portions of it several times and it was the last chapter to finish. My problem with the subject of custom rifles is that I myself customize rifles for tactical use. This is a technical book. How can I be objective in discussing the subject of custom rifles without expressing my own philosophy and opinions? A lot of what you read in this chapter is opinion based on my own understanding and experience in the subject of tactical rifle design. Some of the ideas and opinions will be disagreed with by others. You can read about my philosophies and opinions and throw them out or you can read and think about them before you decide on whether the points I make are valid or not in your own mind. Remember one thing after you are through with this chapter, there are a lot of well made, practical rifles, of exceptional accuracy at reasonable prices out there. If you are in the market for a tactical long range rifle, I advise you to look them all over before making a choice.

Much has been learned about the tactical value of sniping over the past four decades. Only within the past few years has much of this been made aware to many so consequently, the sniper and his rifle have become more popular as a tactical tool now than at any time in its past history. Demand for sniper rifles has encouraged many firearms manufacturers to produce long range tactical rifles. Competing for rifle sales has kept pricing reasonable. Just like the target rifle did in the past, sniper rifle demand is directly responsible for much of the changes in stock designs, improved barrel quality, improved scope sights, improved ammunition and components, and much more. As a result, the entire firearms industry, including the shooter, has benefited.

When deciding on a tactical rifle design, consider what specific activity you are going to use the rifle for. Is the rifle for urban police work? Is it for rugged military field operations? Is it for long and close range tactical sniper competition? Is your rifle going to be used for several purposes including hunting? What weight and size can you handle for each type shooting? What caliber can be used for both long range competition and short range urban police work?

After you decide what you are going to use the rifle most for, look at the sniper rifles available. Some manufacturers come right out and tell you what kind of tactical shooting the rifle was designed for. However, the use of the word "sniper" in the nomenclature is sometimes politically incorrect so the rifle takes on other names, but underneath all of that it may have been originally designed as a sniper rifle. Usually the agency or the actual user of the rifle will give away it's tactical purpose. Many foreign sniper rifles cannot be imported into the US if they carry the word "sniper" in their nomenclature and have to be renamed "target," "long range," "precision varmint with sniper accuracy," "marksman," "sharpshooter," etc. To display products at the annual SHOT SHOW the promoters prefer that your rifle or any product not have any reference to shooting human targets. Norm Chandler was hassled at the SHOT SHOW in Dallas one year when he displayed his "Chandler Sniper" rifle and **Death From Afar** books on a table. Benelli and HK were also at the show with all their combat weapons and so were many other manufacturers and dealers because the way many firearms sell today is "tactical."

Carlos Hathcock sighting down Malcolm Cooper's Arctic Warfare rifle at the 1996 Dallas Shot Show. Norm Chandler kneeling. Lones Wigger, John Plaster, and Malcolm Cooper (holding the Chandler Sniper). (Photo courtesy of the late Guy McCracken Jr.)

As an example, we know the Marine Corps M40 rifle, developed for the Marines during the Vietnam War by Remington Arms, was not originally intended to be a SWAT or varmint (animal type) rifle. However, it was soon brought out as the "Varmint Special" and later the "Police Sniper Special" or PSS. Today's current police model is labeled the Remington 700P. On the other hand, today's synthetic stocked Varmint Specials in .308 Win caliber make excellent police sniper rifles because they have lower cost, medium weight barrels, and acceptable accuracy. The Dallas PD purchased twelve .308 Win. Varmint Specials for their TAC team snipers to replace the old wood stocked Remington 700 BDLs. All of the new 700's have had their Leupold 3.5x10 scopes mounted for 1000 yard shooting by Texas Brigade Armory.

Some designs are better suited for tactical field use and others are better suited for urban tactical use. But most rifles can be used for both. Rifles originally intended for field use are usually lighter in weight, smaller in size, and more rugged. Some rifles were originally designed for target use and later listed as sniper rifles because sales were better or they were being used as sniper rifles. Your tactical rifle may look like a match target or bench rifle that may not be totally suited for rugged field use. These may appear awkward, heavy, have bulky square stocks with odd looking grips, lots of moving parts, and look more like a bench gun or static position urban sniper/counter-sniper rifle, but don't rule these out for field carry. Many have been in use for a while by some of the toughest field rifle users in the world, military Special Operations units. Any rifle can become a tactical long range field rifle if you take one out and use it as such. Put a scope on your "across the course" match bolt rifle and use it in a long range sniper match. The only difference between your rifle and some of the tactical rifles you shoot against is the finish.

Considerations and Features of the Long Range Tactical Rifle

Each feature on a rifle is designed for one or several purposes that should be carefully considered before a choice is made. You will pay for them whether you need them or not. An added feature is not only costly, it adds weight, it is something else that can break or needs adjustment, or can get in the way when the going gets rough. On the other hand, not having a feature may handicap you on specific uses, slow you down, cause you to miss. Here are some of the considerations and features found on many factory and custom tactical sniper rifles:

(1) *Accuracy and precision* is usually the primary concern when selecting a sniper rifle. 1/4 moa at 100 yards is desirable, but not necessary for a lot of shooting. It is good to know that your rifle has that capability so you can only blame yourself for not hitting a target. For a police sniper rifle, 1/2" moa or less is desirable for making precision surgical shots (more on this in the chapter on terminal ballistics). 1 moa rifles may be adequate for some shooting, but for precision shots at long or short range it is unacceptable. You already introduce your own error into the shot, you don't need to add a lot more to it with an inaccurate rifle. Unfortunately, extremely accurate rifles usually cost more.

(2) *Caliber* choice is the 7.62mm NATO/.308 Winchester for all around use. Practical field target engagement distances can be out to 1000 yards in real or training situations. Also expect to engage close moving or stationary targets under 50 yards. Accurate .308 Win ammunition suitable for all tactical purposes is easily obtained at reasonable cost. .300 Win Magnum rifles are heavier and longer, but will do better at the longer ranges. However, it is a little more difficult to make rapid multiple shots on close targets with the heavier recoiling magnum rifle.

(3) *Weight* will be a major consideration especially if you are a military field sniper or are going to participate in sniper matches like the Proskopathlon's (Greek for "Scout's Test") 8 mile first day. Most custom long range tactical rifles today weigh between 12 to 18 lb. Don't let your rifle get excessively heavy with too many add on accessories. Urban police sniper rifles don't need to be heavy, but because the LE sniper will normally not have to traverse rugged terrain for weeks at a time, a heavy rifle is not a burden. A heavy rifle tends to be more stable for most supported positions when sighting and they also resist movement caused by wind. Weight reduces some of the accuracy affects of the shooter jerking the trigger or holding the rifle incorrectly when firing under pressure. Consider the rifle's weight for both carry and shooting.

(4) *Long or short barrels*. Short barreled sniper rifles with 20" heavy or medium barrels are light, handy, and many metropolitan police departments use them. At closer ranges these rifles shoot as accurately as the longer barreled ones as evidenced by the bench and silhouette competitors. However, for long range shooting you will be handicapped by the poorer ballistics delivered by the shorter barrel in any caliber. Ideal barrel length for the .308 Winchester and .300 Winchester Magnum round is 26" to take advantage of the ballistics. Longer barrels are cumbersome and add too much weight for what you gain in increased ballistics. 1-12" twist is preferred for the 168 gr and 175 gr. HPBT Match ammunition. Use 1-10" twist for the .300 Win. Magnum.

(5) *Heavy, medium, or light weight barrel*. Heavy barrels like the Marine and Army contours weigh about 1 to 1.5 lb. more than the medium weight barrels found on the varmint and police type rifles. Heavy barrels heat up slower and will not "warp" as fast as the light barrel and string shots across the target after a few rounds are fired. The heavy match barrel

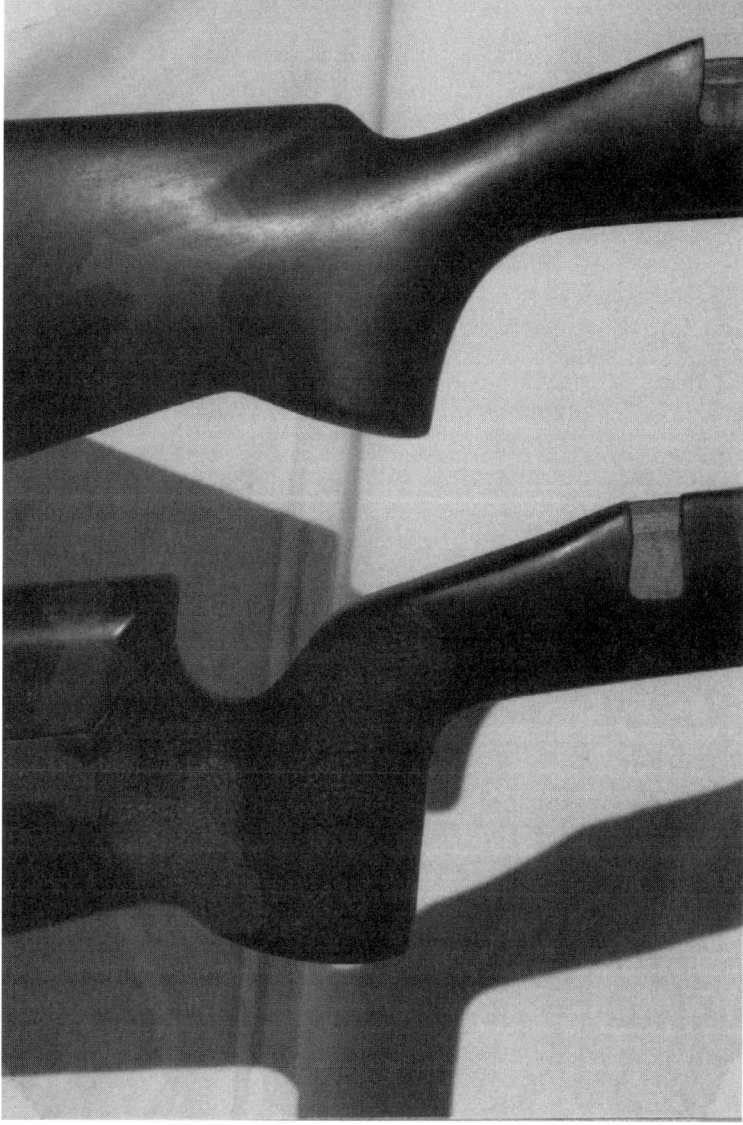

A rifle's "feel" when carrying in the field is determined by its weight, length, and ergonomic design. A rifle's grip/handle determines comfort and ease of firing for the hand as well as ease of carry for long periods of time in normal positions. Angle of pistol grip on AR-15/M16 variants allows for extended periods of field carry with the hand and trigger finger positioned for firing. Center grip is on McMillan's M40A1 General Purpose Hunting stock. Extremely comfortable, the small of the grip is a sporter style for easy field carry. The grip's lower half drops into a steep angle giving full support to the last two fingers and making the hand feel like almost holding a straight pistol grip. The lower stock is the M40A3 pistol grip. Not quite vertical, but very comfortable when shooting from the prone or bench with artificial support. The thumb cutout behind the grip allows the stock to be grasped in almost the same manner as a standard sporter stock allowing for quick positioning to the shoulder with the hand and trigger finger already positioned for firing. (Mike R. Lau)

CHAPTER 4: The Custom Sniper Rifle and Telescopes

increases the rifle's weight to allows steadier holding on target. Barrel vibrations and whip, as well as the felt recoil are also reduced. 26" heavy weight barrels lose about 1/2 lb. when fluted and are becoming more popular as field carry becomes a major consideration in tactical sniping.

(6) ***Control feed or push feed bolt***. Classic Winchester M70s, original Mauser actions, and Ruger 77s, have the claw type extractor that holds the cartridge after it releases from the magazine for chambering. Some feel this is a must have item on their rifles. Unless you are going to shoot the rifle upside down, this is not a necessity. Most modern military rifles and machine guns push the cartridge into the chamber without controlling the live round and do not experience chambering problems.

(7) ***Sako extractor vs. Remington spring extractor***. Both extractors are fine for the rifles they were originally intended for. Machining the bolt head on a Remington 700 to allow installation of a Sako type extractor weakens it by having to remove a lot of metal around one of the locking lugs. Luckily the Remington 700 bolt head is plenty strong because a lot of gunsmiths, including myself, do this for customers wanting it.

(8) ***Synthetic stocks*** have proven to maintain zero better than wood stocks. Military field rifles and police rifles carried in cars are exposed to extreme climate temperatures and humidity year round and should have stocks made of synthetic materials. Synthetic stocks come in many different styles and camouflage or solid colors. Choose one that fits your style

Grip widths and lengths on rifle stocks come in all sizes and shapes. Different shooters have different size hands, finger lengths, and shooting grips. From left, grip of M24/700P with palm swell. Next is Remington 700 ADL Synthetic sporter. Center is the M40A1 grip. Note its tapered width and long length. The two on the right are Remington's BDL wood sporter stock and Remington's Varmint Special. (Mike R. Lau)

Grip on M40A2 (left) and M40A3 (center) have slight palm swells on both sides. Flared grip on M40A1 is wider at the bottom than the other two stocks. (Mike R. Lau)

or preferred shooting and is also comfortable to carry if you intend to do a lot of walking with it. Wood stocks are OK on sporting rifles and they look good.

(9) *Sporter vs. target stock*. For extended field use, sporter style or close to sporter type stocks are more comfortable to carry in the field. This was proven by the Marines in Vietnam when they discarded the target stocks from the M70 rifles and replaced them with sporter stocks. Wide forends with flat square bottoms are hard on the hands if you have to carry the rifle some distance this way or if you shoot with the hand under the forend during position type shooting. The square or slightly rounded corners cut off the blood circulation in the fingers. Wide rounded target type forends don't give this problem and are easier to shoot when position shooting using a glove and sling. When shooting off a bipod or artificial support, like a sand bag, it doesn't matter what the shape of the forend is like. For most tactical field uses, the sporter forend seems to be more comfortable for both shooting and carrying. That is why hunters still choose sporter stocks instead of target stocks.

(10) *Adjustable butt plate* allows the rifle to be adjusted for the specific user until he passes the rifle to another police officer or military sniper. Shooters that don't intend to pass their rifle to someone else may not need an adjustable buttplate. It is also good for those who have a lot of extreme weather and temperature changes and wear heavy clothing one week and a T-shirt the next. A system of adjustment using fiber or plastic spacers is better that a metal adjustable butt plate because it is lighter, has less moving parts, and you don't

CHAPTER 4: The Custom Sniper Rifle and Telescopes

Very important to the feel of the rifle for carrying and position shooting is the shape and width of the forend. From left to right: M24/700P; M40A2, M40A3, M40A1; and the Remington Varmint Special. Wide and thick square forend of the M40A2 was designed to take the stud on the forend tip for attaching the Parker Hale bipod. When using the bipod or shooting prone off a sandbag or backpack, shape of the forend is not critical. Harris bipods require the extra sling swivel stud as on the M24 or 700P stock. (Mike R. Lau)

have that gap between the stock and recoil pad plate. For a target shooter who requires different lengths of pull or cant, the fully adjustable ones are fine. For a tactical competitor or urban tactical shooter, having the buttstock a little shorter than normal works good because the rifle comes up to the shoulder faster in tight confining spaces and it doesn't catch on clothing as often. Many adjustable buttplate devices add a lot of weight to the rifle. The Army's all aluminum model is one that doesn't. It is not a good practice to share rifles especially among police officers because each officer will have a different zero.

78 *The MILITARY and POLICE SNIPER*

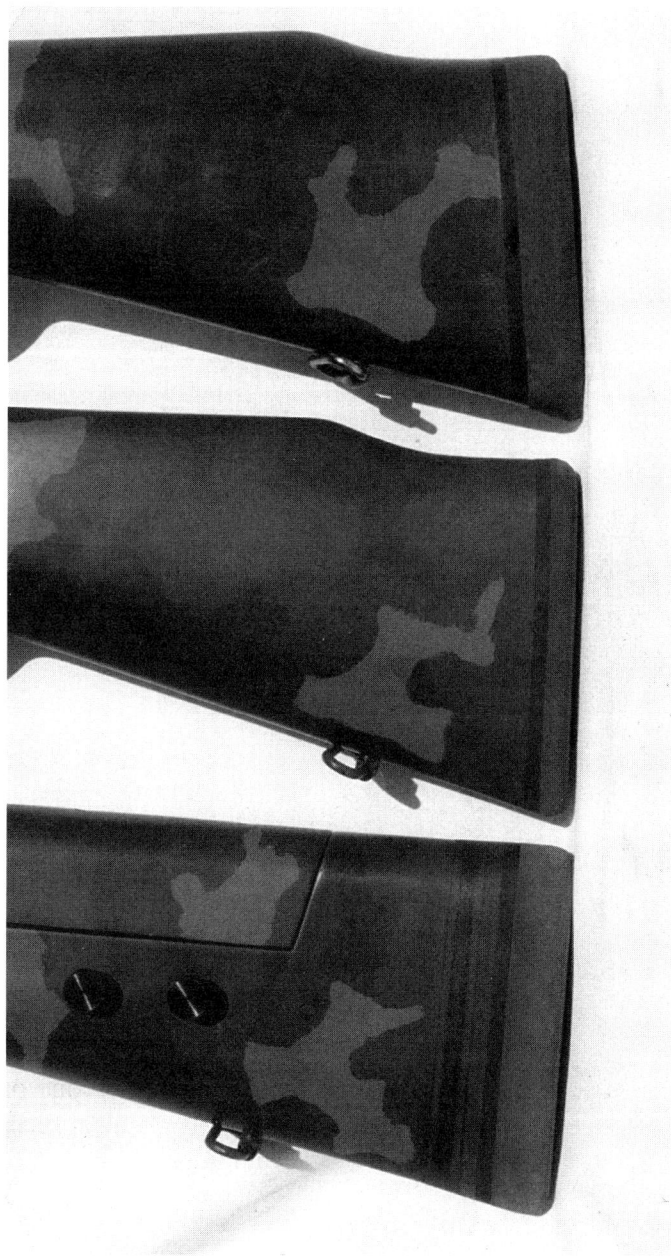

Texas Brigade Armory's M40A1 sniper rifle comes with the standard .60" Pachmayer rubber buttpad as used on the Marine Corps rifle. 1" pads are standard on the M40A2 and A3, and optional on the M40A1. Removable butt spacer system (bottom stock) or three way adjustable buttplate system is also available as an option. (Mike R. Lau)

(11) *Adjustable cheek pieces* on the field sniper rifle are OK if they are easy to adjust and maintain that adjustment with rough use. Many current stock designs are such that they will not allow some to get a good cheek position on the stock and see through the scope correctly. Almost all cheekpieces adjusted correctly for the shooter require placing them back down in their lowest position or removal to allow the cleaning rod to be used from the chamber end. Once you have the cheekpiece adjusted for you, make a small mark on the stock and cheekpiece so you can return it to that height. Like the adjustable buttplate, the adjustable cheekpiece allows you to give the rifle to someone else to adjust for and shoot. Another way to return the cheekpiece to its correct height is with spacers.

(12) *Sling Swivels*. The McMillan stock on the M40A1 has sling swivels that are massive. The **nut** that secures the swivel is a about 2"

CHAPTER 4: The Custom Sniper Rifle and Telescopes

Texas Brigade Armory's M40A1 adjustable integral cheekpiece option. (Mike R. Lau)

long and about 3/4" wide. You can run with the rifle slung across your back. If you plan to do a lot of movement while carrying the rifle, get good sling swivels with the shanks securely attached to the stock.

(13) ***Semi-auto versus bolt action***. Off-the-shelf bolt action rifles are generally more accurate than semi-auto rifles. There are a few exceptions, however, such as Knight's SR-25 and some accurized match M14/M1As and AR-15/M16s. But these are exceptions and require added work and increased cost. Military snipers don't like the auto ejection of the spent cartridge because it gives the sniper's firing position away during firing. The reflection of light from the ejected cartridge case is very easily detected even when not looking in the direction of the sniper's position. Bolt rifles can have longer barrels so they can take advantage of superior ballistics for the same calibers in the semi-auto.

(14) ***Detachable magazines***. These are popular with police. They allow quick reloading for repeat shots. Many gun experts, and military snipers justify not needing detachable magazines with reasons like "we only fire one or two rounds from the FFP" or "if we need firepower from clips or auto weapons then we screwed up by making contact with the enemy and this mission was meant for a larger infantry unit." Gunsmiths usually say that the magazine interferes with the solid bedding of the bolt rifle and weakens the cross sectional area of the receiver by having to remove metal from the side rails, thus reducing accuracy. Whatever the reason for using a detachable magazine is up to you. Robar builds a detachable magazine bolt sniper rifle. Remington's 700P has a detachable magazine option and Steyr's SSG bolt rifle has a detachable magazine as standard. All three have adequate accuracy.

Texas Brigade Armory's M40A Series rifles can be modified for the new Remington 700 Detachable Magazine in magnum or standard calibers. (Mike R. Lau)

(15) **Bipods**. Many sniper rifles are now coming with bipods. The most popular are the ones made by Harris Engineering, Parker Hale, and Versa Pod. The last is a copy of the Parker Hale and is a nice quality unit for a price of around $100. Harris bipods are the most popular because they are light weight and come in numerous extendible lengths and only run around $50-$60. Parker Hale bipods cost around $200 and can be had in either aluminum or steel.

(16) **Muzzle brakes** are sometimes recommended for .300 magnum and larger caliber rifles because of the increased recoil and muzzle whip. A good muzzle brake put on by a competent gunsmith will not disturb accuracy. There are many good ones out there that I have used and these are the VAIS and CFL's. Both of these have holes drilled with a rearward rake to vent gasses slightly to the rear. Muzzle brakes that reduce dust signature by

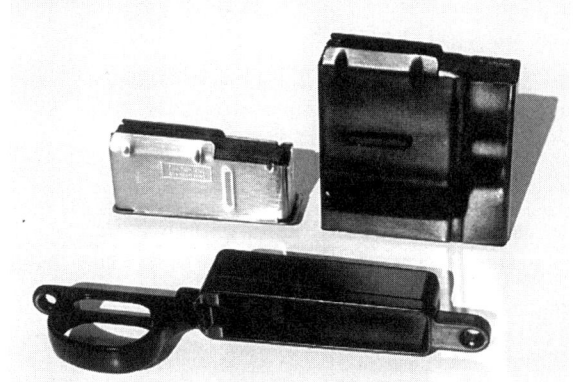

Kwik Klip conversion for long action Remington 700 holds 5 or 10 rounds in standard calibers and 3 and 8 rounds in magnum calibers. A short action clip magazine is also being designed by the manufacturer. (Mike R. Lau)

being closed at the bottom don't seem to give good accuracy because the gas pressure exits unevenly around the bullet as it leaves the muzzle brake.

(17) Make sure your rifle is **maintainable**. If you shoot a lot, you will want easily available

CHAPTER 4: The Custom Sniper Rifle and Telescopes

repair and replacement parts at low cost. This includes the major components of barrel, stock, and action. By having standard parts made by a well known US manufacturer you should not have this problem for a while. Foreign rifles and parts seem to be a little more expensive and harder to come by. Custom machined actions, such as the Stolle, Paramount, and Hall, will hopefully have parts available for at least while the manufacturer is still around. For example, Steyr rifle barrels are not threaded and are put on by inserting the barrel into a heated receiver and then letting the receiver cool. Most gunsmiths in the US are not set up for removing and replacing Steyr's barrel unless it is machined off and the inside of the receiver is threaded.

Rifle Telescopes and Mounting System Considerations

Use only high quality scopes. Don't let the scope be the limiting factor of your weapon system's effectiveness. Some PDs have found out the hard way that the level of consistent accuracy and precision with the weapon system is determined to a large part by the quality of the scope sight. It is in my own opinion, based on experience, that when purchasing a rifle scope, you get what you pay for. Most telescope sight manufacturers appear to have access, if not actually using, the latest and greatest in scope technology. It also appears that the cost for manufacturing a scope sight is very competitive world wide. Therefore, with very few exceptions, the higher the scope's cost the higher the quality.

(1) For the LE sniper, the preferred telescope for the all around urban and rural sniping weapon is the variable powered with a low setting of at least 2X or 3.5X and an upper power of 10 to 14X. Low powered fixed scopes that have excellent light gathering qualities and quick to use reticles are good for the close in rifles. For the tactical field military scope, 10X seems to work fine.

(2) The scope should have a range finding system that is accurate, quick, and easy to use. The range finder should be usable with any type or size targets for a variety of shooting situations.

(3) The scope should be fully adjustable for the maximum windage and elevation you expect to encounter and then some. Windage and elevation should have positive click stops that preferably can be felt and/or heard. The adjustment knobs should have 1/4 or 1/2 MOA adjustments that are marked clearly with direction of impact change indicated. They should be easily read and adjusted by the firer from his natural point of aim, without he or she having to get off the rifle.

(4) If the scope has a parallax or focus adjustment, is should be easily reached and adjusted.

(5) The scope sight should have high quality internal construction that correctly moves the reticle to the corresponding windage and elevation graduation marks on the dials. The inner reticle tube should return to exact zero all the time with no refiring, even after it has

From left to right: Leupold Mark 4 M1, Leupold Mark 4 M3, Leupold 6.5x20X - 30mm tube with 50mm objective, Leupold 6.5x20X -1". M1 has 72 minutes of elevation and 45 minutes of windage in 1/4 moa increments. M3 has Bullet Drop Compensator (BDC) dial for 1000 yards or meters with 1 moa elevation adjustments and 1/2 moa windage adjustments. 6.5x20s have excellent light gathering optics and 20X setting makes pin point aiming at all distances easy. All scopes available with the mil-dot reticle.
(Mike R. Lau)

been cranked on all day at practice. High quality scopes, like Leupold's, have proven to do just that.

(6) In addition to being able to withstand rough handling, the scopes internals should be rugged enough to handle extreme weather conditions. Problems with scopes during extreme weather changes are moisture leaking into the scope body to cause internal fogging, breaking of reticles, and slipping of range finding decals from glass lenses.

(7) The scope should have a matte finish.

(8) **Scope mounts**. A good scope mount is of course a necessity. It should not allow the scope to come loose or bend the scope in any manner during mounting. A very popular one is the one or two piece dovetail mounts. These are fine on hunting rifles and small caliber centerfire rifles, but are not strong enough for the heavier recoiling .308 or .300 magnum rifles. I have seen the rear scope ring pop out of the screws on the rear portion of the base. A better choice would be the dual dove tail design. If used on the close-in rifle with open sights, the mount should allow for quick removal of the scope. Preferred fixed scope mounts are the Unertl Marine Corps type, Leupold's Mark 4, MGWs, and others that are of heavy all steel construction.

Stay with the proven quality scopes and you will not go wrong. You read **Precision Shooting Magazine** to find out whose barrel, what action, what cartridge, and what bullet dominates the competition world. Do the same with the LE and military snipers. Find out what equipment they like and why. Read about the sniper matches and see what is being used, how it is being used, and why it works.

Leupold

The tactical scope market in the US right now is dominated by Leupold. The reason for this is extreme high quality, availabil-

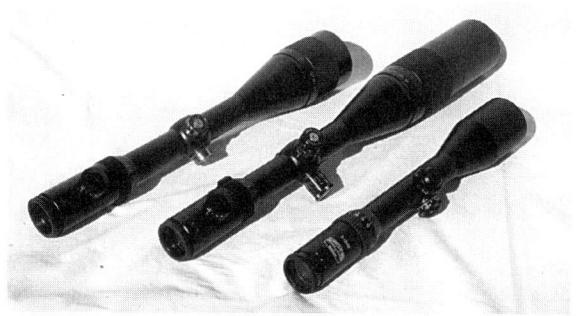

Left to right: NightForce 3.5-15X - 56mm, NightForce 8-32X - 56mm, and Schmidt & Bender 3-12X - 50mm. All three have excellent high quality optics and 30mm tubes. NightForce scopes have reticles that light up. (Mike R. Lau)

Bausch and Lomb Tactical Elite 10X on Sean Little's TBA M40A3. Telescope has all the desirable features including mil-dot reticle, large adjustment knobs and easy to reach focus on eyepiece. Optics and adjustments are first rate. (Mike R. Lau)

ity, and numerous models and reticles to choose from. They stay ahead of the competition by bringing out new models to satisfy customer preferences at competitive prices. They have the largest selection of tactical scopes for tactical shooting so people turn their first. Leupold is like a giant custom scope shop that bends to the customer's needs. Their tactical scopes meet all of the desirable qualities for a tactical scope as listed above and if they don't, Leupold will come out with the change soon.

New for 1998 is Leupold's Vari-X III 3.5x10x40mm Long Range M3 scope. The scope is like the Mark 4 M3 with the BDC

CHAPTER 4: The Custom Sniper Rifle and Telescopes

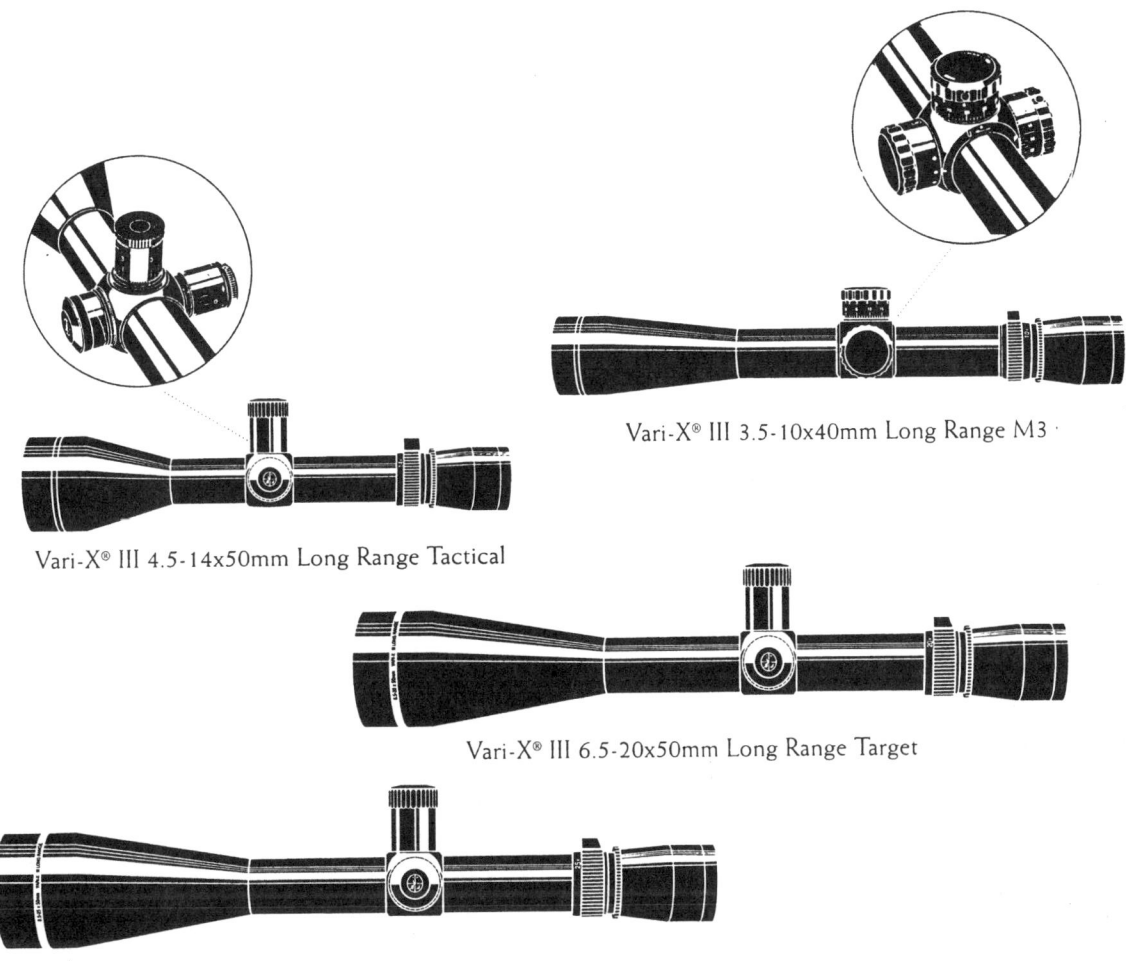

Fig.4.1. Leupold's new models for 1998 are the 3.5X10 Variable M3 and selected Vari-X III models with the focus moved to the turret housing. Note that the BDC knob on the M3 is now graduated for the 1 moa click adjustments.

and the parallax focus except it is a variable powered model. The lower powers will now give the LE and military sniper greater ability to shoot in dimmer light, engage close-in targets, engage moving and short exposure targets easier. In addition, the new BDC dials for .308, .300 Win Mag, .223, and .30-06, now have graduation tick marks for the 1 moa elevation increments. Leupold has also moved the parallax/focus adjustment from the objective bell to the turret housing on three Vari-X III models.

If you own a Leupold scope and it does not have mil-dots, Premier Reticles can install this type reticle in your scope. They are an authorized Leupold distributor and specialize in reticle installations in Leupold scopes. They are a quality name and their logo on a Leupold scope is held in high esteem by tactical and sport shooters alike: Premier Reticles, 920 Breckinridge Lane, Winchester, VA 22601.

A few other good tactical scope manufacturers out there are Redfield, Weaver, Burris, Tasco, NightForce, Bausch and Lomb, Springfield Armory, Swarovski, Schmidt and Bender, Springfield Armory, and U.S. Optics. When you look at these other manufacturers' line-up of scopes, they have several to choose from and offer the same quality and some of the same features or more than the Leupolds, but their choices are not as wide and some lean more toward the commercial sporting use rather than tactical. Springfield Armory offers a good selection of tactical scopes. When they saw that tactical shooters were going to mil-dots, they made a good decision by adding mil-dots to the reticle while still keeping their quick range finding brackets and level.

The Custom Sniper Rifle

Just like the competitive rifle shooter, the discriminating tactical shooter demands the ultimate in accuracy and utility. To meet this requirement the shooter usually turns to the custom rifle builder. Custom rifles cost more than large factory manufacturers' rifles, but then the shooter expects a lot more. Custom rifles like that produced by Texas Brigade Armory, Iron Brigade Armory, AWC, Robar, McMillan, L.O.D, and others, cost between $1500 to $4000.

Texas Brigade Armory customizes the popular Remington 700 action into a very practical line of long range tactical rifles that have all the functionality of higher priced models at lesser cost. My rifles are not of unique design using components I designed myself. They are built from already proven and popular existing components that others have spent a lot of time and labor to develop. TBA's products are designed from rifles that are proven to be practical and have already withstood the test of combat and time. I consider them to be "classic" in the sense that I believe after all the hoopla of radical new designs, fads, and innovations, the simple "classic" rifle will prevail. As a custom rifle builder, I generally see the experienced field tactical shooter and competitive target shooter, who also want a long range hunting rifle, going toward the less complex and easy to carry rifles like the M40A1, M40A3, and M24. These rifles also seem to be the choice of the ex-military, active military, and police sniper. I see the long range prone tactical shooter that does not intend to do a lot of extended field carry in rugged terrain asking for the heavier target type rifle with options such as the hand rail, three way adjustable buttplate, straighter grip, wide flat bottom forend, and adjustable cheekpiece.

The primary reason most shooters come to me is that they want an exceptionally accurate rifle at reasonable cost and with practicality in rifle design. They want it to shoot and carry equally well for all types of shooting whether it be in the close confines of the urban environment, slogging through rough terrain in harsh climates for weeks on end, or for long range tactical shooting. They want to be able to use the rifle on targets as well as for hunting. They also want it to look good. Even after they have put the rifle through hard use and it's all dinged up, they want to be able to sit down and lay it in their lap and look at it's graceful lines and aesthetic beauty. A rifle is more than a tool, it is an art form that possesses more than the ability to shoot. It should have character and personality.

Like most small custom shops, Texas Brigade Armory can only put out several rifles a month from its shop. I am the only gunsmith and machinist most of the time. I usually do not employ more than one or two part timers at any one time. And these have been trained by myself so I know what kind of quality they produce. If they do not meet my standards I do not keep them around long. One of my assistants is Sean Little, who is very familiar with the USMC M40A1 because he is a Marine Scout/Sniper. I cannot guarantee delivery on time. I cannot deliver large contracts for agencies. What I can deliver is quality and accuracy that will meet or exceed your expectations! Texas Brigade Armory builds the custom long range tactical rifle just like the bench rest custom rifle builder: to the customer's requirements and satisfaction, one at a time.

A Few Secrets of the Small Custom Shop

To give the reader an idea of what goes into the building of a custom bolt rifle, I will describe to you a few of the assembly proce-

CHAPTER 4: The Custom Sniper Rifle and Telescopes

dures that Texas Brigade Armory uses. The first step in building a high quality rifle is to use only the best quality components available. Use of poor quality materials will cause a lot of problems including damaging the reputation of the custom gunsmith and his business.

Barrels

Many persons buying firearms with stainless steel barrels today are mainly interested in the metal's ability to resist rust and corrosion. Discriminating shooters use stainless barrels not because it resists corrosion better, but because it is more uniform in hardness throughout the bore when compared to chrome moly steel. This is a result of the mixing of the alloy in the manufacturing process. A bullet going through a plain chrome moly steel barrel will hit hard and soft spots in the metal causing some deformation of the bullet which will reduce accuracy. I prefer to use mainly Hart, Kreiger, K&P, and Obemeyer barrels. Consistency of quality is what these makers turn out. I don't specify that I want a match grade, super premium, best barrel they can turn out. These makers only produce one grade barrel. I don't have to measure these maker's bore straightness or uniformity of groove and land diameter because I know they already did it. I get the same high precision every time so I don't even bother to think about it. The way I determine the quality of the barrel made is when I machine it, when I shoot it, and when the customer tells me how it shot.

Lathe Operations on the Barrel

TBA has developed some of its own methods for truing the receiver, bolt, chamber, and barrel threads. I carefully studied the methods of several well known precision barrel put-on-smiths in the country. I then examined the best of their precision lathe operations and experimented with these methods to try to improve on them. Some of these methods could not be improved upon. For the other operations, I was able to increase precision by breaking them down and increasing precision at a specific step within. A couple of operations required standard tool modifications to increase precision in alignment. I developed a few of my own operations and rearranged normal sequences of operations that increases the mating precision between the major components of receiver, bolt, and barrel. There is a drawback, some of these steps are awfully time consuming and require additional set up. An accountant for a large manufacturer would pull his hair out if he saw how long it takes me just to set up the barrel in the lathe before I even start cutting the threads and chamber.

When I first set up a K&P barrel in the lathe, I put the small shank, that Ken Johnson leaves on the breech end, into the chuck. I then clean the inside of the tailstock before putting in the 60 deg center which I replace frequently if scored. I clean the muzzle's 60 deg taper that Ken used last at his shop and also clean into the bore. I hand cut the 60 deg taper again with a Clymer 60 deg center cutter in the bore and then put the barrel on the tailstock center. I go back to the breech end in the chuck and indicate the outside of the breech end to within .001 to .0005. Then I come back to the muzzle end and lightly apply the steady rest to the tapered barrel. When I measure the barrel run out in relation to the bore with the 60 deg center in it, the outside of the barrel will run-out around .002 to .003. I back the tailstock center away and recut the 60 deg center with the Clymer held between the tailstock center and the inside of the muzzle. I readjust the steady rest and remove the tailstock center and will now get between .001 to .0005 run-out on the outside of the barrel in relation to the center of the bore. I recheck this by measuring the run-out on the inside of the bore and it usually measures around .0001. Ken cuts these barrels very precise, inside and out. I return the tailstock center into the muzzle and then cut a short and shallow straight shank section along the end of the excess muzzle barrel length. The barrel is taken out of the lathe. The straight shank on the muzzle end is inserted into the chuck with the shoulder of the cut shank butted up against the chuck jaws. Now the barrel won't move away from the tailstock or cutting tool and the chuck won't be clamping down on a tapered section of the barrel. I repeat the same procedure for aligning the breech end as with the muzzle before cutting the threads and chamber. If a barrel does not have an untapered section or the section that the steady rest is to run on is not within .0005, I turn a short section near the breech end after pre-centering the bore to the lathe. This concentric section is used for the steady rest and will cause the barrel not to create run-out when it is held only by the steady rest at the breech end. To keep the 60 degree center from wearing the taper in the bore quickly, I only take shallow and small cuts at low speed during each lathe operation and use a lot of oil. I recut the 60 degree centers after each operation and remeasure and adjust the barrel to stay within the same run-out inside the bore as when I first started.

Besides being favorable to uniform barrel vibrations and uniform heating/warping, you can see that the more perfectly concentric barrels are easier to center in the lathe. You can imagine how much more work it would be to align a barrel in the lathe that measured several thousandths out of concentricity on the outside of the barrel to the bore's center. I have found that the four mentioned barrel makers turn the outside of the barrel to very close tolerances in relation to the bore's axis. You know whoever turned the barrel down on the outside was meticulous about the precision on the inside.

A lot of my customers are Law Enforcement types who are not allowed to shoot handloads or loadings other than what their department specifies for duty use. As a result, the most popular chamberings in my rifles are for the .308 Win with the 168 gr. HPBT, followed by the 175 gr. HPBT. The length of the Remington 700 short action forces the ammunition producing companies, as well as handloaders, to keep the overall loaded length of these rounds to 2.80". We all know that throat length is critical to the rifle's accuracy. The custom rifle builder can cut the headspace and throat length specifically for the 168 gr. and 175 gr. match loadings. Factory rifle chambers I have measured show throat leads from .15" to .17" as a normal average!. Yes, that puts the 168 gr. HPBT ogive more than 1/10" from the rifling! However, some rifles that showed this much throat lead still shot within 3/4 moa. I have found that the most accurate throat lead length is at .010". Putting the bullet ogive right into the rifling or only a few thousandths off seems to reduce accuracy somewhat. It may be because variations in case length and bullet seating depth may be jamming some bullets into the rifling causing inconsistent pressures. My chamber reamers have been specially modified by Clymer to decrease throat length to my desired minimum specification. You can use a Stoney Point gauge to measure OAL of the 168 or 175 gr. bullet to see how far off the rifling you are. For example, if the gauge pushes the bullet out to an OAL of 2.85." and your factory ammo or handload is 2.80" then you are .050" off the rifling. TBA barrels are crowned 90 degrees to keep the hole in the bore perfectly square to the muzzle. An 11 degree crown can be cut at an angle to the center axis of the bore causing the bore exit to be oblong so is not a preferred method.

Lathe and Milling Machine Operations on the Receiver

Receivers on TBA's M40A series rifles are the preferred Remington 700 actions. These cylindrical receivers bed uniformly into the stock and has a proven accuracy record. I use some of the common methods of truing the action as done by other well known gun-

CHAPTER 4: The Custom Sniper Rifle and Telescopes

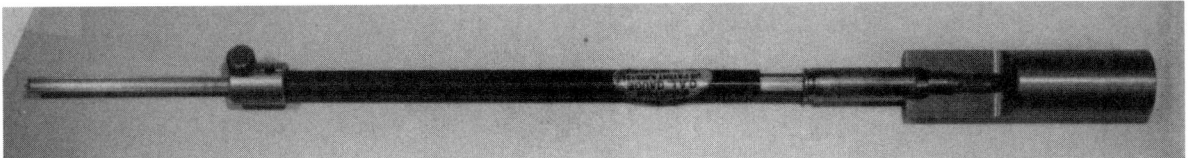

Stoney Point's OAL Gauge is a must have gadget for handloaders and gunsmiths. Tool consists of a sized case that has its base threaded for the tool's outer body. A thin rod on the inside on the outer body goes through the case to push a bullet into the rifling at the chamber end. The rod is locked down with the brass screw and the tool with bullet and case is remove from the chamber. OAL of cartridge is measured. Sectioned rifle barrel chamber is not part of tool kit. (Mike R. Lau)

Tubb or Chandler recoil lug is optional on any TBA Remington 700 rifle. (Mike R. Lau)

smiths. I have developed my own method to increase the precision when truing the receiver face to the inside threads and when truing the face to the threads and shoulder of the barrel. I modified an existing tool that is already available on the open market to do this operation. Methods for truing the bolt face, lugs, and lug contact areas in the receiver are standard lathe operations followed by good old fashion lapping on top of that to remove any tool marks. Anyway, that is just the beginning, but you get the idea of how meticulous machining operations can become if precision is to be maintained. I keep the tread fit between the barrel and receiver tight, but not where I cannot screw the barrel on with some effort by hand. I also reduce the tolerances on the breech counterbore which helps to keep the bolt centered to the bore. I will not go further into detail on my barrel and receiver lathe operations because at this time I consider them trade secrets that took me a while to develop.

TBA's Remington 700 receivers are clip slotted like the Remington 40X receivers. This allows the receiver to handle a Unertl type USMC mount with lugs on the underside which are handfitted to the clip slot cutouts. The scope base is machined from steel with Redfield rings pinned and brazed to the base. The underside of the base is epoxy bedded to fit the contour of the receiver and then the rings are lapped. This unit is so tight that even if three of the four screws came loose, the scope mount will not shift at all. TBA can build this scope mount to fit on a Remington 700 for any scope, including the NightForce.

A lot of good custom gunsmiths put on barrels using their own methods and special tooling, but they are just as meticulous. Some gunsmiths may think this is excessive, but I have found that the rifles shoot better if I take the time and effort. Throughout the entire rebarreling process I know exactly what tolerance I allowed on each of the steps. Tolerances have to be kept to a minimum because they add up when two pieces are mated. Some of the larger custom shops, and of course, the big factories cannot afford to have this must attention to detail. Yield goes down tremendously when barreling a receiver in this manner. I would rather have low yield during machining than have to rework the barrel and action later when it won't shoot.

Redfield Palma rear sight and Lyman front globe sight are available as an option on TBA rifles. Warner, Paramount, and MCS rear sights as well as equivalent quality front globe sights are also available. (Mike R. Lau)

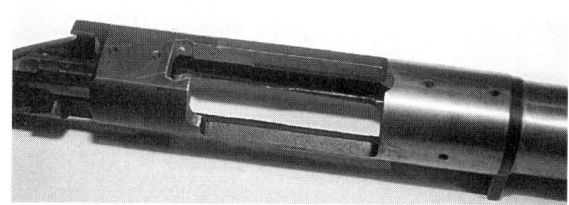

Clip slot is cut on all TBA Remington 700/M40A conversions for mounting of the Unertl type scope mount or for charger clip use. (Mike R. Lau)

Stocks and Bedding

TBA uses the McMillan fiberglass tactical stocks on the M40A series. These stocks already have a proven record of stability, strength, and accuracy. They come in a variety of camouflage and solid colors which are molded in epoxy and not painted. When you order a stock from McMillan, you can specify "sniper fill" which is to fill the buttstock with epoxy/fiberglass to increase the strength of the stock. This also adds weight, so by leaving it out, the stock is lighter. McMillan can also produce an extra light weight stock of full dimension, weighing about 1 lb. or possibly less if I remember, by making the epoxy less dense. This is done by adding a special filler material made of very small round plastic balls to replace epoxy. It is only used as the filler material along the inside length of the stock and not to the outer "shell" of solid fiberglass and epoxy. TBA's McMillan stocks are finished with Pachmayer Presentation or Decelerator rubber pads and Wichita Engineering sling swivels, standard "Marine Proof" hardware.

TBA rifles are bedded with 1/2 inch aluminum pillars to the McMillan stocks using Devcon. I have used another brand in the past, but have found that it is not as strong as Devcon. It has too much stainless steel powder in it which seemed to make the compound weaker by reducing the amount of epoxy. When that other brand hardened, the epoxy had a crumbling effect when you cut into it, whereas Devcon appears to be much harder. Regular two-part epoxy without stainless steel particles is not used for bedding because it is not as hard as Devcon. However, it is used to seal the unbedded portions of the stock. A thin layer of Devcon separates the aluminum pillars from the receiver and floorplate/trigger guard assembly. 65 inch pounds on the stock screws appears to be ideal and this torque has already been proven to be right as both the Army and Marine Corps use this specification.

Barrels are bedded free floating except for 1-1/2 inches forward of the receiver. Bedding a short length under the 26 inch barrel gives the barrel additional support to it's heavy weight. For a shorter barrel this may not be a problem, but for a long 26" barrel the receiver can flex a few thousandths of an inch. Imagine what other stresses it is causing to the receiver during firing. The section of the barrel

CHAPTER 4: The Custom Sniper Rifle and Telescopes

that is bedded forward of the receiver should be parallel to the bore. A tapered barrel section will allow the bedding fit to get loose if the rifle sets back even slightly in the stock after the first few rounds are fired.

Floorplate and trigger guard of the TBA M40A series rifles are the modified Winchester 70 steel parts, just like on the Marine Corps rifle. Modifying the parts are a little tedious, but are less expensive than the aftermarket one piece steel BDL types. They also maintain the classic appeal of the Marine rifle.

On TBA's M24 rifles, the action is bedded over the aluminum bedding block in the H-S Precision stock. This will take out the play between the receiver and the stock along the sides.

The Rifle's Finish

Finish on the steel components of tactical rifles are usually matte black or olive green. This can be applied in many different ways with different materials, but because the barrel is usually stainless steel, you cannot parkerize or blue it. Issue Marine Corps M40A1 barreled actions have a black oxide finish and are painted over by the Marines to match the environment of their mission. The black oxide is applied with a heated tank similar to hot blueing. All metal parts being finished this way, usually require a hydrochloric (HCL) acid pre-dip in order to etch or "gas" the outer surface of the metal. A very thin oxide layer of black color is formed on the surface of the metal that is fairly durable and looks really nice. If the finish is scratched by an object that leaves its color on the black oxide, you can sometimes rub the surface discoloration off and get back the black finish. If the oxide is actually scraped off, then you have to match the oxide's flat black color with paint. The only way to get back the black oxide is by refinishing the metal parts. Extra care must be taken to prevent the HCL and the hot black oxide solution from seeping into the bore. These chemicals will etch the metal surface like parkerizing does and can damage the bore and the sharp edge of the crown. To slick up the action surfaces where the bolt slides in and out, the black oxide should be removed which requires honing the action. This is tedious work. Black oxide is actually a finish for stainless steel and will not really provide weather protection for the chrome moly receiver and other steel parts. Because of the deep etching process to obtain black oxide, I do not recommend refinishing any rifle this way for the second time.

To get away from the black oxide finish, many large manufacturers and custom gun shops paint the metal parts. Remington uses their own trade named black paint on the M24, Robar uses Roguard, and L.O.D. uses an olive green paint of their own formula. Many of the paints in use on firearms today are commercial high tech polymer resin or epoxy paints that are similar to Sherwin Williams' Polane or Dupont's Imron. These paints usually require a base coat primer and can be thermally set with 130 degree heat to hasten the curing and adhesion. Similar paints are being used in the automobile, computer cabinet, and furniture industry so they are tough. Over the counter Krylon (Division of Sherwin Williams Company) produces a line of epoxy camouflage paints in flat black, OD green, and brown. These paints are advertised as made for camouflaging outdoor sporting equipment. They go on easily and come off hard. Brownell's has brought out a moly-teflon black matte paint that sets with oven cure at 350 deg for 30 minutes. I am familiar with all of these finishes because I have used almost all of them including black oxide. I personally like the moly-teflon finish because it is easy to put on, cures faster than the Dupont and Sherwin Williams, and is almost as tough. It is not as thick as the polymer resin finishes because it does not require a primer undercoat so it is easy to touch up. It also does not get shiny like the polymer resin paints after you rub the finish a few times or carry your rifle in a gun case. The best reason why I like the moly-teflon coating is because I can leave it in the receiver. The teflon actually imbeds itself into the metal pores when the metal is heated to 350 degrees. Leaving the coating in the receiver actually makes the action slicker and yet provides a weather proof finish. Even if the black moly is worn off inside the receiver, the teflon is still there. TBA is set up to do any of the finishes described above as well hot blueing, parkerizing, and electroless nickel plating.

Final Assembly and Testing

TBA's .308 Remington custom rifles are checked for 5 round capacity and proper feeding through the internal magazine after final assembly. .300 Win Mag rifles hold three rounds. All triggers on TBA's custom rifles are adjusted to 2.0 lbs. with no discernible creep or overtravel. All TBA rifles are test fired with a minimum of ten rounds to check safety and functioning. Accuracy testing is also done at the same time and more rounds will be fired if necessary. Trigger pull quality is evaluated during accuracy firing and readjusted if necessary. Headspace is also rechecked on each rifle to insure it did not increase after test firing due to threads of barrel and receiver setting tighter.

Other Services Performed by the Small Custom Shop

A complete and experienced custom shop should be familiar with accurizing most of the popular firearms. Besides accurizing bolt action rifles, TBA performs full upgrades of M-14/M1As to National Match specs using Reid Coffield's "Ultra Strength" 5 pillar bedding method with the single rear lug. Other match conditioning procedures include replacing the standard barrel with a medium or heavy weight Kreiger or Douglas/Barnett match barrel, opening of the flash suppressor, welding the front ferrule to gas cylinder, NM sights, trigger job, and making all stock clearances. In addition, TBA upgrades AR-15/M16s for NM use by replacing the barrel, trigger parts, sights, handguards, and tightening improvements. Complete upgrade and accuracy work on NM, Practical Pistol, and Carry/Conceal pistols and revolvers are also done in the shop.

Gunsmithing Tip: 1000 yd Come Up. Elevation required for .308 Win caliber rifles to zero at 1000 yards requires between 35 to 45 MOA depending on muzzle velocity, barrel length, temperature, ammunition, and the way you shoot the rifle. Most of the commercial 1" tube telescopes have 35 to 45 MOA *maximum* adjustment for elevation. Don't count clicks when checking this because there

Texas Brigade Armory's Designated Marksman (DM) Rifle is a completely accurized conversion of the Springfield Armory M1A and is similar to the DM rifle used by the Marine Corps to increase sniper capability at the Fleet Security Force squad level. Stock is the McMillan M-14 and scope is Leupold's 3.5x10. Comb height can be increased with a leather cheekpiece made specifically for the M-14 by Turner Saddlery which can also be purchased from Iron Brigade Armory. (Mike R. Lau)

will be a few clicks on the extreme ends of elevation and depression that won't move the reticle. Use a boresighter to check this. Many of the scope mounting systems will mechanically center the scope so that you have about half of the elevation available which is about 25 to 35 minutes with about 20 MOA of depression (which you don't need). For a 1000 yard zero you need at least 40 minutes elevation for the .308 Win round. Many scope mount bases can be tilted to increase the scope's useable elevation by shimming the rear and epoxying under the front. Dual bases cannot be tilted as easily and shimming only the rear base will bend your scope. A couple of manufacturers have bases with a built in tilt angle machined into them. Or go to a custom gun builder, like Texas Brigade Armory, who will machine and mount a base for your scope. Also, when you are zeroed for near extreme limits of the elevation knob, sometimes you can only put a few clicks of windage on because the scope's inner tube has bumped the inside of the main outer tube at the top or bottom. When you force the windage knob to move the reticle to one side in this condition, you can damage the scope. The reticle will also follow the direction of the curvature of the inside of the main tube.

A solution to limited elevation and windage when using 1" tube scopes is to use the mil dot on the reticle's wire as an aiming point. The best solution is to go to a 30mm tube scope. Most 30mm tube scopes already have over 65 to 90 MOA elevation adjustments and their mounts already allow them to use the majority of this elevation.

Gunsmithing Tip: Leupold M3 Scope Mounting. The BDC cam in the Leupold M3 scope for any caliber allows you to elevate from 0 to 1000 yards or meters with one revolution of the dial. This means your scope mount must position your M3 scope fairly precise, so when you zero at 100 yards, you can relocate the "1" on the dial over the tick mark on the turret housing. If the scope mount does not position your M3 scope correctly, you will not be able to put the "1" on the BDC dial over the tick mark. The scope's angle is either pointed too high or too low for the dial and you are only able to sweep a portion of the single revolution of the cam. If this is the case, then your 30mm base and rings are not designed for the M3. If you put the Leupold M1 scope in an M3 mount, you will automatically lose about 15 moa to depression which is OK because the M1 has 65 moa elevation total.

Gunsmithing Tip: Old and New Remington 700 Receiver Heights. Occasionally I get a customer that tells me their one or two piece scope mounts don't fit their 700 receiver correctly and the scope won't zero. What I usually find is that the scope mount is not the correct one for the receiver. In 1974, Remington Arms Co. raised the height of the 700's receiver bridge by .017 inches to allow room for the anti-bind rail that makes bolt

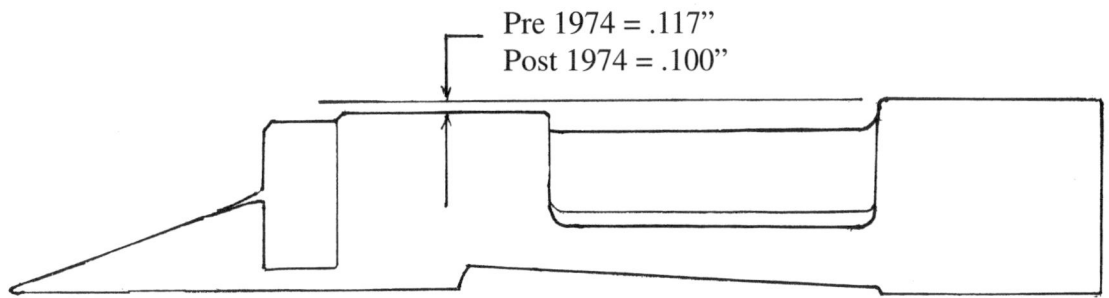

Fig.4.2. Remington changed the height of the 700's bridge in 1974 to accommodate the bolt guide rail in the receiver. Scope mount manufacturers followed up with base height dimension changes. (Mike R. Lau)

operation smoother. The receiver ring was not raised or enlarged to come up the same amount. As a result, the scope mount manufacturers had to reduce the rear base or rear portion of the one piece base to account for the increased bridge height. Using an old pre-1974 scope base on a new receiver will cause the back end of the scope to be to high causing the rifle to shoot low and some rifles will not even come up to 100 yard zero. Using a new, shallower rear base, on an older pre-1974 Model 700 will bend your scope when it is mounted.

Quality of Work

TBA is totally dependent on the quality of its work. Besides precision fitting the barrel to the receiver and carefully bedding it to the stock, all other work has to be of high quality because we deal directly with the customer. Every scratch, chip, gap, dent, missed spot, crooked cut, bad extraction, stiff bolt opening, heavy trigger, can be examined and pointed out at the leisure of the customer. If the rifle doesn't shoot the way the customer expected it to, or the workmanship doesn't look good, then TBA pays by having to re-work the rifle at our own expense or worst yet we lose a customer and everyone else he tells about us. We deal directly with the customer and the customer knows exactly who did the work, so he knows exactly who to complain to. I am the customer complaint department. The customer is Quality Control.

The Custom Sniper Rifle Design

All custom tactical rifles by TBA use the M40A1 barreled action built on the Remington 700. The Marine Corps Unertl type scope mount can be fitted for any scope. TBA also offers as options: any other popular caliber, different barrel lengths, different twist rate, other stock colors, and a fully adjustable buttplate or spacer system.

My Personal Favorite

My interest has always been in military weapons, but I was always disappointed in the accuracy of military rifles. Even the M21 with ART was not accurate enough and I never liked the inconsistent accuracy between one M21 to the next. I always lose interest in rifles that are not consistently accurate. I need to know that my rifle will shoot that nice ragged one hole 5 shot group at 100 yards if I do my part. A heavy target type rifle with its shiny stainless steel barrel or its awkwardly looking non-classic shaped stock is OK and has its purpose, but I want my rifle to look military. I like the feel of a rifle that carries easily in the field, but with the accuracy that could compete with the best high power competition rifles available. Then, in the 1980's, I discovered the USMC M40A1. The rifle had everything I was looking for. It was awesome looking with its heavy black barrel, camouflaged classic shaped stock, and a ruggedness that had the seal of approval: "Marine Proof." Best of all, it could be made as accurate as a competition target rifle because it was made from similar quality components. Hart barrel, McMillan stock, Remington 700 action, Winchester 70 steel floorplate and trigger guard, Unertl type scope mount. It was also appealing because it had nostalgia and history associated with it. The same kind of nostalgia and history you feel when you hold a Colt Single Action Army revolver or an M-1 Garand rifle in your hands. When you sight the M40A1's mil-dot reticle onto a target, you can't help imagining what it must have felt like to be LCPL Tom Rutter looking across no man's land in Beirut. You watch on the evening news as Americans and foreigners push their way through the security gate of the U.S. Embassy compound in Monrovia, Liberia. Then you get a close up of the Marines on the roof of the Embassy building as they observe across the street for potential problems. You get all excited when you no-

CHAPTER 4: The Custom Sniper Rifle and Telescopes

For all around use, the TBA M40A1 is the classic standard field grade sniper rifle. It is TBA's most popular model weighing only 13.75 lb. with Mark 4 M1 scope. (Mike R. Lau)

TBA's classic heavy rifle is the M40A2 SEAL type. It has a heavier type stock with an adjustable metal saddle cheekpiece. The deep forend was originally designed to accommodate the extension stud on the front end for the Parker Hale Bipod. It is a fine prone or static position rifle because of it's weight and larger size. It is a very comfortable stock when firing the .300 Win Magnum caliber as it absorbs recoil well. Rifle in .300 Win Mag with Harris bipod and Leupold 10X Mark 4 scope weighs about 15.8 lbs. (Mike R. Lau)

The improved field grade sniper rifle is TBA's M40A3. The McMillan stock has a pistol type grip and slightly flat forend. It also comes with an optional integral adjustable cheekpiece. It is a good compromise between the sporter M40A1 style stock and the heavier tactical rifles as it combines some of the best features of both. A lot of shooters are going to this stock because the combination straight and curved grip feels comfortable when shooting from the bench or very low prone position. The slightly curved underside and thumb cutout on the top side of the grip allows for long periods of comfortable field "ready" carry with finger near the trigger while moving. Rifle weighs 14.75 lbs. Fluted barrel reduces weight by 1/2 lb. (Mike R. Lau)

tice one of the Marines is holding an M40A1. You know the M40A1 has been field tested in Grenada, Panama, Somalia, Haiti, Beirut, Bosnia, Saudi Arabia, and a lot of other hardly known places like Monrovia.

I don't worry about a vine getting caught on some fragile part or if I accidentally knock the rifle against a rock or tree. I don't worry about an adjustable cheekpiece or buttplate coming loose in the field or getting filled with mud because there are none. I don't worry about the buttstock cracking or breaking off if I slip or suddenly have to get down and use the stock to break my fall while wearing a full Alice pack. I like the looks and feel of the checkered Remington bolt handle and the way it gracefully sweeps back. I never tire of the dark non-reflective matte finish or the dull sheen of the beautiful combination of camouflage colors on the stock. The rifle doesn't glitter nor is it gaudy. It has a warm luster and soft glow about it. I don't worry about the rifle's finish getting scratched or nicked when I am in the field for days stalking and moving constantly. It is a field grade rifle and is meant to take wear and abuse. A rifle that is in pristine shape tells me the user doesn't use it a lot. Look at Kent Gooch's M24 or Sam Chesnut's TBA M40A1, or Sean Little's USMC M40A1. Nicks, scratches, wear on the finish, all tell a story of where the rifle has been and what it has done. Their rifles have character and personality. They are used all the time for real purposes and also win matches.

I like the way my 13.5 lb. M40A1, with the Leupold Mk 4 M1 scope, steadies onto the target whether from the prone in the grass or from a static firing position on top a building. Once in position, it is like an artillery piece that doesn't move. All I have to do is adjust the sight and press the trigger. I like the way the classic grip of the M40A1 fits my hand whether I am shooting off the bench or firing at a moving target offhand. I like the feel of the slim rounded forend when I walk up the steep slope of a ridge looking for game, or when I have to run 300 yards to get to my next firing position. When I sling the rifle across my back to free my hands, the rounded stock doesn't cut into my back. Yes, that classic rifle is called "classic" because it will always be the favored and never go out of style regardless of how many fads or trends in rifle design come and go. Yes, I like the way it shoots, the way it carries, and the way it looks. Today, the M40A1 is still my personal favorite, just like it was many years ago.

Author behind USMC M40A1. (Mike R. Lau)

CHAPTER 5: TERMINAL BALLISTICS

This section concerns the placement and effects of bullets on the human body. DO NOT SKIM OVER or DELETE this section from your training. It is fundamental to all tactical shooting, whether it be rifle, pistol, shotgun, assault rifle, or submachine gun. It is one of the most important aspects to the understanding of tactical shooting whether you are a military sniper, police sniper, or civilian competitive tactical shooter.

Military and police sniping deals with the shooting of human beings, persons, men and women, and sometimes children. We take away their identity, make it politically correct, and easier to kill them by labeling them "enemy soldiers," "felonious suspects," "terrorists," "criminals," or "target." It is the duty of the American soldier to use deadly force as soon as possible in order to keep the enemy from destroying our own troops. It is the policy of the police officer to *not* use deadly force until the situation makes it necessary, usually after alternate solutions fail. The police sniper will use deadly force on a suspect to save the lives of, or prevent bodily harm to hostages, bystanders, fellow officers, and him or herself. For the military sniper, wounding enemy personnel about to fire on friendly troops, is acceptable if the enemy cannot continue to be a threat. For a military sniper, just wounding a drug lord that has ordered the killing of high government officials in a foreign country, does not accomplish the mission. The difference between the 8 ring and 10 ring on the 600 yard prone slow fire target is two points. The difference between shooting a criminal gunman in the brain stem or shooting him in the lower abdomen from 86 yards away, can mean death to a hostage or fellow officer. There are instances where the situation is so volatile and serious that the sniper must shoot the suspect at the first clear opportunity even if there is no immediate threat to life at the precise moment. In either situation, the sniper's purpose is to shoot to kill to prevent the suspect from doing anything else.

Real world tactical shooting, whether with rifle, pistol, or shotgun, involves shooting at human beings. It is hoped that you never have the need to put a human being in your sights and press the trigger. For the competitive tactical shooter, learning about terminal ballistics gives him or her a better understanding of what the military and police sniper has to do in real life situations. Competitive tactical shooting should require the use of targets that have body shapes and heads. (In practical pistol shooting, when persons complained about targets having heads, the targets sometimes have to be replaced with rectangles.) Competitive sniper matches not only use body silhouettes, but targets may have the vital shot

City of Dallas Police SWAT is considered one of the best units in the country. They handle several times more the number of SWAT type call outs than most other same size large cities in the US, yet the number of situations being resolved with the use of deadly force is one of the lowest in the country. Dallas SWAT also handles a large number (over 600 a year) of high risk narcotics/drug raids and arrests, yet has one of the lowest number of incidents of injuries to officers and felons alike. (Mike R. Lau)

placement areas outlined on the target. Hits in these areas count for more points. Hits outside these areas may count for less or no points even if there is a hit somewhere else on the target. Almost anyone can put a bullet through a full size upper body silhouette out to 400 yards even without correcting for a 5 mph wind and jerking the trigger. However, if the only shot that counted had to be in a 4" circle in the center of the chest, your would have to shoot with precision. Now you would have to correct for the wind, find the natural point of aim near the mid-upper center of mass, hold the crosshairs to a specific point on the target, and squeeze the trigger. If you did not know where the instant incapacitation areas on the target are then you will have difficulty imagining where these areas are. As you will see in the following chapters, knowledge of effective shot placement with respect to terminal ballistics is important in many of the tactical shooting situations and is referred to continuously. Military and police snipers already know terminal ballistics, and thousands of others have learned the same material in the past. Learning about where to effectively place shots on the human body will not make you a cold-blooded killer. If you find the subject disgusting and repulsive, then stay with shooting bullseye targets. It is important to know how to make an effective body shot. A police or military sniper kills to save his own life or that of another.

CHAPTER 5: Terminal Ballistics

Surgical Shooting Techniques

The target is the human being and the shot is not to center of mass, but on a very small, select area of the human body. Both head and chest shots will give varying degrees of *instant incapacitation*. Shots to other parts of the body may cause **temporary immobilization** and may eventually result in death. The Marines have classified body shot effects into four different types: (1) the instant kill shot, (2) the kill shot, (3) the emanate death shot, and (4) the disabling shot. They call this **Surgical Shooting**.

The Instant Kill Shot

If the terrorist gunman, holding a hostage, has his finger on the trigger, the trigger finger will relax and not jerk, when he is hit by the bullet. To get this result, the bullet must sever the gunman's *medulla oblongata* (brain stem) from the base of the brain. This junction area, where the brain and the spine join, controls body motor function or body movement. In the top of the *medulla oblongata* is the *pons* which transmits the motor information to the body and coordinates left and right side movement. On each side of the *pons* are the two halves of the *cerebellum* which control body balance. Below the *pons*, in the lower portion of the brain stem, is the *medulla* which controls breathing and heart rate.

The aiming point for the ***instant kill shot*** is a two inch by four inch area running through the eyes and ears and a two inch wide band running down the nose when viewed from the front of the face. A shot in this area will usually result in ***flaccid relaxation or instant paralysis*** and the body will fall like a "sack of potatoes." A rifle bullet traveling at high velocity (usually over 1800 fps) *may* cause flaccid relaxation even if it is within 2 inches of the medulla oblongata due to ***hydrostatic shock***.

The Kill Shot

A shot elsewhere in the brain area will cause incapacitation, but the body may experience muscle tightening with involuntary violent jerking or even flipping before death.

Many times the suspect does not give you a clear shot or the correct head positioning and you may be forced to adjust your aim or wait. In an actual situation, a bank robber came out of the building with a hostage, heading for the escape car. Officers arranged to have a police car down the street sound the siren yelp. The gunman turned his head to look and the sniper killed him. The bullet creased the hostage's face. That's how close they were.

You can usually tell how effective your head shot was by the way the suspect falls. If he goes straight down, limp, or pitches forward, flaccid relaxation occurred. If the suspect falls to the side, most likely you have partially incapacitated him. Always be ready to fire a second shot in this case to protect others.

The Emanate Death Shot

Sometimes you cannot get a good head shot on a suspect because he may be partially hidden by a barricade, moving, or there is a high risk of hitting a hostage or bystander if you miss. Another reason for not taking the

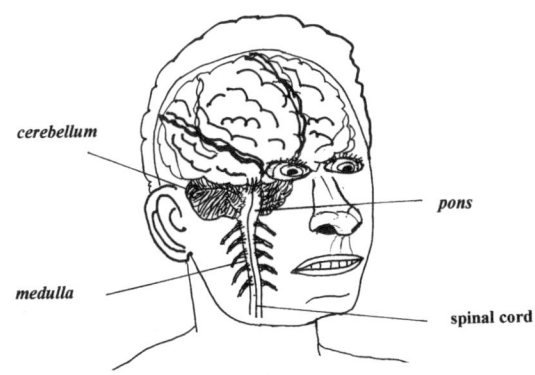

Fig. 5.1. The Brain Stem (Medulla Oblongata) is about an inch in diameter and connects the spinal column to the brain. Severing or pulverizing the area of the cerebellum, pons, or medulla will cause instant incapacitation or "flaccid paralysis."

Fig. 5.2. Your aiming point on the target when facing you is along a line between the tip of the nose to the center of the bridge of the nose. A hit anywhere between the eyes up to the eyebrows and within a an inch of either side of a line running down the center of the nose will cause instant incapacitation. The eye and nose areas have less protection offered by teeth or the bone. If you are above the target or he is looking down, aim for the top of the forehead or hairline. If the target is looking up with his chin up aim for the bottom of the nose.

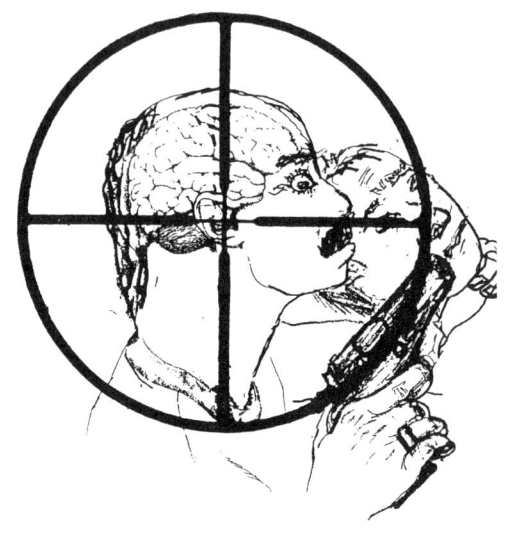

Fig.5.3. A profile target's aiming point is the ear canal or the temple. Many pathologists and doctors say aiming for the center of mass of the head will get the required results. In front of the ear is the cheek bone and below and to the front is the jaw. There is a hard bone right behind the ear that could "soften" the effect of the bullet. Be specific in your aiming point.

head shot is because you don't feel confident enough to accurately place the shot. Your own skill, your weapon's accuracy, or if the distance is a little too far, may be the reason. It is important that you decide when not to make the head shot. If this happens, the body shot that will cause incapacitation is the heart shot or mid sternum shot. If the heart is missed, oxygenated blood in the brain can function from 30 to 60 seconds. Coordination with a second sniper to take an immediate second shot is usually considered in most law enforcement sniper situations.

The Disabling Shot

This shot is made in a non-vital body area, and whether intentional or not, immobilizes your target. Most think of a shot in the leg or shoulder as such areas, but these could also produce death if not treated promptly.

Tissue and Organ Damage

When a bullet impacts the human body it will stretch and tear the body tissue that is directly in it's path. Soft point or hollow point bullets, designed to expand, will create a wound channel the same size as the deformed bullet. This path of damage is called the ***primary wound channel***. If the bullet breaks up in the body tissue or upon impact with bone (***structural shot***), the fragments (including bone) may cause *several* primary wound cavities. A full metal jacket bullet that does not break up will penetrate the body deeper because of its retained energy.

As the primary wound cavity is created by the bullet, a secondary wound cavity is created by the pressure wave around the moving bullet. Since fluids cannot compress, this pressure is hydraulic. Muscle fiber will absorb some of the pressure and there will be

CHAPTER 5: Terminal Ballistics

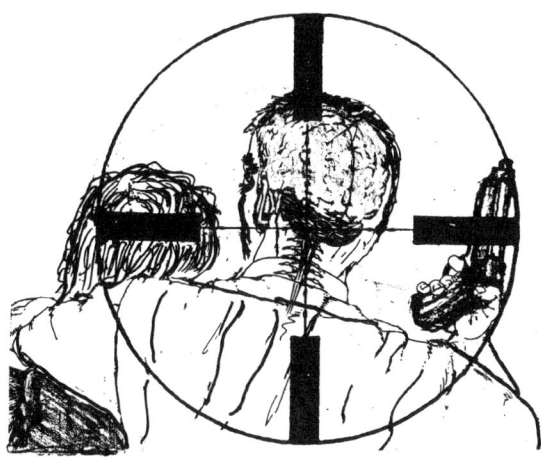

Fig.5.4. For a target facing away from you, aim about 1 inch below the base of the skull where the neck begins. If you aim above the horizontal centerline of the skull the bullet may deflect upward slightly on the curvature of the head and your target may not be instantly incapacitated.

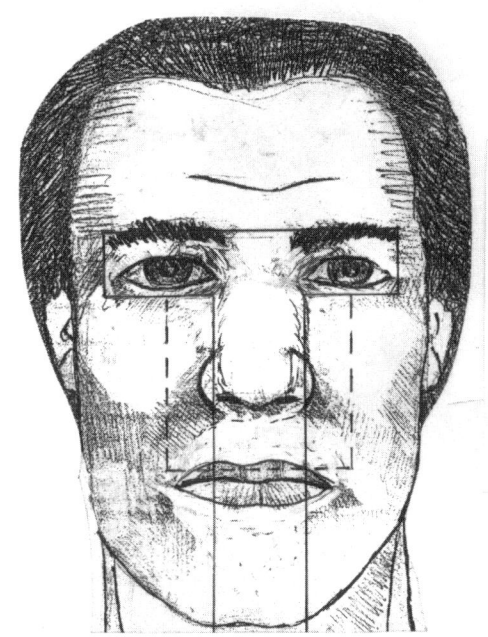

*Fig.5.6. The **FBI "T"** is represented by the solid line that travels up the center of the target's face and extends to the edges of the ocular openings of the skull. This represents the speculated actual area of the medulla, pons, and cerebellum. The dotted lines represent the **"functional medulla shot"**. A shot placed in this area serve the same purpose as a direct hit on the medulla due to the temporary cavity formed by the hydrostatic shock of the high velocity bullet.*

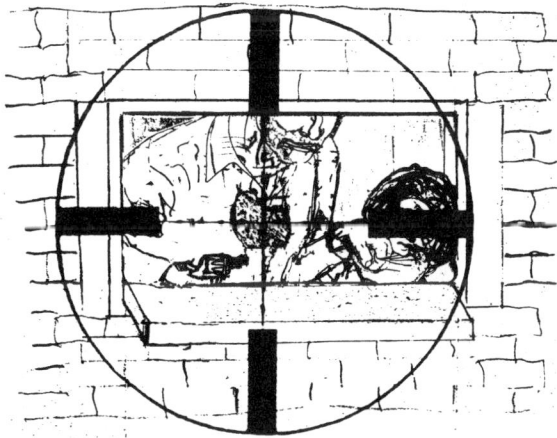

Fig.5.5. When you cannot make the head shot, your next point of aim is the mid-sternum. When the bullet hits the chest and expands or breaks up, the projectile and fragments of the bullet if any, and also from the chest bones will cause massive internal damage to the lungs, heart, and arteries. The bullet or fragments may also sever the spine. The criminal suspect may not be instantly incapacitated and may be able to move the upper arms for 1 or 2 minutes afterwards because of blood supply still in the brain. This is the reason why the law officer continues to fire even after the target is down.

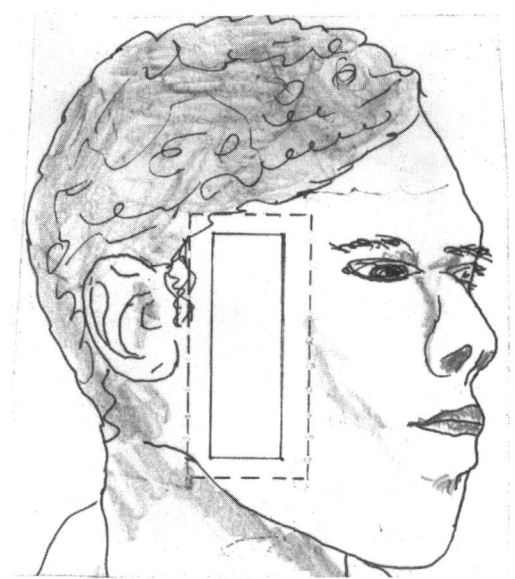

Fig.5.7. A slightly angled profile shot using the FBI "T". The bullet will have to shatter and penetrate the jaw bone, cheek bone, or teeth before reaching the vital area of the cerebellum and medulla.

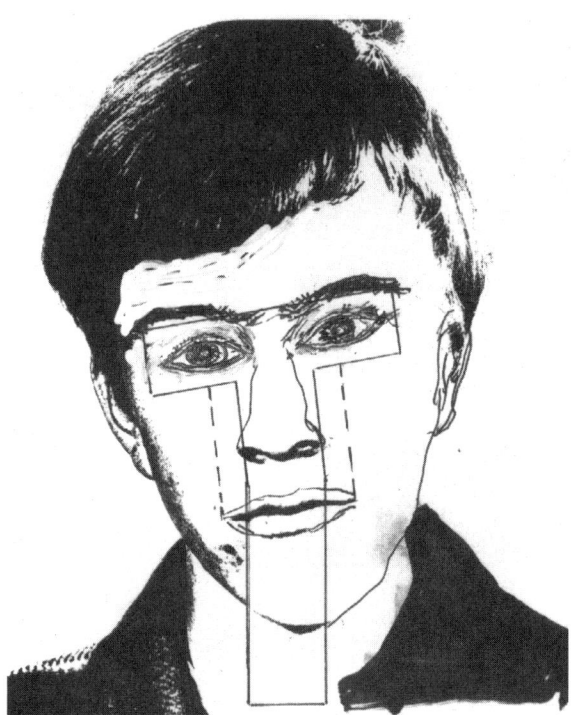

Fig.5.8. Another consideration for the sniper is that the suspect will tilt his or her head. This alters the visual location of the top of the brain stem in the neck area and the "T" is no longer straight.

Three reasons why the Dallas SWAT has an enviable law enforcement record is (from left to right) Sr. Cpl. Sam Lepere, Sgt. Bob Newton, and Sr. Cpl. Tony Black. These three veteran LE officers provide the bulk of the weapons and tactical training to the Dallas SWAT members. Each one has been trained as an instructor in almost every conceivable LE weapons and tactics school in the U.S. which includes everything from pistols, shotguns, sub-machine guns, assault rifle, and long range rifle. In addition they are certified NRA instructors in most rifle and pistol courses. All three have been or are currently certified NRA competitive shooters in rifle or handgun categories. These three put their tactical officers through demanding and continuous training that maintain their skill levels to peak performance and which result in one of the best SWAT units in the U.S.. (Mike R. Lau)

varying degrees of bruising and tissue damage. Organs, such as the liver and kidneys, are less flexible and cannot absorb as much of the pressure which results in greater injury to them. Body tissue and fluids are displaced temporarily and then return to it's position around the primary cavity. Damage to tissue in this secondary channel can be none or severe depending on the force of the pressure wave or type of organ nearby. Projectiles traveling at high velocity create greater pressure waves and consequently greater tissue damage. 5.56mm bullets have a tendency to be deflected by bone and will take another path following the surface of the bone until deflected again.

When the 5.56mm SS109 (M855) 62 gr. bullet was being considered by NATO, there was some concern that the lethality of the new bullet would be reduced because of the lower velocity and greater stability (non-tumbling effect) caused by the fast 1 in 7" twist barrel. A study was conducted by Dr. Martin L. Fackler at the Wound Ballistics Laboratory of Letterman Army Institute of Research, San Francisco, California, to determine if the 62 gr. bullet would exhibit the same lethality as the 55 gr. projectile. Each type bullet was fired into gelatin blocks, which simulate body tissue mass, with the 7.62mm NATO Ball round as control. The results showed that the 62 gr. bullet was just as lethal. The extensive tissue damage caused by the small 5.56mm/.223 bullet is not only caused by high velocity and pressure, but the bullet tumbling and then breaking up. As the bullet turns 90 degrees in body tissue, it bends, flattens slightly, and then breaks up at the crimping cannelure, causing the rear portion of the bullet to fragment into

CHAPTER 5: Terminal Ballistics

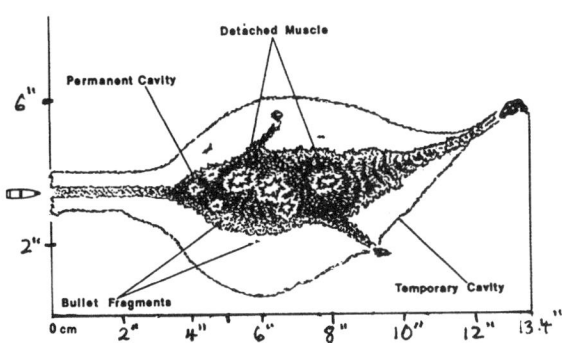

Fig.5.9. Wound profile of 5.56mm M193 Ball.
Vel: 3094 fps. 55 gr. FMJ-BT.
Final Wt. of Bullet: 35 gr.
36% Bullet Fragmentation

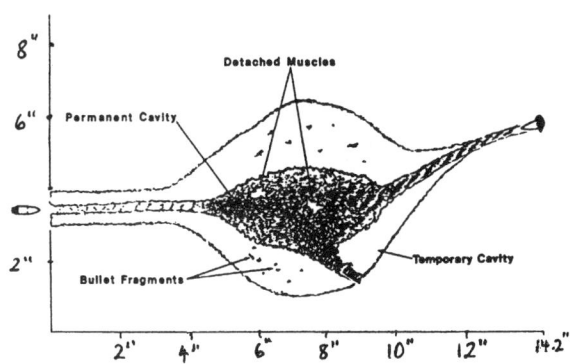

Fig.5.10. Wound profile of 5.56mm M855 Ball.
Vel: 3034 fps. 62 gr. FMJ-BT.
Final Wt. of Bullet: 31 gr.
50% Bullet Fragmentation

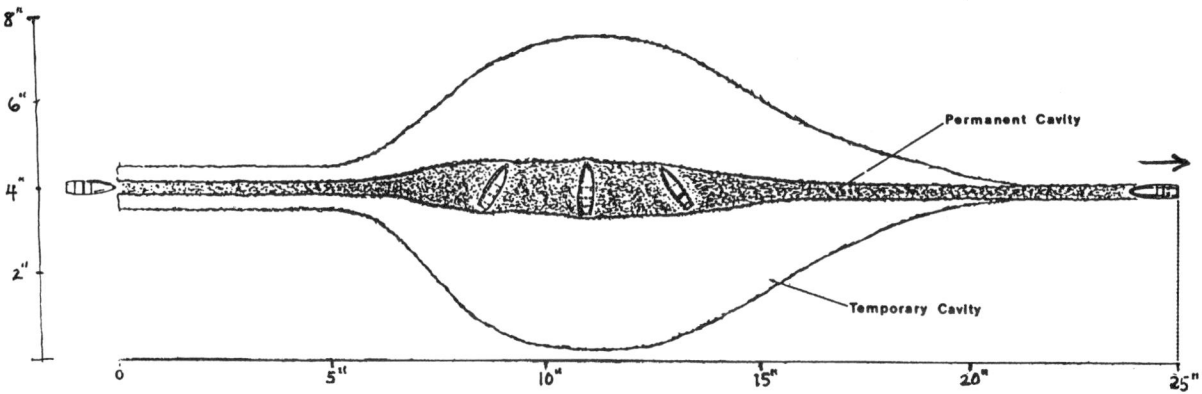

Fig.5.11. Wound profile 7.62mm NATO M80 Ball. Vel: 2830 fps. 150 gr. FMJ

many pieces. At the lower velocities beyond 200 yards, the breaking up of the bullet is much less or does not occur.

Shooting Through Glass: A Terminal Ballistic

With the increase in hostage rescue and anti-terrorist training by the military and police, general considerations should be made concerning taking out a target on the other side of glass. The Marine Corps conducted firing tests through different types of glass to determine if a target holding a hostage, with other hostages or bystanders nearby, could be engaged by the sniper. Types of glass included automobile windshields, plastic laminated safety glass, and plate glass from 1/8 inch to 1 inch in thickness. Bullets were fired at 90 degrees and 45 degrees to the glass. Targets were located from 1 to 12 feet from the glass. Results and conclusions of the testing showed that:

(1) The jacket of the M118 ball would explode into fragments upon hitting glass. The fragments would follow an erratic path in differing directions on the other side of the glass. The bullet core would follow the original path, but would tumble after penetrating the glass.

(2) No matter what angle the bullet hit the glass, the path of the bullet core was not

Deer with nice rack taken by Mike Tucker with his TBA M40A1 at 150 yards. Small dark spot on neck above barrel is where Federal Match 168gr. HPBT entered. Other side of neck showed same small exit wound. Second bullet entered the left shoulder and exit wound of greater size can be seen on right shoulder. Mike said internal damage to deer's major organs in heart/lung area were extensive. Sierra does not recommend that any of its HPBT MatchKing bullets be used for hunting due to poor expansion characteristics. (Mike Tucker)

affected up to 5 feet beyond the point of initial impact.

(3) When glass is hit by a bullet at an angle, large amounts glass fragments become secondary projectiles on the opposite side and travel perpendicular to the flat surface of the glass. The remaining glass fragments followed the bullet core in a cone shaped pattern along the path of the bullet.

(4) Large fragments of plastic from the laminated safety glass embedded itself in the target that was one foot from the glass. In addition, there was damage to the target caused by bullet jacket fragments.

(5) If the target is from one to two feet from the glass, there was very little, if any, secondary fragmentation impacting on hostages or bystanders to the target's immediate left or right.

(6) As the distance of the target from the glass increased, the cone of glass fragments gets wider causing a higher probability of friendly personnel to be hit.

In tests conducted by others some other general comments can be made:

(7) 5.56mm/.223 caliber bullets are not reliable for hitting targets on the other side of glass and almost all types break-up.

(8) 7.62mm/.308 caliber bullets are more effective than smaller calibers, and full metal jackets are more effective than some hollow point or soft point ammunition.

(9) Some tempered glass may absorb most of the bullet's energy to cause it to be ineffective in causing major injury or lethality to subject target.

(10) If not a hazard to hostages or bystanders, two snipers can take a shot simultaneously at the target. One of the bullets will break the glass and the second should take out the target.

In addition, a test was conducted at Ft. Meade in 1984 to examine the effects of the sniper firing through glass located one yard in front of himself. A target was placed 100 yards from the sniper. Both 7.62mm M118 and 7mm Magnum rounds were fired at the target. One round hit the target in the abdomen and another in the right hip. These were considered lucky shots and proved that you should break the glass, like in the movies, or open the window, before taking the shot.

Other Glass Shooting Considerations

1. Sliding glass doors and windows that are opened means you have to fire through two layers of glass. Shooting through double paned storm windows and doors that are opened means even more glass.

2. Shooting at a target on the other side of an automobile (Beverly Hills, CA, shootout), may mean shooting through two layers of glass at different angles to each other. A moving automobile makes it even harder.

3. Window tinting materials acts like tape and will hold glass together.

4. Windows in high rise buildings are very strong or thick to protect people from falling out. Learn about these from glass distributors in your hometown.

5. Deflected bullets, lead cores, and steel penetrators still have high energy and pose hazards to persons unseen in nearby rooms.

6. In aiming at a target on the other side of glass, consider the 90 degree glass explosion on the other side. Aiming directly at the target may injure the hostage with glass if hostage is closer toward you than target. You don't want hostages to receive glass fragments because glass cannot be seen in x-rays.

7. Synthetic windows may sometimes be easier to penetrate, but may result in viewing distortion. On some bus windows, bullet may punch a larger diameter hole, maybe 2", but the resulting cracks make it difficult to see through.

Shooting at a target behind glass should be carefully considered if there are risks involved. However, if the sniper decides there is no other choice at the moment he should take the shot, considering the peculiarities of shooting through glass. Utilizing two snipers shooting simultaneously would be more effective. Snipers should receive training in shooting through glass so they can learn for themselves as to what to expect and can make good tactical decisions before taking the shot in a real situation.

In 1993, Quantico's Marine Corps Scout/Sniper Advanced Class (1-93) conducted a glass shooting experiment using M40A1s and different types of ammunition. Firing consisted of two and three snipers firing simultaneously at one or two Styrofoam targets or gelatin blocks on the other side of different types of glass, including curved automobile windshields. The results supported the original findings in that bullet and glass fragments, or the lead core of the first bullet to strike the glass, was found to hit the target effectively. It also demonstrated that shooting at the targets with more than one rifle simultaneously increased the probability of the target being hit effectively. What was most interesting was the use of Federal's .308 caliber 165 gr. soft point Trophy Bonded (bullet which was brought out around the time. The bullet's lead core and copper jacket are "fusion bonded" together. Bullets fired through glass hit the target with almost all of it's original weight, compared to the M118 Special Ball which fragmented. The bonded bullet by design is accurate and the soft point maximizes energy transfer to the target.

"When a situation gets to where you have to take down the suspect and all you see of your target is a foot, then aim for center of mass. If you shoot the foot, a leg may appear. Aim for center of mass. If you hit the leg, the body may appear. Aim for center of mass," Tony Black, Dallas Police Tactical Unit. (Mike R. Lau)

CHAPTER 6: INTERNAL BALLISTICS, ACCURACY, PRECISION, AND DEVELOPMENT OF THE 7.62x51mm SNIPER CARTRIDGE

There are two other types of ballistics, besides terminal, that the sniper must understand, to be effective in target engagement, *internal* and *external*. *External ballistics* are concerned with the projectile in flight and the many factors that affect it on it's way to the target. We will study these in the next several chapters and look at the different ways we can control these to maximize *accuracy* and *precision* in our shooting. Mr. Webster says *accuracy* and *precision* are synonymous. However, sometimes we refer to them differently. *Accuracy* at the target is affected by the *precision* of our ammunition, and mechanical design and assembly of the rifle. At the target, we sometimes refer to *accuracy* as hitting the center or the exact aiming point on the target. *Precision* is how tight a group we fire regardless of whether it hits the exact aiming point. Whatever way you want to describe the two terms, our primary concern is to hit what we aim at and do this consistently under any condition and in any situation.

INTERNAL BALLISTICS and it's Affect on Accuracy and Precision

Internal ballistics concerns what goes on inside the rifle before the bullet exits the muzzle. The ammunition consistency, speed of the firing pin, chamber pressure, velocity of the bullet traveling through the bore, and recoil, are some of these. For the most part, the sniper is required to use the rifle and ammunition that his unit or agency provides so he may have little or no control over internal ballistics. Many LE snipers purchase their own weapons to get higher precision and accuracy. Fortunately, the Army, Marine Corps, and many police departments in the US have excellent weapons and ammunition for their snipers and counter-snipers.

The DEA agent's TBA M40A1 rifle (left) was fired at the Desert Marksman Range in Palmdale, California. He was aiming at the bottom of the 1000 yard target. The rifle on the right is a .300 Win Mag built by TBA for Sam Lepere. Both Sam and the DEA agent were testing long range handloads at the time. (Sam Lepere)

Of course, to get the best internal ballistics, high quality match grade ammunition should be used. However, sometimes even that ammunition may not give you what you need by way of external and terminal ballistics. In choosing ammunition, there may be trade-offs between internal, external, and terminal ballistics that must be made, as you cannot have it all. Effective long range accuracy usually requires heavy bullets at high velocity to reduce wind resistance, retain supersonic veloc-

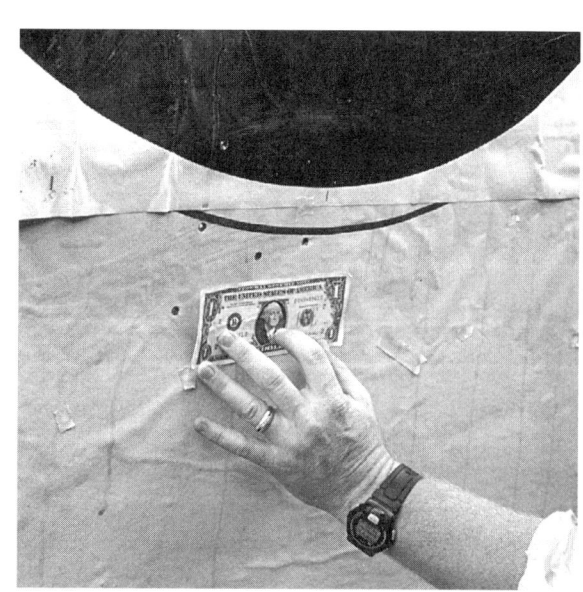

This 1000 yard 5 shot group, measures 5-1/2" across. Group was fired by a Los Angeles County DEA agent with a Texas Brigade Armory M40A1 in .308 Win caliber. Ammunition was handloaded using Federal cases and 210 primer. Powder charge was 41.0 grains Varget behind a 185 gr. Berger VLD. (Sam Lepere)

ity, and have enough energy to do damage. On the other hand, the rifle cannot be excessively heavy or have excessive recoil. Hollow point match bullets are usually more accurate than full metal jacket or soft point bullets. Full metal jackets are more reliable in penetrating the body, equipment, glass, and some barrier materials.

Determining Accuracy of Ammunition

Measuring bullet run out and weighing components can be fun and the information gathered can be useful. However, the military and law enforcement sniper is usually stuck with the factory loaded ammo their organizations select and issue to them so make the best of it. If we find, however, ammunition that will not shoot as well as we expect, we can always request a new lot or different manufacturer and hope for the best.

The way most of us test factory ammunition is to shoot it at 100 yards from a bench or prone with support. It is practical and convenient and a good place to start as it will eliminate a lot of problems associated with accuracy. You don't have to mess with unknown wind, elevation, and other weather factors. You can also shoot off a bench with sandbags and pay more attention to the way you get down on the rifle and press that trigger. Keep in mind, however, that accuracy at 100 yards does not mean proportionally the same accuracy at 1000 yards. A good example is 168gr. Federal Match round. From a good rifle, the cartridge will mostly shoot 5 round groups inside 1/2 MOA @ 100 yd. A few of these groups will be under or near 1/4 MOA at 100 yards. At the same time, shooting some not so good lots of M118 Match or Special Ball from the same rifle cannot be made to shoot inside 1.2 MOA @100 yd. Yet, when both types rounds are fired at 900 yards, the bad shooting M118 bullets shoot better than the Federal match. This is due to the effects of atmospheric conditions on the external ballistics of the round. The heavier 172 gr. M118 bullet is superior to the 168 gr. HPBT bullet at bucking the wind and retaining velocity at the longer ranges.

Many police departments around the country don't have facilities to shoot beyond 100 or 200 yards anyway and some don't really see the need to. On the other hand, many Sheriff's departments and State Highway and Public Safety agencies have the need to practice long range because they cover a lot of open territory. Many of these organizations also don't have easily accessible long range shooting facilities.

In selecting off the shelf ammunition for your department make sure you buy enough of a single lot each time your budget allows. There are accuracy differences between lot numbers and this becomes obviously apparent with extremely accurate ammunition like the Federal's Gold Medal Match. I have seen one lot shoot extremely fine groups in the teens consistently at 100 yd. Another lot will shoot only between .24" to .33" groups in the same rifle. Although for a police sniper rifle this is nothing to worry about, it can cause sleepless nights to the sniper who competes in sniper matches or is paranoid about accuracy. It is possible for ammunition lot differences to cause all your rifles to have totally different zeros and elevation adjustments.

Target accuracy can be measured in many different ways. A very accurate method that is used by the military and firearms/ammunition manufacturers is the *Mean Radius*

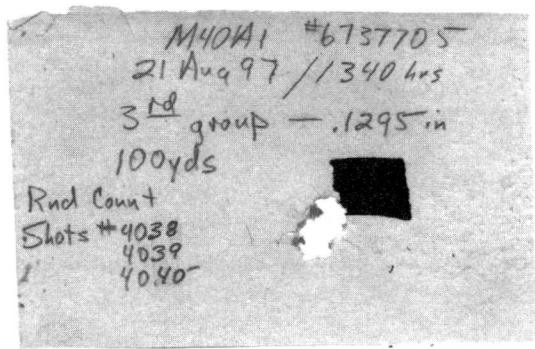

Typical group fired by Sam Chesnut at 100 yards from Texas Brigade Armory M40A1 in .308 Win caliber. Three shot group measures .1295" (Mike R. Lau)

(MR). This method of measuring requires finding the center of the group (*centroid*) and then measuring the distance from the centroid to each shot. Normally 5 to 10 ten shot groups are fired so you can get a good average of the distances to the centroid. MR calculation will reduce the effects of the obvious flyers in the groups, so resulting radius size of the group represents the majority of shots. A second method is to just measure the group across the two widest holes at the outside edge and subtract the bullet diameter. This is basically the Extreme Spread (ES) method of measuring the group size: center to center of the two widest shots. Most of the time we just discount flyers and write our excuses next to the hole: "jerked the trigger" or "called shot." ES is not statistically accurate because you are just measuring the distance between your two worst shots regardless of where the majority of the other shots are. However, ES has more meaning to us because the size of the group fired at 100 yards is basically our rifle/ammunition's real world accuracy expressed in MOA and accounts for our own skill too. Unaccounted for flyers in the group means you or your rifle system have problems that need correcting. If not corrected, then expect that kind of accuracy in the field.

The more groups fired at 100 yards, the better data you will receive on rifle and ammunition accuracy. We all know that 10 shot groups are usually larger than 5 shot groups which are usually larger than 3 shot groups. Normally several 10 shot groups are fired to reduce statistical probability error, but it may not be practical or economic to do so. A sniper's most important shot is the first one out of the cold bore, so 3 or 5 round shot groups are adequate. Let the rifle cool down between shots. Make sure you wipe out all solvent or oil from the bore before firing or your first shot will be high, even if it's freezing out there.

Any 3, 5, or 10 shot groups, can be converted to one of the other, without firing additional rounds. For example, if you want to convert a three shot group measuring .37" ES to it's equivalent 5 shot group ES, then divide the ES by .637 which equals .58" for the 5 shot group. Another example: Suppose you wanted to compare accuracy of similar type ammunition made by two different manufacturers. You fired Remington's .308 Win Match for accuracy using 10 shot groups to get an average ES. Later you tested Winchester's .308 Match ammunition by firing 5 shot groups to get the average ES. Comparison can be made by converting one of these to the other's equivalent group size using a multiplying factor from the table below.

Table 6.1. Conversion Factors for Obtaining Extreme Spread (ES) from Mean Radius (MR)

To obtain ES for:
3 shot groups	Multiply MR by	2.65
5 shot groups	Multiply MR by	2.90
10 shot groups	Multiply MR by	3.08

Table 6.2. Conversion Factors to Obtain Equivalent ES from Different Size Shot Groups Without Firing Tests.

If the number of shots in your group is:	divide the ES by	To get equivalent ES if you were to shoot this group:
3 shots	0.637	5 shot
5 shots	0.812	10 shot

Multiply the larger group's ES by the same factor if you would like to see what the equivalent ES would be if you were to fire the smaller group. These factors have been statistically proven with large sampling of fired groups by ballisticians working for large ammmunition manufacturers. If you fire only one three shot group to calculate an equvalent 5 shot group you may not be statistically accurate. However, it will give you an idea of what to expect, so have at it. Our time and ammunition is limited.

(Note: The late Creighton Audette asked Lake City AAP what relationship they saw between ES and MR. LCAAP random sampled the mean radius and the extreme spread of 126 ten shot groups of 7.62mm tests. ES was about 3.3 times the MR. If you are interested in some of the other methods of measuring and testing ammunition see Creighton Audette's 5 part article on "Testing Rifles and Ammunition" in *Precision Shoot-*

CHAPTER 6: International Ballistics, Accuracy, Precision, and Development of the 7.62x51mm

Building 45 at Lake City Army Ammunition Plant is where accuracy testing takes place for all ammunition produced by the facility. (Mike R. Lau)

ing Magazine, Oct, Nov, Dec 1994; Jan, Feb 1995. Explanation on how to do the calculations plus comprehensive, but easy to understand, treatise on internal ballistics, barrel and action vibrations, and more.)

Potential Accuracy of the Sniper Rifle and Ammunition

Military and law enforcement snipers have their differences, but they both share the need for very accurate weapons system. For the military sniper, engaging targets at long range is preferred, limited by specific conditions and weapon/ammunition system. In many military situations, the need for instant incapacitation of the human target is usually not necessary. An injury, or even a miss on a target, may not mean the end of the world to the sniper or the mission. He may get a chance to take a second shot or come back another day and try it again. When the Army and Marine Corps request improvements in ammunition with Lake City Army Ammunition Plant (LCAAP) and Army Research, Development, and Engineering Command (ARDEC), the resulting ammunition is measured by it's *"probability of hit"* on human size targets at specified distances. This assumes that the sniper is properly trained and his weapon system can realistically produce the accuracy desired. A more accurate sniper/weapon/ammunition system, allows a target to be engaged at longer ranges with a higher probability of hits. The maximum effective range of the M24 is 800 meters (70% probability), but this does not mean you cannot engage a target at 1050 meters. Military sniper ammunition should be chosen with long range accuracy as primary consideration. Terminal ballistics for military sniping usually means the projectile still has enough power to injure or kill at long range. This also means the bullet may have to penetrate heavy layers of clothing, personal equipment, body armor, and light cover or concealment such as brush.

7.62mm ball ammunition is tested at LCAAP with the old Mann "V" rest and the 1903 Springfield action with 21-7/8" heavy barrel. Ernie Williams rapid fires 10 rounds through the firing device to reduce the effects of wind. Target is blank paper on a roll at 600 yards. Target paper is retrieved and groups are measured after all groups are fired. The Mann rest is adjustable for windage and elevation. Lake City AAP has some of the same machines used to test Caliber .30 ammunition way back in the 1920's at Frankford Arsenal. (Mike R. Lau)

7.62 x 51 mm Match Ammunition

US military snipers have always demanded the most accurate ammunition produced by the military that was <u>*legal for use in combat*</u>. Developed in the late 1950's, Sierra brought out the .308 caliber 168 gr. HPBT match projectile which was loaded by several of the large commercial companies, as well as individuals, for 300 meter International Matches. Known as "Mexican Match" because many of these matches took place south of the border, the US government first addressed the legality of this bullet since much of the ammunition was purchased by the AMUs and MTUs with many active duty military shooters participating in the US team.

However, because the "Mexican Match" ammunition was developed for match use only (and packaged in white "non-military looking" boxes by the commercial manufacturers) the government declared this not to be a major concern.

Around the late 1950's, Frankford Arsenal also produced it's own, in small lots, of International Match 7.62mm ammo, designated T275. This round used the 172 gr. FMJBT bullet. Eventually the 168gr. HPBT gave way to FMJ loadings and the result was the 7.62mm M118 Match (military white box). 1967 M118, made by the Lake City facility, had a 600 yard MR of 1.73", or an approximate group size of 5.6" and it served well for

CHAPTER 6: International Ballistics, Accuracy, Precision, and Development of the 7.62x51mm

Business end of Building 45. Civilian technicians fire 100 round belt of 20 mm tracer in M61 Vulcan Gatling to allow author to photo. Took all of 2 seconds and almost missed getting photo of tracers. Large flat board with center cut out is a safety switch. If the seven piece 20 mm projectiles come apart when exiting the gun, the pieces will hit the board that will short out another board less than an inch behind it. The electrically operated Vulcan will immediately stop firing when this happens. Range goes out to 2400 yards on farthest hill. Observation bunkers along side the range allows technicians to observe ammunition effects at each back stop from a not too comfortable distance. (Mike R. Lau)

US military snipers in Southeast Asia. It is important to note that in the 1970's, accuracy of newly manufactured M118 began to wane and this was evident in the scores of many high power rifle match competitors who used it. This was one of the reasons why NM ammo was reworked by rifle teams. Going full circle, many shooters began loading the 168 gr. HPBT bullet once again. Demand by the military rifle teams, caused LCAAP to develop the prototype 7.62mm "Special Match" which later became XM852 match ammunition in the early 1980's. Many competitor's scores improved and the superior accuracy of the M852 over the M118 Match soon became apparent.

At the 1983 National Matches, Sgt. Larry Tedders, USMC, set a new national record of 200-15x on the 1000 yd. Wimbledon course using the 168 gr. HPBT in a long barreled custom bolt action rifle. Since many military snipers also participated in rifle competition, the accuracy of the M852 over the M118 did not go unnoticed by them. With it's hollow point bullet, however, the M852 was still considered illegal for combat use and boxes were marked "NOT FOR COMBAT USE." By the mid-1980's, M118 was down graded from "Match" ammunition to "Special Ball" (manila overlabel and later brown box). Accuracy with the downgraded M118 Special Ball in M40A1s was nothing to brag about, with extreme spreads grouping over 1 MOA at 100 yards. The Corps' original accuracy acceptability requirement of 1.5 MOA accuracy at 600 and 1000 yards for the M40A1 was becoming difficult to obtain with the M118 Special Ball. Cpl. Little told me that some M118 ammo he has fired in M40A1s may impact as much as 2 MOA lower than the corresponding range on the Unertl's elevation cam at the greater distances. Sgt. Sam Chesnut and SFC Kent Gooch have also encountered SB lots that could not be used because it was so inaccurate.

Early commercial white box "Mexican Match" .308 Winchester ammunition loaded with the Sierra 168 gr. HPBT bullet. Ammunition was loaded for Army AMUs and Marine MTUs and cartridges in Western box has headstamp of "W C C 6 0", while the Remington cartridges have "R A 60 (NATO cross symbol)." Bullets were long loaded to decrease bullet jump. (Mike R. Lau)

Frankford Arsenal 7.62mm International Match, T275. Cartridges are loaded with the 172 gr. boattail FMJ bullet. Note velocity of 2250 to 2300 fps which was determined to give the best accuracy for this bullet at 300 meters. Headstamp is "FA 56 (NATO cross symbol)." (Mike R. LAU)

Desire to use M852 for combat grew during the 1980's because of the increased use and importance of the US military sniper's role in places such as Grenada, Panama, and Beirut. Now that Lake City AAP was producing M852, this ammunition could be requisitioned and issued through normal military supply channels by combat units. So the legality of using hollow point bullets in warfare once more became a legal issue.

To provide an answer to this legal dilemma, studies were conducted by the Army's Judge Advocate General's Office (JAG) at the request of the US Army's Special Operations Command. (United States SOCOM, is a joint organization made up of the Army Special Forces, Army Rangers, Navy SEALS, and Air Force and Army special operations aviation. This combined force was designed to eliminate the miscoordination and interservice rivalries that caused problems in past operations such as in Grenada. Marine Expeditionary Unit, Special Operations Capable, MEUSOC, are not part of USSOCOM, but recently all Marine infantry battalions are now being trained for special operations.) A reply memorandum from Army JAG to Army SOCOM, dated 12 Oct 1990, made some very interesting arguments and conclusions that must be included when discussing military sniper ammunition. To make their case, JAG lawyers examined article 23e of the Annex to the Hague Convention IV of 1907, which prohibits "arms, projectiles, or material of a nature to cause superfluous injury" or "unnecessary suffering," as later defined. Also examined was the

CHAPTER 6: International Ballistics, Accuracy, Precision, and Development of the 7.62x51mm

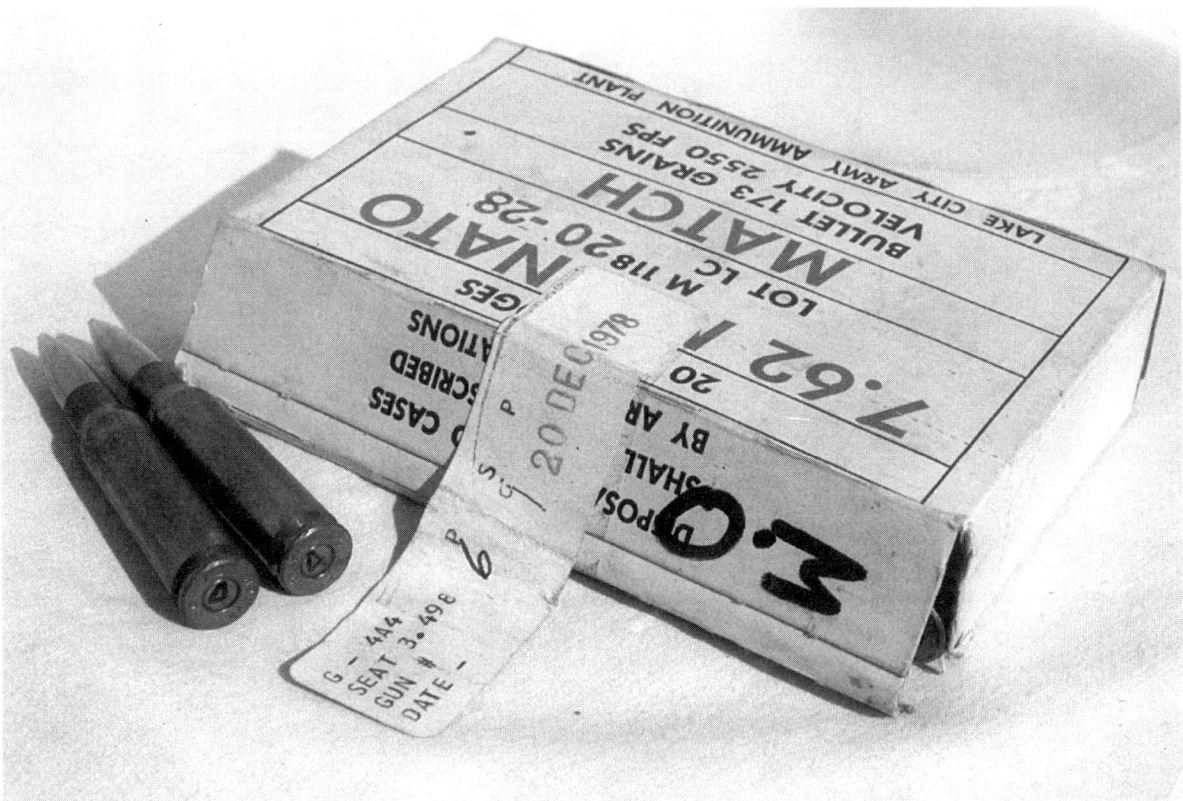

"Quantico Mexican Match". This box of 7.62mm M118 Match came from Marine Scout/Sniper Jim Furgeson for my collection. Jim says that Quantico has it's own loading room, the only one authorized in the Corps at the time, but by no means the only one. Large lot quantities of 1970's Match Ball had their bullets pulled with one collet type press. A second press would then reseat a HPBT Match bullet into the case. Sometimes the powder was pooled and measured individually before going back into the case. And then there were times when the cartridge cases were reloaded with new commercial canister powder of a "top secret" loading that would push the HPBT bullet to a higher velocity. Only the two persons in the loading room and the senior rifle team members knew what the powder charge was. Box in photo is one of these loadings and was made for or by CWO4 Bob (Gunner) Goller. Note label has G and 4 printed on it and a numeral "4" stamped on the primer. Headstamp is "LC 72 MATCH". (Mike R. Lau)

1899 Hague Declaration Concerning Expanding Bullets which prohibits the use of bullets with a "hard envelope which does not entirely cover its core or is pierced with incisions." Legal experts then provided reasons why these provisions were not violated. An argument made was that the "open tip" or "hollow point", of specifically the Sierra 168 gr. and 180 gr. MatchKing, was small and designed to provide better aerodynamics for long range accuracy and that they do not have the expansion characteristics that a normal soft point or hollow point hunting bullet would exhibit. It was noted that Sierra recommends that the MatchKing should not be used for hunting. In wound ballistics tests, conducted on material similar to human tissue and examination of persons actually hit by these bullets, it was determined that the MatchKing bullets did break up and fragment after entering the tissue. However, tissue damage was not to the same degree as that caused by the hunting bullets or by several *foreign NATO ball ammunition*. Examples specifically mentioned were the 7.62mm NATO bullet used by the Federal Republic of Germany and 5.56mm NATO ammunition used by Sweden. Tests were also conducted with 168 gr. bullets that had the nose

Top left: white box 1967 M118 Match; bottom left: 1983 M118 Special Ball with manila overlabel on white Match box; top right: 1992 brown box M118 Special Ball; lower right: 1992 brown box M852 Match. (Mike R. Lau)

closed up, called "closed tip," but these could not maintain the same accuracy as the normal "open tip" MatchKings. An important argument stated that recent wound ballistics research showed that almost all jacketed pointed military bullets have the tendency to break up on impact with body tissue, bone, and clothing/equipment worn by the soldier. This point being made to say that most pointed military bullets could cause needless suffering and injury and, therefore, all pointed bullets violate the article of the Hague Convention. However, since most of the world's nations have adopted pointed bullets for military use because of it's ballistics advantages, the pointed bullet was a "necessity for war" and outweighed the added injury or suffering it may cause. Finally, the legal people argued that the "principle of discrimination," a fundamental law of warfare, applied here. Methods used in combat should, when possible, distinguish between enemy targets vs. non-combatant targets such as wounded and sick personnel, medical personnel, and civilians. It was pointed out that Army and Marine snipers in Vietnam expended 1.3 rounds of ammunition for every verified enemy soldier killed at an average range of 600 yards. This is in contrast to 200,000 rounds of small arms ammunition expended per enemy killed when total usage was considered. Summarizing this argument, the legal experts stated that "the 7.62mm open tip MatchKing bullet provides maximum accuracy at very long ranges." Its tendency to break up after hitting a target is no more than ball ammunition and "is not a design characteristic" nor a reason for using it. "The military necessity for its use — its ability to offer maximum accuracy at very long ranges — is complemented by the high degree of discriminate fire it offers in the hands of a trained sniper." Conclusion: the 168 gr. and 180 gr. Sierra MatchKing HPBT bullets were in effect legal for combat use against enemy personnel. This conclusion was

CHAPTER 6: International Ballistics, Accuracy, Precision, and Development of the 7.62x51mm 115

concurred by Air Force, Navy, and Marine JAG, the Army General Counsel, and the Department of State.

So there we have it! Our military snipers can now use the M852 in combat for increased accuracy and it showed up in the Gulf War. But hold on, it doesn't end here! A couple of problems arise when using the M852 ammunition in Corps' M40A1s or Army's M24. First of all, it is known to many long range shooters that M852 Match, with it's 168 gr. HPBT bullet, may not always be as accurate as M118 SB at ranges past 800 yards. This is especially noticeable when shooting M852 in barrels 24 inches or shorter, in colder temperatures, or with high crosswind conditions. In my own experience with chronographing M852 and M118 ammunition, I also noticed that the first, cold barrel shot, is sometimes 15 to 25 fps slower than the average of the following shots. With shorter barrels and colder temperatures, both the 168 gr. HPBT and 173 gr. FMJ projectiles approach transonic velocities when nearing 1000 yards. Although the 168 gr. may start out faster at the muzzle than the 173, the lighter bullet sheds its retained velocity quicker and is slower than the heavier M118 bullet at all distances beyond 100 yards. When the projectiles approach transonic speeds it has a tendency to lose some stability and can be affected more by wind. This may be why we sometimes see M852 or 168 gr. Federal Match keyhole on the 1000 yd target and yet we still hear the loud "crack" of the still supersonic bullet when pulling targets in the butts. However, the 168 gr. may sometimes still print better extreme spreads than the 173 gr. at the longer ranges because the 168 gr. HPBT ammo is inherently more accurate if the cross wind conditions don't exceed 7 to 10 mph.

The second problem with using M852 in the M40A1 and M24 is that the trajectory is not the same as the M118. It will not correspond to the elevation cam in the Marine's Unertl 10X scope or the M3A's M118 BDC elevation knob. So what's the solution? Design a new projectile for the 7.62mm cartridge that has similar velocity, trajectory, and ballistic coefficient as the M118 SB, but with the accuracy of the Sierra HPBT MatchKing.

Improved M118 for LONG RANGE

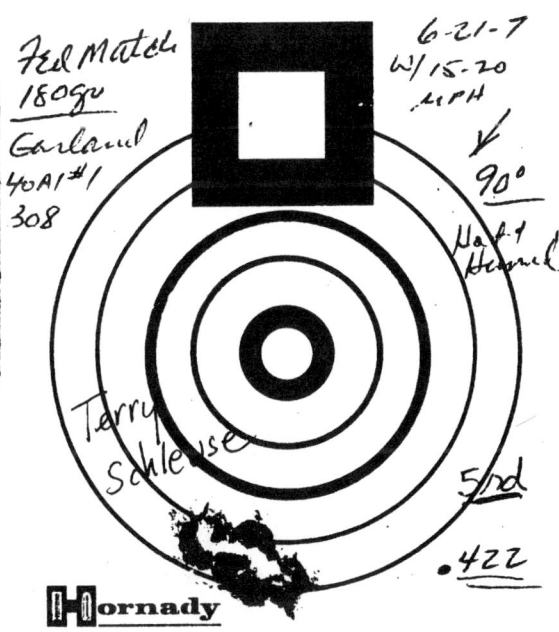

Federal 180 gr. HPBT Gold Medal Match made for in small quantity lot for SOCOM in 1994 and is not available commercially. Target on left was fired by Terry Schleuse using one of his three TBA M40A1 rifles. (Mike R. Lau)

Lake City Army Ammunition Plant has been operated under contract by Winchester/Olin since 1985. From left to right: Bill Melton, Plant Engineer; CPT Rory Tegtmeier, Plant Executive Officer; Dale T. Pollard, Contract Operations Officer; and LTC Michael W. Lambrecht, Plant Commander. (Mike R. Lau)

Sniping

In November of 1993, Marine Corps Systems Command issued a memo requesting that the accuracy of M118 Special Ball for the M40A1 was to have an objective accuracy of 1 MOA (10.47" ES) with a threshold accuracy of 1.5 MOA (approx. 15" ES) at 1000 yards. In the following year, US Army Research, Development, and Engineering Command (ARDEC), at Picatinny Arsenal, Dover, NJ, and the Lake City Army Ammunition Plant (LCAAP) began informal tests to develop a new 7.62mm cartridge that would have greater accuracy at 1000 yards than either the M118 Special Ball or M852 Match ammunition currently being produced. Under the supervision of Paul Riggs, (ARDEC), new ammunition was developed at LCAAP with the plant's contract operators of Winchester-Olin.

The new ammunition was referred to as "M118 Long Range," but nomenclature, packing boxes, and cans continue to designate the experimental ammo as "Special Ball." From here on, to prevent confusion, the new sniper ammunition will be referred to as "M118 LR" and the previously used M118 Special Ball as "M118 SB." In 1995, Winchester employees at LCAAP submitted cost estimates for necessary tooling improvements and initial production of M118 LR begun. Sierra developed a new .308 caliber 175 gr. HPBT MatchKing bullet for this project and an initial lot run of 60,000 bullets were produced. This new bullet is supposed to have a similar ballistic coefficient as the 173 gr. FMJ projectile in order to have a similar trajectory and velocity. Propellant charge of lot tested (LC-95M301S468) is 44.2 grains of Winchester-Olin's WC 750, a ball powder that is similar in appearance to WC 748. Produced by LCAAP, the primer is the same as that for

CHAPTER 6: International Ballistics, Accuracy, Precision, and Development of the 7.62x51mm

match ammunition as well as the use of the red primer sealant. Like all match ammo produced by LCAAP, a black tar substance is used inside the case mouth for waterproofing and the bullet is uncrimped. There is no body cannelure like that on the M852 case and the case mouth is annealed. Besides the hollow point bullet, the LR ammo is identified by a new headstamp with "LR" for "Long Range" along with the familiar "LC" and the last two digits of the year of manufacture. Cartridges are packed in a 20 round tan cardboard box with a plastic divider and a white overlabel is pasted to the front of the box. This new label is unique with the cartridge designation "M118 Special Ball," part number, and lot number, printed over a barely discernible Marine Corps Globe/Anchor emblem with "USMC" in the background. The words "NOT FOR COMBAT USE" is NOT on the label or box. Other changes being made to increase match quality of the cartridge will be to reduce case wall and neck thickness variations, more uniform centering of the flash hole, and removal of the flash hole burr. New tooling at LCAAP will also reduce bullet runout, case weight variations, cartridge overall length variations, and tighten headspace. New powder charging plates have been installed with a better sweep design to decrease powder charge variances. I have measured the powder charges from several randomly selected M118 LR cartridges which showed very uniform powder charge weights. Muzzle velocities of LR samples, measured by myself with an Oehler 35P, revealed insignificant velocity variations and was almost as uniform as 168 gr. Federal Match and my own handloads with individually measured powder charges. Chronographing M118 LR in my test rifles with 24" and 26" barrels showed average muzzle velocity of 2601 fps and 2710 fps respectively, when temperatures were around 60 degrees. On colder days, first shot out of a cold barrel went as low as 2524 from the 24" barrel. One interesting observation was that LC 79 M118 occasionally recorded 2550 fps from the 24" barrel when the temperature was 60 degrees. Both M118 LR and M118 SB have similar trajectories, with only very insignificant MPI changes. This is due mainly to the lower Ballistic Coefficients (BC) of the SB's 172 gr. bullet (.495/.483/.463) versus the higher BC of the LR's 175 gr. (.496/.485/.485).

Accuracy test set up at LCAAP for the new M118 LR consists of 6 new test fixtures utilizing Hart stainless steel barrels and Remington 700 receivers. These test fixtures are more like an actual M40A1 instead of the heavy Mann barrel fixture using the same 1903 Springfield receivers that have been in use since the dawn of time and still in use today at Lake City to test other small arms ammunition.

Preliminary accuracy tests comparing M852, M118 SB, and M118 LR, were conducted throughout 1995 at Quantico, VA, Ft. Benning, GA, and at Crane, Indiana. Both fixed mount systems and individual riflemen tested accuracy of the ammo and the tests results proved very interesting. The following

Mark Hudson, technical engineer at Lake City Army Ammunition Plant, prepares to assemble six M40A1 test fixtures for accuracy testing the new Marine Corps M118 Long Range sniping ammunition. The Hart barrels were assembled to Remington 700 actions for new machine rest test fixtures. (Mike R. Lau)

is only a summary of these results as much data was generated and would be too lengthy to reprint at this time. Most of the standard deviations calculated for comparison were acceptable and there were significant differences noted between the ammunition types in most of the specific tests.

1000 yard Accuracy Test: Crane, Indiana, 31 July 1995. Wind: 0-5 mph, Temp: 86-95 deg. (report was prepared by William McCombs, Paul Yester, Jr., J.D. Smith, and John Yarbor)

Rifle: M40A1 SN224211

	Radial ES	Hor ES	Vert ES	MR
M852	23.49	10.11	21.10	8.20
M118 SB	22.47	10.40	20.73	7.81
M118 LR	13.21	6.73	12.29	4.82

Rifle: M40A1 SN221195

	Radial ES	Hor ES	Vert ES	MR
M852	23.32	12.31	20.67	7.94
M118 SB	30.87	16.71	28.41	10.34
M118 LR	16.94	12.27	14.43	6.46

Rifle: Barreled Action SN 10408

	Radial ES	Hor ES	Vert ES	MR
M852	20.96	10.00	18.31	7.33
M118 SB	17.25	7.70	15.71	5.90
M118 LR	12.02	6.29	11.63	4.26

Rifle: Barreled Action SN 10407

	Radial ES	Hor ES	Vert ES	MR
M852	19.71	4.81	19.09	6.41
M118 SB	19.91	9.49	18.30	6.80
M118 LR	11.88	7.53	10.00	4.04

1000 yard Accuracy Test at Quantico, Virginia (date and temperature conditions of test not given in memorandum which was prepared by John G. Mardo):

2 fixed mounted rifles, combined data:

	Radial ES
M852	14.08
M118 SB	15.67
M118 LR	12.18

4 individual riflemen test, combined data:

	Radial ES
M852	N/A
M118 SB	22.36
M118 LR	18.54

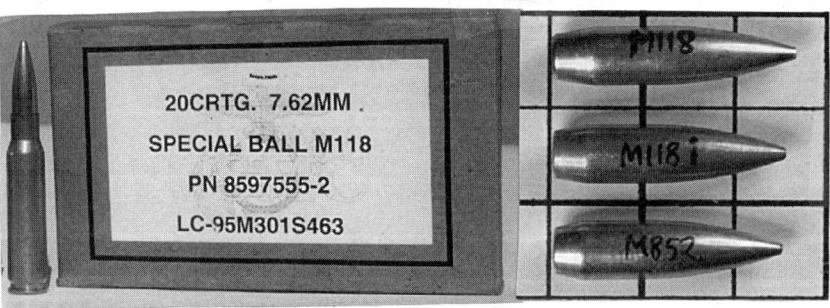

Headstamp of new 7.62mm M118 Long Range sniper ammunition. Photo was taken from unfinished Lake City AAP case without flash hole. Projectiles used in Army and Marine Corps sniping ammunition: 173 gr. FMJBT for M118 SB; 175 gr. HPBT made by Sierra and used in M118 LR; 168 gr. Sierra HPBT used in M852. Note that M118i is synonymous for M118 LR where "i" designates "improved". Further tests using commercial powders and bullet coatings are continuing. (Mike R. Lau)

CHAPTER 6: International Ballistics, Accuracy, Precision, and Development of the 7.62x51mm

1000 yard Accuracy Test at Ft. Benning, GA, using 2 rifles, combined data, no mean radius given:

	Radial ES
M852	15.80
M118 SB	21.11
M118 LR	13.51

Probability of hit on 10, 15, 20 inch target at 1000 yards range are as follows:

	10" TGT	15" TGT	20" TGT
M852			
Crane	.683	.903	.978
Quantico- fixed	.894	.992	1.000
Quantico- mksm	N/A	N/A	N/A
Ft. Benning	.858	.975	.997
M118 SB			
Crane	.631	.866	.961
Quantico-fixed	.850	.978	.998
Quantico-mksm	.592	.867	.972
Ft. Benning	.678	.910	.982
M118 LR			
Crane	.917	.995	1.000
Quantico-fixed	.945	.977	1.000
Quantico-mksm	.759	.955	.995
Ft. Benning	.934	.995	1.000

The following conclusions and comments were noted in the summary of both tests conducted:

1. Test personnel noted key-holing of M852 bullets on the 1000 yard target, but none for the other two types.

2. At 300 and 600 yards, the accuracy of the M118 LR was statistically better, or the same in some cases, when compared to M852 and M118 SB. At 1000 yards, average weighted ES from all sites showed M118 LR: 12.75", for M118 SB: 20.55", and for M852: 18.71". This represented a 38% accuracy improvement of LR over SB and 32% improvement over M852.

3. The difference in average mean point of impact (MPI) along a horizontal line of M118 LR and SB, and M118 LR and M852 was no greater than 1.20 MOA at 300 and 600 yards. At 1000 yards the MPI between M118 LR and M118 SB was the same. Significant elevation adjustments (not specified) were needed when switching from M852 to M118 LR or SB.

4. Data obtained on velocity indicated that the retained velocity between LR and SB was no more than 32 f/s average difference with the M118 LR being higher at all ranges. M118 showed an average of 50 f/s more retained velocity at 300 yd., 97 f/s at 600 yd., and 161 f/s at 800 yd. than the M852. (Author's note: The Crane report gave muzzle velocities and BC of 2691/.493 for M852, 2677/.541 for M118 LR, and 2664/.535 for M118 SB, which would indicate 26" barrels may have been used to get initial velocities.)

5. Based on Doppler/Pejsa analysis, it was conjectured that both the M118 LR and M118 SB projectiles are less susceptible to destabilizing factors when the bullet enters the transonic region when compared to the 168 gr. MatchKing bullet.

This new ammunition appears to be just what the Corps is looking for and will improve the Scout/Sniper's probability of hits out to 1000 yards and beyond. Continuing development and testing with the M118 LR is being conducted today to include studies on different propellants which are less temperature sensitive and gives more uniform velocities to decrease vertical spread. Also being studied is the use of NECO's moly coating and Kincaid's Danzac on the 175 gr. Sierra HPBT to increase accuracy and prolong barrel life in the M40A1.

From left to right: Michael Lodge, Senior Engineer, 7.62mm Ammunition Production, LCAAP; Mike Lau; Paul Riggs, and Orest Hrycak, Pistol Ammunition Production, ARDEC. (Mike R. Lau)

CHAPTER 7:
THE .308 WINCHESTER, 5.56mm/.223, .338 LAPUA, CALIBER .50 BMG, and MORE

Commercial .308 Winchester Ammunition

Today's standard police sniper round is the .308 Winchester and the most popular loading is the match 168 gr. HPBT. The manufacturer to beat is Federal Cartridge Company's .308 Winchester Gold Medal Match. This is the standard that some manufacturers compare to when they produce their own line of .308 Win match ammo. Federal's .308 GMM in 168 gr. HPBT loading is so accurate out to 600 yards that it will beat the pants off some good handloads. Some have told me that they've spent a lot of time match prepping cases and using competition quality handloading tools only to find out that the Federal .308 Match shot better or was equal.

I can remember years ago, when the only way to get super accurate ammunition that would exceed your own skill and the advances in weaponry was to handload. Now, thanks to companies like Federal, ammunition accuracy standards have been pushed to new heights and at competitive prices. As a result, the entire ammunition industry, including the military, is having to up their standards of accuracy and quality. What separates match ammo from other ammunition is consistency. Consistency in components, assembly, and performance. This is something that Federal, Remington, Winchester, Black Hills, and others have mastered to become the leaders in high quality ammunition. When shooter's prefer factory ammo over handloads you know something has changed.

Not too long ago I evaluated the new M118 LR for Paul Riggs. I began by comparing the M118 LR to currently manufactured commercial .308 match ammunition to see what the commercial companies were doing that LCAAP was not doing when it came down to component quality and assembly. As a result, myself and a team of experienced shooters and reloaders that assisted me, got to examine a lot of commercial match ammunition and components. We also did extensive reloading and firing tests to determine the feasibility of substituting commercial propellants in the Lake City LR case to improve it's accuracy. One of the things we learned was that there are a lot of ammunition makers out there that have very high standards of component quality and accuracy. Two of these companies, Norma of Sweden and Lapua of Finland, are already well known to the serious competitive shooting community, but because they are foreign manufacturers and their ammunition is sometimes not easily available, they don't have the popularity of companies like Federal, Winchester, and Remington, with the non-competitive community. However, we are not concerned with reloading or just match ammunition. We are concerned with the preferred commercial ammunition that is currently being used by the law enforcement community.

Considerations for selecting sniper ammunition for law enforcement use is similar to military ammunition with some additional considerations.

(1) It must be accurate and of high quality.

(2) It must be reliable and function in all your weapons that are chambered for it.

(3) The loading has to give you the desired terminal ballistics with considerations as to hazards to hostages and bystanders by excessive penetration, ricochet, excessive power.

(4) It should be purchased from a reputable manufacturer.

(5) It should be purchased in bulk for long use consistency and economy.

LE sniper ammunition, like military ammunition, must be able to withstand the excessive heat of the police cruiser and not give excessive pressure or extremely increased velocity when subject to high temperatures. On the other hand, like military ammunition, the propellant must not breakdown in extreme cold which also gives excessive pressure. Unlike military ammunition, LE ammunition does not have to be able to have the long storage life of decades. LE ammunition is shot up or replaced frequently by department policy to insure the officer always has new and reliable ammunition. Most of the large and small ammunition manufacturers in the US meet all of the above criteria.

The 168 gr. HPBT Match cartridge.

This .308 Win loading is the choice of most law enforcement agencies for use in sniper/counter-sniper operations. Popularity of the 168 gr. HPBT loading is due to it's accuracy and easy availability by several reputable manufacturers including Remington, Winchester, Federal, Black Hills, Hornady, and IMI. All of these manufacturers' ammunition appear to be of consistent high quality from lot to lot with only minor differences in accuracy noted. As discussed before, the HPBT match bullet made by Sierra is not designed for quick expansion like the soft point bullet and is not recommended for hunting game. It has a good BC and will give the penetration and terminal ballistics required for the LE target and at most LE sniper distances. At close ranges the bullet has been known to pass through the target with considerable remaining velocity and energy that could pose a hazard to civilians and other officers. Care must be exercised when setting up a perimeter to eliminate crossfire hazards. LE snipers are very cognizant of ammo lot accuracy because of the extensive practice they do and availability of weapons that give extreme accuracy. When Sam Chesnut consistently shoots .18 moa groups with one lot and the next lot goes .24 moa, he notices the difference.

The 175 gr. HPBT Match Cartridge.

This loading has only recently received a lot of popularity because of the military use of the M118 LR and it's increased performance at long range over the 168 gr. Beyond 600 yards, the 175 gr. exhibits ballistics similar to the M118's 173 gr. FMJ bullet by maintaining supersonic speeds longer and bucks wind better, but with improved accuracy. Most law enforcement agencies, however, continue to

CHAPTER 7: The .308 Winchester, 5.56mm/.223, .338 Lapua, Caliber .50 BMG, and More

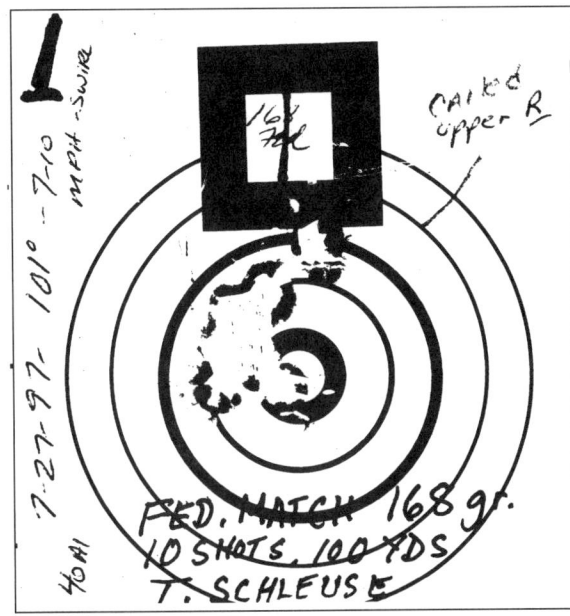
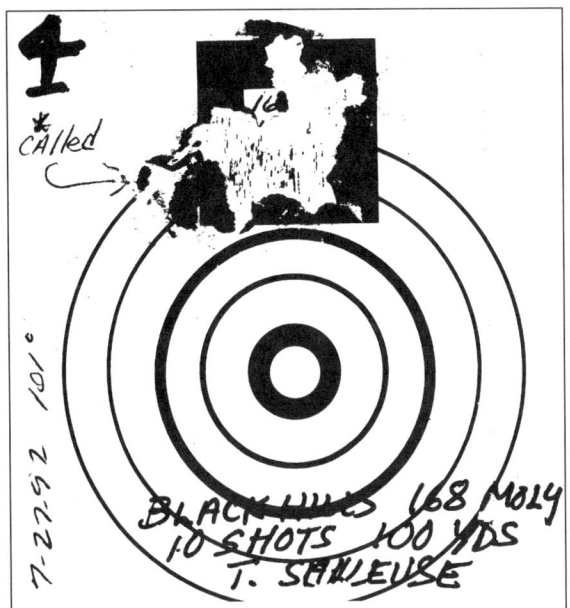
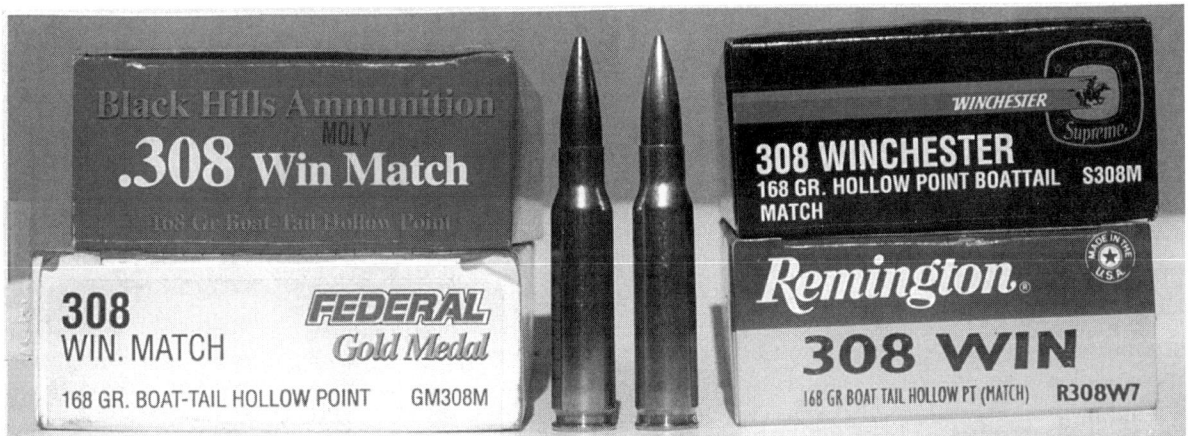

Four different makes of .308 Winchester 168 gr. HPBT Match ammo. (Mike R. Lau)

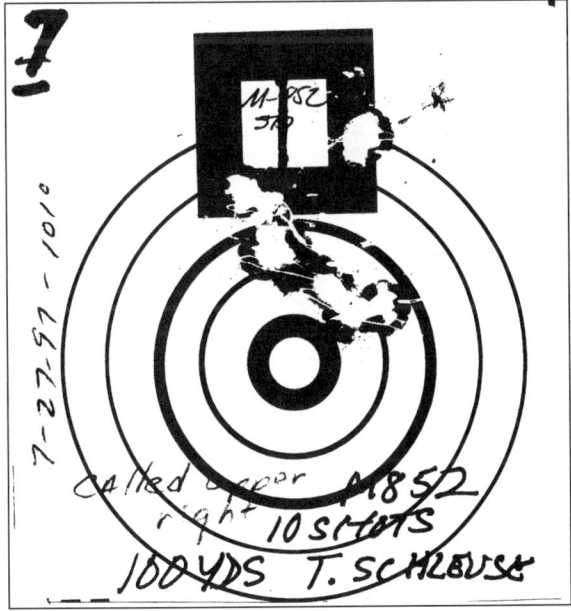

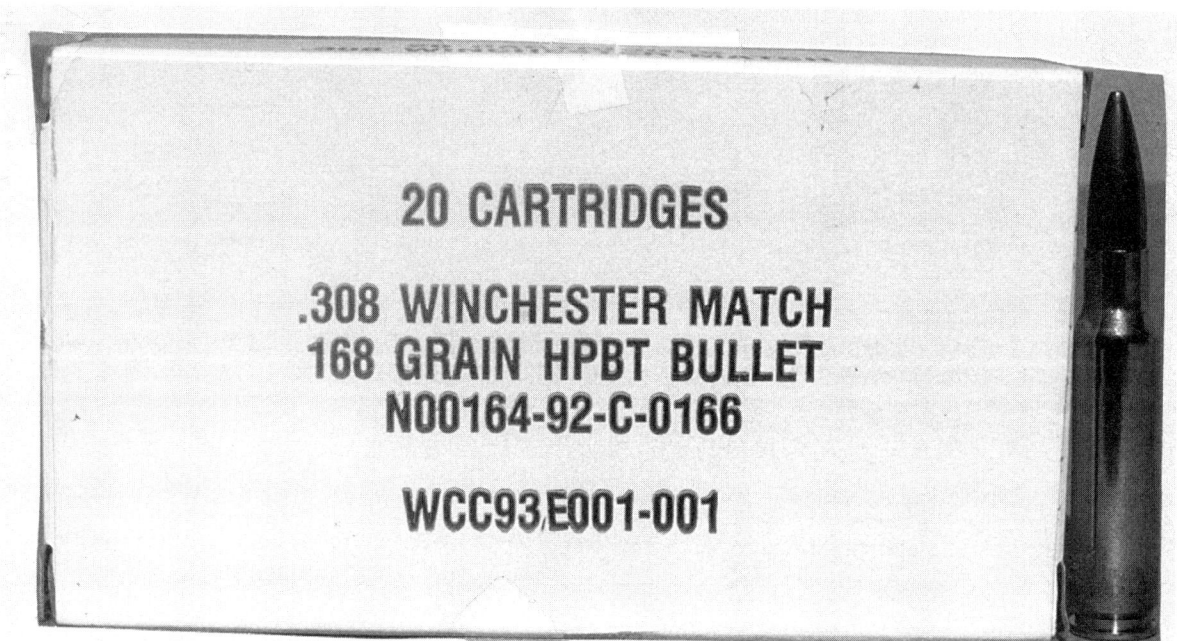

Military contract loading of Federal's 168gr. HPBT in 1992. (Mike R. Lau)

Winchester Ranger 168 gr. HPBT Match. This is the same loading as the commercial match so the Ranger tactical loading in .308 Win is not boxed this way any longer. Their Law Enforcement catalog now recommends the Supreme Ballistic Silvertip Boattail as an alternative for the Sierra or Hornady 168 gr. HPBT match loadings. This new bullet combines Winchester's Silvertip soft point into the Nosler Ballistic Tip design with the Lubalox coating (looks dark like Moly bullets except has Silvertip). This is an excellent idea, as it gives the LE sniper a bullet that will give the accuracy of a hollow point boattail match round, but with the desirable terminal ballistics of a soft point. (Mike R. Lau)

CHAPTER 7: The .308 Winchester, 5.56mm/.223, .338 Lapua, Caliber .50 BMG, and More

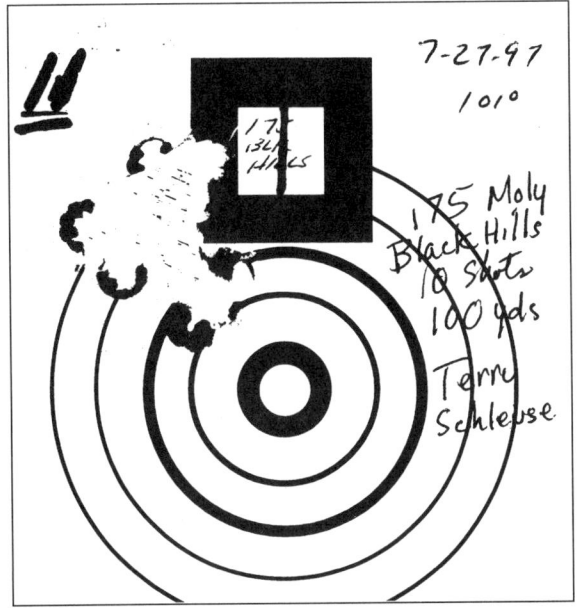

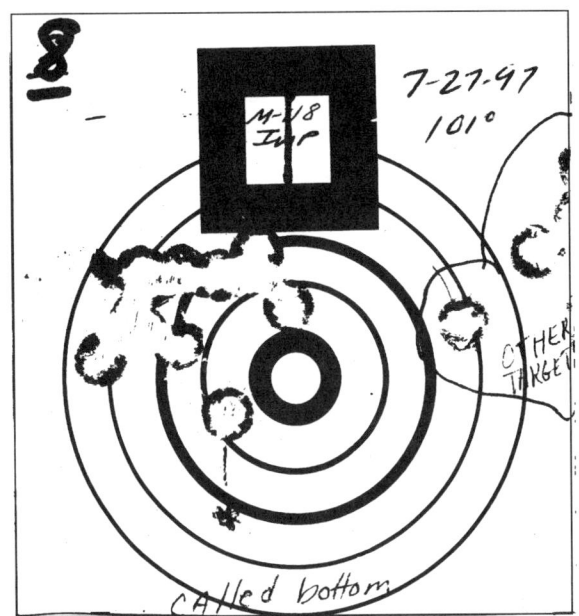

The 175 gr. HPBT manufacturers. Federal Match 175 gr. appears to put you closer to rifling by .01" for the same OAL because of longer ogive (bullet is wider toward the tip than 168 gr.). Accuracy at 100 yds is about the same as the 168 gr. (Mike R. Lau)

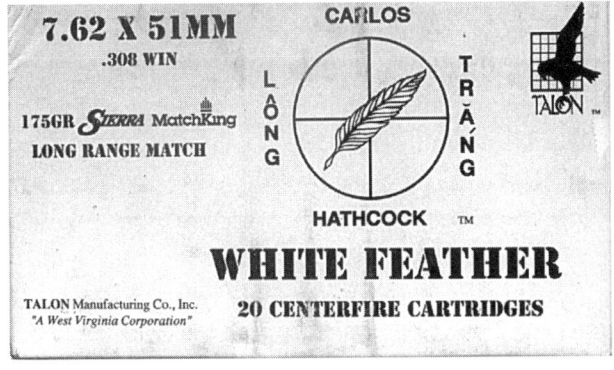

Federal Premium 165 gr. SP. This was the Arlington, Texas, Police Department sniper ammunition before they switched over to the 168 gr. Federal Match. (Mike R. Lau)

use the 168 gr. loading because there is no need to change. Long range competitive tactical shooters choose the 175 gr. HPBT factory load if they don't handload once they see the advantages of the heavier bullet. Black Hills, Federal, and Talon Manufacturing are producing this load. Black Hills also produces a moly coat loading.

Federal's 165 gr. Soft Point Classic or Premium and other .308 Win Ammo

This is still a very popular cartridge with many law enforcement agencies. The soft point bullet gives better terminal ballistics as far as expansion goes. Its reduced penetration makes it safer to use in urban environments than the hollow point match bullet. One drawback is that the ammunition is not as accurate as the .308 Win match ammo. Federal's 165 gr. bullet moves out from the 24" barrel fairly quickly at 2700 fps. It will impact about 1/2 to 3/4 moa higher than the 168 gr. HPBT using the same zero.

There are other .308 Winchester loadings worth mentioning that, although developed specifically for sporting or training uses, they can be used for special tactical shooting situations. I already mentioned the use of Federal's Trophy Bonded bullet being excellent for shooting through glass because of its reduced fragmentation yet complete expansion (see chapter on terminal ballistics). Another loading that has a reputation for accuracy yet gives good terminal ballistics is Federal's 150 gr. Nosler Ballistic Tip. Winchester has also combined with Nosler to market a 168 gr. Supreme Ballistic Silvertip loading which is their Silvertip loading with "Lubalox" coating, but the bullet is designed like Nosler's Ballistic Tip. Both Winchester's and Federal's Ballistic Tip loadings are designed to move the center of gravity towards the rear of the bullet similar to the HPBT match. The light and slick pointed insert in the bullet's nose reduces drag and increases long range accuracy, yet still allows for maximum expansion at most ranges. For maximum shock and terminal ballistics affects there are high speed light weight loadings such as Corbon's 125 gr. JHP whose velocity is listed at 3150 fps.

Accuracy of Commercial .308 Winchester Ammunition

Terry Schleuse is a new shooter and handloader. Ever since he owned a TBA M40A1 in .308 Win a couple of years ago, Terry has developed a tremendous interest in shooting and handloading because of the accuracy of the rifle. I asked Terry to do an accuracy test for me with several factory .308 Win ammo I had on hand. In order to not have him shoot a lot of groups and only select the best, I only gave him 10 rounds of each type and he was told not to rezero for each loading. The next several targets is what he came back with. Terry only uses a bipod *without* a sock or rest under the buttstock. The groups fired by Terry are not statistically valid and he is neither a top competitor or a bench rest shooter. He would be typical of most anyone shooting under normal conditions trying his best. The groups he shot are valid in that they are typical of what one person would shoot with one rifle on any one day under one condition. A military or police sniper cannot fire a number

CHAPTER 7: The .308 Winchester, 5.56mm/.223, .338 Lapua, Caliber .50 BMG, and More

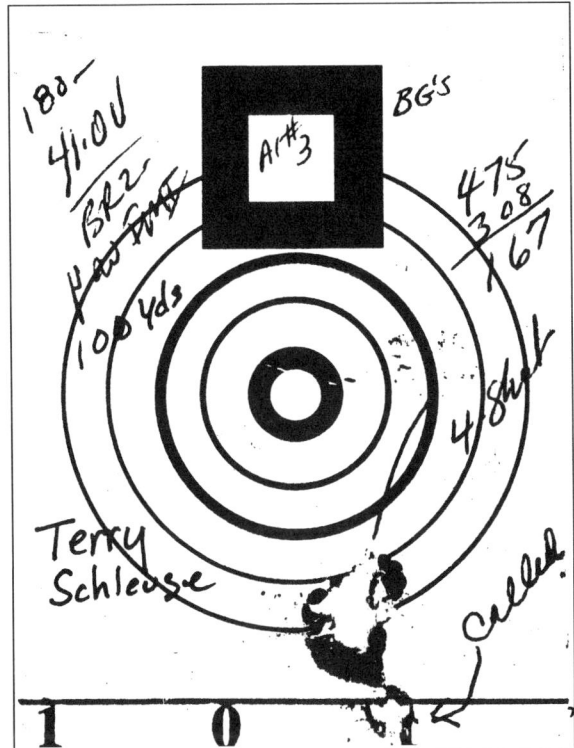
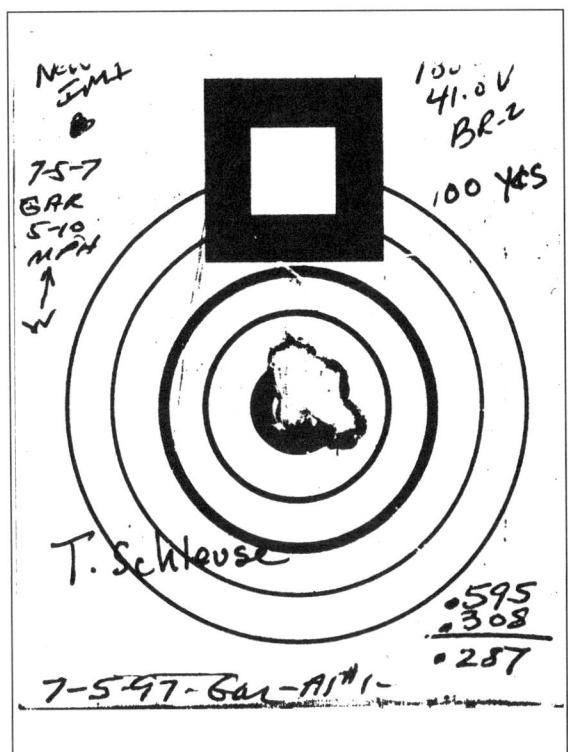

These two 5-shot groups were shot by Terry using handloads out of the same TBA rifle. IMI (Israeli Military Industries) brass was used. (Terry Schleuse)

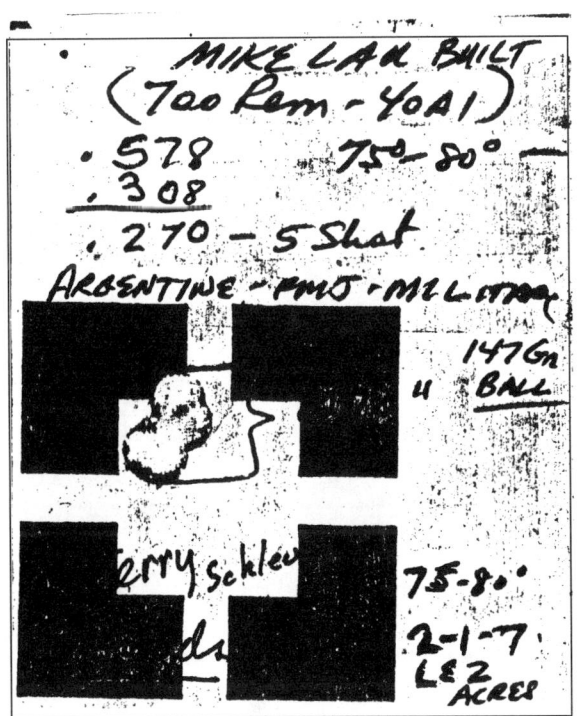

Terry shot this group with surplus Argentine 7.62mm ball ammo purchased from a surplus dealer that advertises in the Shotgun News. Other shooters that have bought this ammo had the same results. It is phenomenally accurate for very cheap ammo. (Mike R. Lau).

of groups at the hostile target and select the best group. What you shoot the first time out is what you get.

5.56mm/.223 CARTRIDGE for SNIPING

Designed as a replacement for the military 7.62 x 51mm cartridge, the 5.56mm (.223 Rem) M193 cartridge allowed the soldier to carry more ammunition in a lighter rifle. Engagement of enemy personnel is effective to around 460 meters with iron sights and provides a reasonable chance of a first round hit with the desired lethality. Critics of the M193 round made a point that the 55 grain projectile could not penetrate armor like the 7.62mm AP rounds. Its penetrating effectiveness was actually only effective to 400 yds. To increase its effectiveness, a prototype 5.56mm round with a 68 gr. heavy ball bullet was developed by Federal Cartridge Company in the late 1960's. Some were sent to Vietnam for sniper use. Meanwhile, Frankford Arsenal began to do it's own development with the XM287 using a similar 68 gr FMJBT. When fired from a 20" barrel with 1-9" twist, the projectile moved along at 2960 fps and had a trajectory, wind deflection, and penetration similar to the 7.62mm M80 Ball round. The XM287 had a problem, it would not be stabilized in the 1-12" twist found in the M16A1s.

When the US adopted the M16A1 as a standard rifle in the late 1960's, NATO set out to find a round that would pass its small caliber cartridge test. By the mid-1970's the US had developed another round designated the XM777 which uses a small *steel penetrator* inside the jacket in front of the lead core. The bullet weighed 54 gr. and was stabilized by the M16A1's barrels, but was not up to NATO Standards.

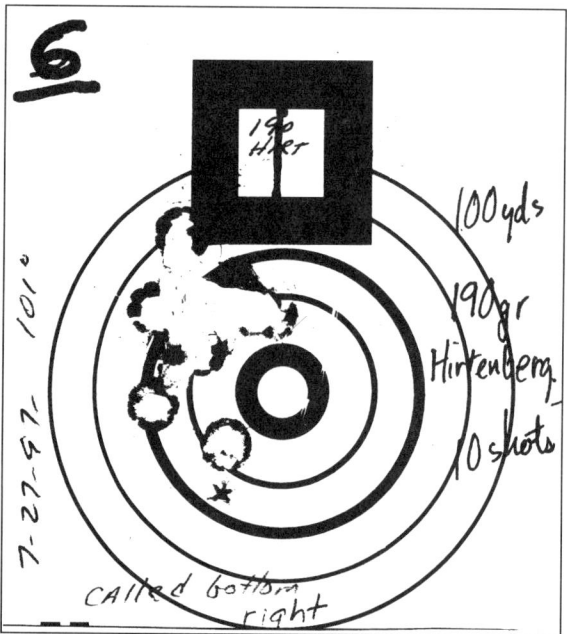

190 gr. Hollow Point Hirtenberger (bottom left) is Austrian sniper ammunition. Originally made for the British Army and for use in the Accuracy International AW (Arctic Warfare) rifle in 7.62x51mm. Box has English overlabels. Contract with Hirtenberger was supposedly canceled. Author fired some and 190 gr. loading exhibits high pressure in normal rifles and has stiff recoil even in heavy rifle. Lapua box is the 167 gr. Scenar (HPBT) loading and is very accurate. Last time I looked, Lapua had 14 different .308 Win loadings including a subsonic 200 gr. load and an extra high velocity 170 gr load. (Mike R. Lau)

CHAPTER 7: The .308 Winchester, 5.56mm/.223, .338 Lapua, Caliber .50 BMG, and More 129

Left target was fired by author at 100 yards. First round out of 5 round group went high to illustrate what happens with cold bore shot when Shooters Choice solvent is not wiped out of bore. On right is 5 shot group fired by Sean Little from a TBA M40A3 with the 168 gr. HPBT Federal Match. Rifle went to the Richardson, Texas, Police Department. (Mike R. Lau)

Still popular with many is the old .30-06 Springfield cartridge. Until replaced by the .308 as the military cartridge, many of the US sniper rifles from WWI to Vietnam were chambered for this venerable classic. (Mike R. Lau)

From left to right: Belgium SS109, U.S. M855 with steel penetrator insert, 5.56mm Ball reloaded for subsonic sniper rifle use during the Vietnam War. (Mike R. Lau)

NATO found the alliance's new standard round in the Belgium made SS109. Produced by Fabrique Nationale (FN), the SS109's 62 gr. bullet has a longer and harder steel penetrator than the XM777. NATO tests require that the projectile had to penetrate, at 573 meters, a mild steel plate of 3.5mm (.138 in) thickness with a carbon content of between .1 % to .2 % and a Rockwell hardness of B55-70. The SS109 projectile will penetrate the old standard US steel helmet at 1100 meters. The M193 55 gr. bullet and the 54 gr. XM777 AP bullet will not penetrate the steel plate at 573 meters. The US adopted the SS109 bullet in our standard M855 Ball.

When the XM855 first came out, the Marine Corps conducted machine rest accuracy tests of 32 M16A1E1s refitted with 1-7" twist barrels. At 93 meters, accuracy ran around 2.5 MOA which was acceptable. The firing of M193 Ball in the 1-7" barrels was also acceptable (2.21 MOA) out to 500 yards. At farther ranges, wind deflection and dispersion were unacceptable with the M193. Firing the M855 in the 12" twist of older M16A1s and AR-15s is not recommended because the instability of the bullet will reduce it's accuracy and penetration.

For most LE tactical situations the 55 gr. M193 or 62 gr. M855 is adequate for close range support and will deliver good terminal ballistics if not firing at a target on the other side of glass. Penetration is not as great as the .308 Win loadings as the small high velocity bullets usually break up when it hits an object. For greater terminal ballistics effects with less danger to other officers and bystanders a good choice is Corbon's 55 gr. JHP, Hornady's 55 gr. V-Max, or Winchester's Supreme Ballistic Silvertip. This last loading is a 50 gr.

At least one company is putting out the M855 62 grain Penetrator for LE use. Winchester's Ranger Tactical Loading now only consists of the M855. (Winchester)

projectile moving out at 3410 fps from a 24" barrel. Sometimes department's can be creative and use specialized ammunition to end armed suspect situations. Frangible target ammunition was used to shoot the pistol out of the hand of a suspect who threatened to shoot himself with little danger of ricochet. Winchester markets their Ranger Frangible line which, besides handgun ammunition, includes two loadings in 5.56mm. Both loads fire a 33 gr. frangible bullet with a MV of 3200 and 3600 fps from a 24" barrel. Bullets are a composition of tungsten, nylon, and copper and completely breaks up when it hit a hard target.

The .300 Winchester Magnum

As we move into the 21st Century, the threat to world security and terrorism does not appear to be going away. Our weapons, equipment, and tactics get even more sophisticated and so does the enemy's. Terrorists are getting more organized, plan operations, and use sophisticated equipment including body armor. His weapons, security, and barriers, keeps you from getting closer. When you try to close the target distance, the situation becomes more hazardous to you. He can see you as easily as you can see him and his weapons are just as lethal. On the battlefield, the Army, Marine, or SEAL sniper faces the same threat: the probability of hitting a target works both ways.

The .300 Winchester Magnum is slowly gaining popularity as military and LE snipers find a need to make longer shots possible. The .300 Win Magnum will increase your probability of hitting a target beyond 800 meters and also have plenty of retained energy to be lethal. The .300 Winchester Magnum also has proven accuracy at long range.

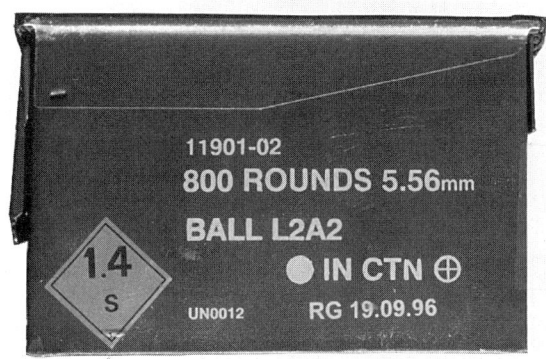

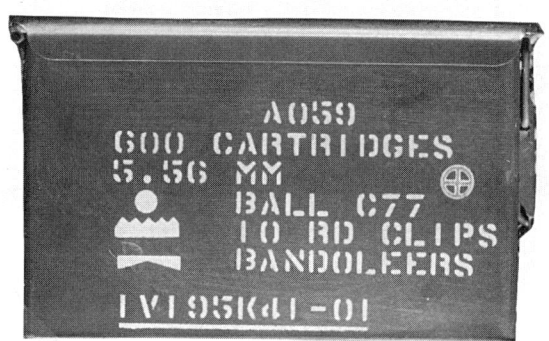

British and Canadian 5.56mm NATO. RORG is Royal Ordnance Factory, Radway Green, Crewe, England. IVI is Valcartier Industries, Inc., Courcelette, Quebec, Canada. NATO ammunition is supposed to be interchangeable, however, the Brits say our M855 is too hot for their L85A1 rifles. The Brit's L2A2 5.56mm NATO round will not function reliably in our M16A2. (Mike R. Lau)

CHAPTER 7: The .308 Winchester, 5.56mm/.223, .338 Lapua, Caliber .50 BMG, and More 133

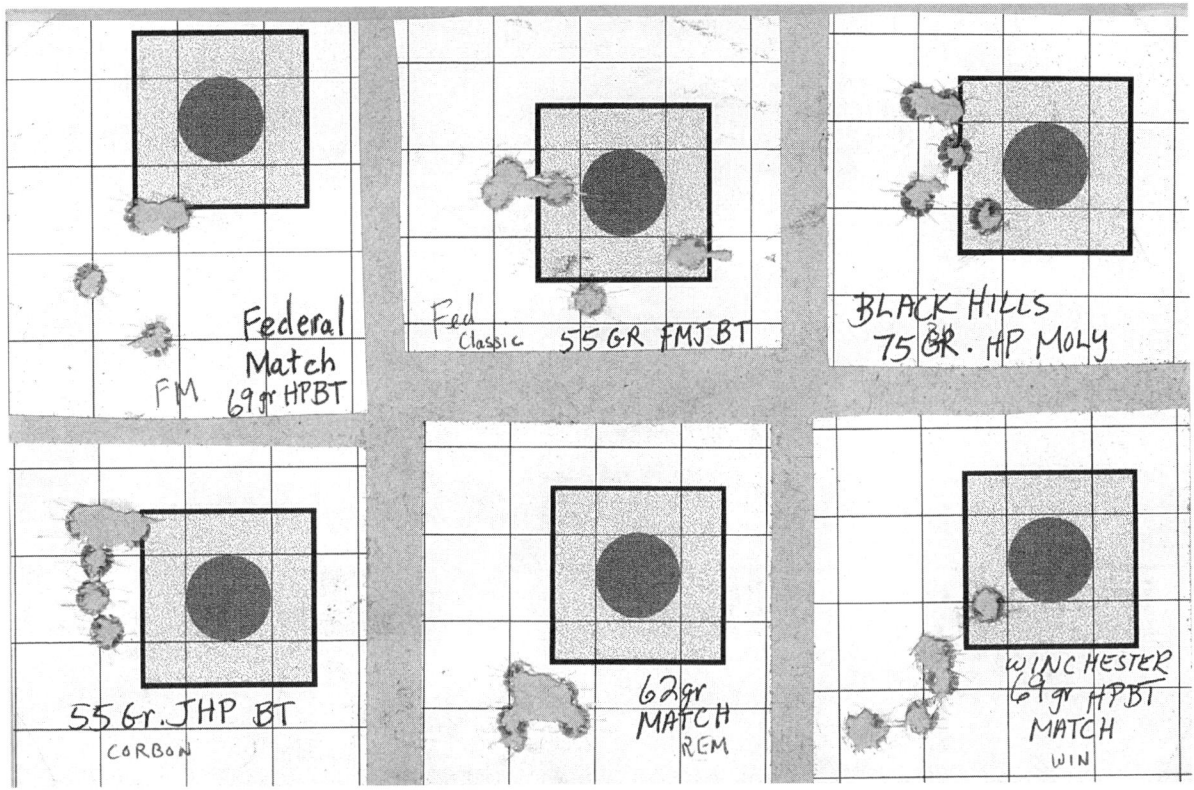

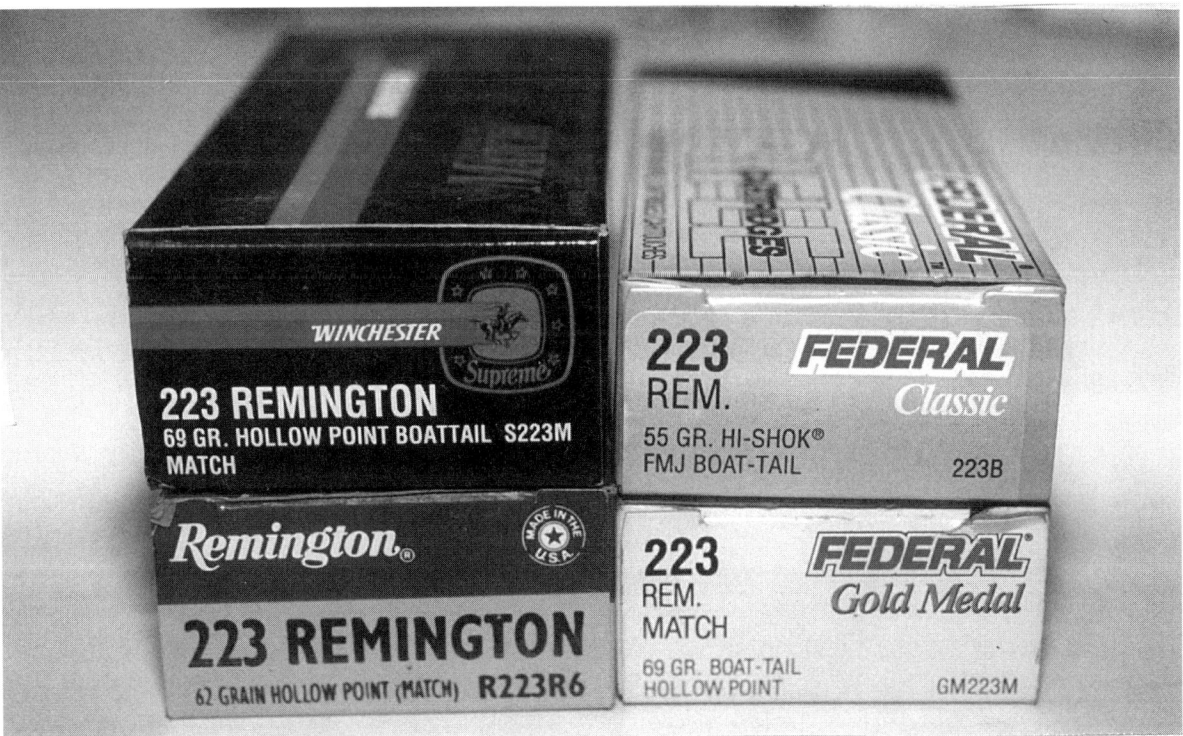

Keith Scullins volunteered to test some commercial .223/5.56mm ammunition with his brand new Savage 112 Varmint Stainless with a Burris 8-32X. The Savage has a 1-9" twist, 26" barrel and gives excellent out-of-the box accuracy. Targets were fired at 100 yards. Keith used a bench rest and sandbag under butt. Other shooters report that the Black Hills 75 gr. Moly loading is extremely accurate in the 1-8" or 1-8.25" twist barrels in the AR-15 target variants. (Mike R. Lau)

Black Hills Moly coated 75 gr. HPBT Match loading is considered by many to be the most accurate factory loading in .223 caliber for match grade AR15/M16 variants with proper 1-8", 1-8.25" and 1-9" twist barrels.

Corbon's 55 gr. JHP is designed for the LE tactical shooter in mind. Its large hollow point provides maximum terminal ballistics effects with minimum penetration and danger to bystanders and hostages. Ammunition is also very accurate for a light high velocity bullet not listed as a match loading and can be used effectively in the older slow twist AR-15s and M16A1s.

The Navy SEALs have an assortment of .300 Win Magnums. The Army has also converted a few of their M24s to .300 Win Mag. There appears to be plenty of .300 Magnum rifles in the hands of trained snipers, but no match ammo was available by any of the major manufacturers. Until now, hunting type soft point ammunition was the only .300 Win Mag factory ammunition available to LE. The best types were the 180 gr. to 200 gr. softpoint boattails and the Nosler Ballistic Tips, but not anymore! Federal recently came out with their new .300 Win Mag Gold Medal Match loaded with the excellent Sierra 190 gr HPBT MatchKing bullet.

Federal Cartridge Company had been producing a similar load for the Naval Special Warfare Forces (SEALs) for a few years now. A public release report on the Navy's .300 Win Mag "Match" ammo was prepared by John Yarbor, Commander, Naval Surface Warfare Center Division, Crane, Indiana, (NAVSURFWARCENDIV Crane), and released back in 1994, details Federal's involvement. The complete report which include 5 data tables, were published in ***Precision Shooting's Special Edition No. 3, Volume 1***, November 1995, and are summarized here.

The Navy was looking for a long range anti-personnel rifle cartridge that could perform better than the 7.62x51mm. Two of the cartridges investigated were the .338/416 (.338 Lapua) and the .30/338 Magnum. Neither of these were selected at the time because they were inadequate for military operations. The .300 Win Mag cartridge was eventually chosen because it was (1) commercially produced in economic quantities, (2) commercial weapons could easily be modified for military use, and (3) the cartridge was proven in long range match competition.

Specifications on the first cartridge lot were to be within SAAMI specs, except that the bullet was to be seated out with an OAL of 3.50 inches to increase case capacity and decrease bullet jump. (Remington 700 long actions can handle cartridges as long as 3.65 inches in the internal magazine box). Accuracy desired was 10" extreme spread (ES) at 1000 yards, and not to exceed 15" ES. Average ES was not to exceed 8.0" at 600 yards or 3.5" at 300 yards. The projectile had to be between 180 to 190 gr. and meet Law of War obligations, meaning full metal jacket.

In 1987, Federal Cartridge was awarded the first contract. They recommended the 180 grain HPBT Sierra MatchKing with the open point "spun" to close it. Due to range constraints, ES was changed to 2.3" at 200 yards. Accuracy of the First Article Lot and

Production Lot gave 2.3" and 1.615" average ES at 200 yd. Muzzle velocity was 2923 f/s and 2941 f/s. After the ammunition was issued, personnel from the Navy Match Rifle Team, and also from Naval Special Warfare, reported the bullets showed instability at 800 yards or about the time the projectile nears sonic speeds. Targets at 1000 yards showed bullets keyholing. It was believed that by closing the open tip, the bullet length was shortened and there was an offset of the symmetry and center of gravity which would cause the bullet to become unstable near the speed of sound. On top of all of this, Navy JAG said the bullet still looked like an open tip so it was illegal to use the new .300 Win Mag ammo in combat.

To fix the problem, some of the First Production Lot was reloaded with the 185 gr. Lapua FMJ bullet. Velocity was changed to 2930 f/s at 70 deg. Accuracy of the modified lot was 9.1" ES at 800 yd and 15.4" at 1000 yd. The Lapua bullets also impacted 14" higher at 1000 yd.

In 1990, a contract was made with Hunting Shack Manufacturing of Missoula, MT, to produce .300 Win Mag ammo for the Navy using the Lapua 185 gr. FMJ and IMR 4350. Three lots were produced, giving muzzle velocities of 2987 f/s, 3008 f/s, and 3009 f/s. Average ES at 200 yd. were 1.66", 1.47" and 1.37". Although the 200 yd ES figures looked good, users reported accuracy problems at the longer ranges.

In 1990, Army and Navy JAG authorized the use of the Sierra 168 and 180 gr HPBT MatchKing bullet for use in combat. Because of this decision and the accuracy problems experienced previously with other bullets, it was decided to use Sierra's 190 gr. HPBT MatchKing bullet in the .300 Mag. A complete evaluation of several propellants were to be studied to insure chamber pressure and velocity were within SAAMI specifications before any further testing and commercial contracting. Powders evaluated were Hercules Reloder 19 and 22, Accurate Arms 3100, IMR 4350, and IMR 7828. IMR 4350 was rejected because desired velocities could not be achieved within SAAMI chamber pressure limits. AA3100 exhibited excessive chamber pressures at the SAAMI upper limit of 125 deg F, but was fired for accuracy anyway. Re 19 and Re 22 showed the least changes in velocity over the temperature range of -40 deg F to 125 deg F. Re 22 showed the least chamber pressure variations over the same temperature range. The results of the propellant test were:

Table 7.1. Accuracy Performance of Navy .300 Win Magnum for Selected Propellant

Propellant Load	200 yd Avg ES	600 yd Avg ES	1000 yd Avg ES
Re 19	1.62"	5.75"	10.79"
Re 22	1.63'	7.20"	13.37"
Re 22	1.79"	5.56"	10.68"
Re 22	2.40"	6.24"	14.02"
AA3100	2.08"	7.65"	13.32"
IMR7828	1.86"	7.46"	12.18"
IMR7828	2.61"	7.71"	13.48"

As a result of this testing, performance specification HS/2024/C91/621 defined the acceptable components as: (1) 190 gr. HPBT match projectile, (2) Re 19 or Re 22 as propellant, (3) commercial 70/30 alloy brass, (4) commercial match primer, (5) 3000 f/s nominal velocity, (6) be within SAAMI pressure at -40 deg F, 70 deg F, and 125 deg F, (6) avg. ES not greater than 2.25" @ 200 yards, with no individual group exceeding 3.5". A test lot of 20,000 rounds, produced in 1992 and constructed by NAVSURFWARCENDIV Crane, verified the new performance specification.

Federal Cartridge Co. was awarded the new contract, N00164-92-C-137, and produced 4 lots under the new performance specification. The new lots passed all Navy test requirements: 28 day temperature and humidity, transportation vibration, 4 day tempera-

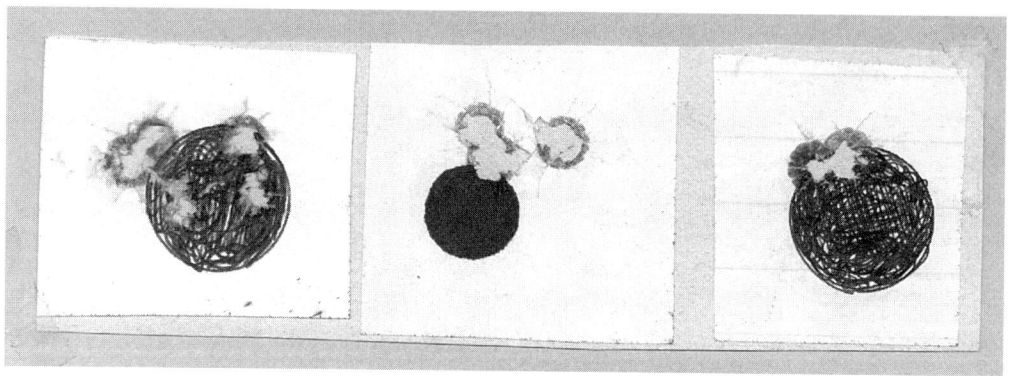

A 5 shot and two 3 shot groups fired from 100 yards with a TBA .300 Win Magnum by Jesse Duffey. His loads were 73 grs. Hercules Reloder 22 and the 200 grain Sierra HPBT MatchKing bullet seated .01" off rifling. Velocity was 3000 fps and Jesse reported that the load only needed 12 moa come up at 700 yards from the 100 yards zero. (Jesse Duffey)

Hunting Shack Manufacturing ammunition for Navy Seal's McMillan 86 .300 Win Mag rifle. (Photo courtesy Stephen Leung)

ture and humidity, 40 ft. and 5 ft. drop, rough handling, toxic fumes, sound pressure, extreme temperatures, and temperature shock and humidity. Reports from the field by match rifle shooters and snipers using the new lots were also favorable.

In 1997, Federal Cartridge Co. introduced their newest entry to the Gold Medal Match series: the .300 Win Mag loaded with Sierra's 190 gr HPBT MatchKing. Unlike the Navy's ammunition with an OAL of 3.50", commercial factory cartridges have an OAL of 3.30". I have shot several boxes of both the commercial and Navy ammunition and found them to be very accurate. Where you had once shot 1.5 to 2 MOA groups with factory soft point ammunition you will now get around .75 to 1 MOA ES. With a custom made heavy barreled tactical rifle, with trued Rem 700 action, minimum chamber, and reduced throat lead, expect .4 MOA or better. Occasionally you will get a 1/4" three shot group at 100 yd., but not as easily as with the .308 Win, because of the increased recoil and muzzle whip. A plus with using the .300 Win Mag is it's more effective terminal ballistics at the longer ranges. The best thing about the .300 Win Mag is the heavier bullet's high ballistic coefficient of .533. This makes it easier to shoot in wind and even if you make an error in wind speed and direction your probability of hitting the target is greater than with the .308.

The bottom line on the use of the .300 Win Mag for sniping is summarized by John Yarbor:

"The .300 Win Mag is the caliber of choice to fill the operational capability divergence between the 7.62x51mm and the .50 Caliber. The 7.62x51mm is effective to ranges

CHAPTER 7: The .308 Winchester, 5.56mm/.223, .338 Lapua, Caliber .50 BMG, and More

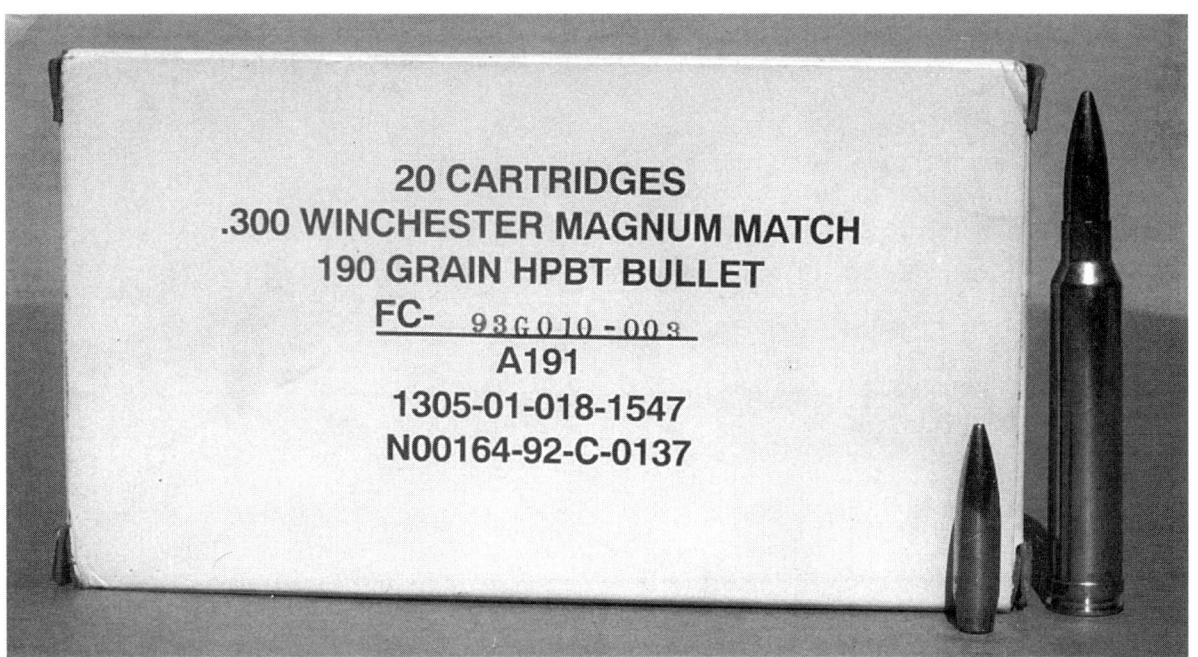

One of the final accepted Federal Cartridge Co. lots with the new acceptance loading. Note that the 190 gr. Sierra HPBT bullet is loaded longer than SAAMI specs and will seat near the rifling in SEAL .300 Mag rifles. (Mike R. Lau)

around 800 yards and the .50 caliber could be employed at ranges between 800 and 1200 yards. However, due to accuracy and muzzle signature characteristics of the .50 caliber, it is not desirable to employ this system at these ranges for anti-personnel purposes. The .300 Win Mag provides pin-point accuracy at these ranges (and lesser ranges) with a signature slightly larger than the 7.62mm."

The .338 Lapua Magnum

This cartridge has it's beginnings as far back as 1983 when Research Armament Company was experimenting with prototype .338 caliber cartridges to fire a 250 grain bullet at 3000 f/s. The Navy contracted for a .338 rifle and the final cartridge design was necking down the British .416 Rigby cartridge to handle .338 bullets. The .338 Lapua was designed with the intent to bridge the gap between the 7.62x51mm and the caliber .50. The Navy decided not to go with the .338 rifle, but pursued the .300 Win Mag as explained earlier. Unlike other magnum rifle cartridges we are familiar with, the .338 Lapua is non-belted.

Lapua brought out the cartridge commercially as the .338 Lapua Magnum or as the 8.6x70mm. Lapua claims that the cartridge is effective out to 1500 meters which actually puts it into the caliber .50's effective range. Replacing the Cal. .50 rifle with the .338 Lapua is reasonable as the accuracy of the Cal. .50 M8 API is about 12" maximum mean radius (MR) average at 600 yards when fired from the Browning Machine Gun. This is approximately equivalent to 39" ES groups. On the other hand, the .338 Lapua is capable of match target accuracy from a well built rifle like Accuracy International's Super Magnum. This makes the .338 a better choice for engaging enemy personnel at long range than the Cal. .50 rifles.

Table 7.2. Summary of the Quantity Lots and Accuracy Results During the Development of .300 Winchester Magnum Ammunition for the Navy Special Warfare Forces.

Lot Number	Mfg.	Bullet	Vel @ 70 deg F fps	Extreme Spread Obtained			Comment
				200 yd Max 2.3"	800 yd Max 8.0"	1000 yd Max 15.0"	
FC-88A001A001	Federal	180 Sierra HPBT	2923	2.30			w/closed tip
FC-88F001-001	Federal	180 Sierra HPBT	2941	1.61			w/closed tip
FC-88F001-001	Federal	185 Lapua FMJ	2930		9.10	15.40	reassembled w/ Lapua
HSM91E010-001	Hntg Shack	185 Lapua FMJ	2987	1.66			excessive pressure
HSM91E010-002	Hntg Shack	185 Lapua FMJ	3008	1.47			excessive pressure
HSM91G010-003	Hntg Shack	185 Lapua FMJ	3009	1.37			excessive pressure
NSW92D010-001	NavWarCen	190 Sierra HPBT	2937	2.00			propelent acceptance Lot
FC93A010A001	Federal	190 Sierra HPBT	3001	1.30			1st article with new propellant
FC-93B010-001	Federal	190 Sierra HPBT	2997	1.47			regular production lot
FC-93B010-002	Federal	190 Sierra HPBT	3001	1.97			regular production lot
FC-93B010-003	Federal	190 Sierra HPBT	2975	1.72			regular production lot

Note: 200 yd MR changed to 2.25" with Federal Cartridge Lots loaded with 190 gr. Sierra HPBT.

The standard loading uses Lapua's patented FMJ *Lock Base* bullet. This bullet has a very odd short boattail with the base pinched to almost closed up over the lead core. Some call this bullet base "rebated". Accuracy standard for this load requires a 5 shot group, at 300 meters, not to exceed 174mm when the height and width are added together. This is about a group size of about 3.4 inches at 328 yards or about 1 MOA. Another loading is designed for hunting with their trade named *Fourex* projectile. The 260 gr. bullet is a partition type, with a soft point and lead core in the front section, and a hollow base in the rear. The contact to rifling is sort of like an artillery round with rotating bands except the bands are not pre-grooved. A third loading available is the AP round which has a slightly different trajectory from the ball loading.

Several rifles are now chambered for this cartridge including Accuracy International's Super Magnum, Mauser's SR 93, Keppler (German) Model KS II, ERMA Precision Rifle Model SR100, Dakota .416 Rigby chambered for .338, and Heym Magnum. For a rifle action to handle the large Lapua round, an internal magazine length of 3.7" is required along with a wider girth. A bolt body diameter of .70", like on a Mauser type bolt is OK, but counterbored or interrupted lug bolt heads should be wider to be really safe. Chamber pressure of the .338 Lapua is 58,000 psi. Some magnum actions that could be used for building custom .338 Lapua rifles would be the Brno 602, Brevex Mauser, Magnum Mauser, Weatherby Mark V, and the A-Square Hannibal. This last rifle is based around the M1917 Enfield action and A-Square already chambers it for .416 Rigby, .460 Weatherby, etc.. Most of the rifles already chambered for the .338 Lapua are more than adequate and some are already tactically designed. Other advantages the .338 Lapua rifle would have over the Cal. .50 rifle would be the difference in ammunition weight and rifle weight. AI's Super Magnum weighs 15.43 lb. while the Barrett M82A1 weighs 29.69 lb. The .338 Lapua does not have the blast signature of the .50 caliber nor the recoil.

Table 7.4. Free Recoil Energy of Sniper Weapons

Rifle w/scope	Wt. Lb.	Free Recoil Energy Ft. Lb.
M16A2	8.75	3.8
US Army M24	14.95	9.1
USMC M40A1	14.50	9.4
Rem 700 PSS	9.50	15.4
Rem 700 Sporter (.308/22")	7.25	18.1

CHAPTER 7: The .308 Winchester, 5.56mm/.223, .338 Lapua, Caliber .50 BMG, and More

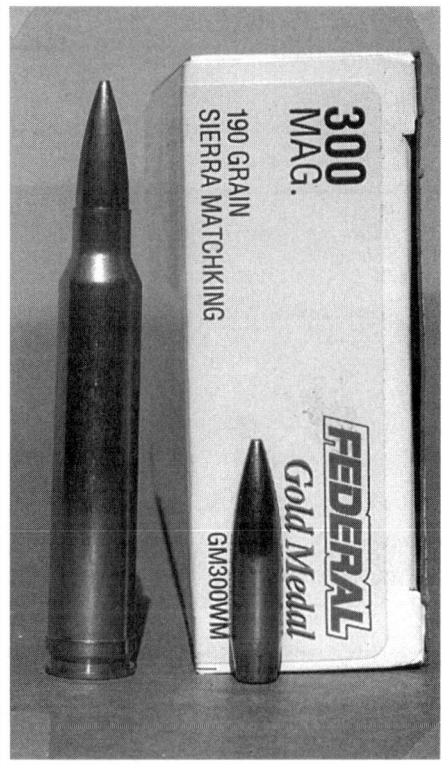

Andres Escobar, International Bodyguard Associates, with TBA .300 Win Magnum M40A2. Andy shoots the new Federal Gold Medal Match loaded with the Sierra 190 gr. HPBT MatchKing bullet. (Photo courtesy Andres Escobar)

AI Super Mag .338	15.43	28.7
.378 Weatherby Tac Rifle	15.43	42.5
Barrett M82A1	29.69	103.2

Note: Some of the recoil energy of the Barrett Cal. .50 is absorbed by the ground through the bipod and also by the recoil spring and bolt. Consequently it doesn't feel like 103 lb. of recoil. .378 Weatherby Tactical rifle is hypothetical using AI's Super Magnum weight.

The .338 could do more than fill in the gap between the 7.62x51mm and Cal. .50, it could replace the .50. However, there is one drawback, a more destructive .338 round needs to be developed to replace the .50's role in engaging light armor and other heavy equipment. Look for the .338 Lapua to become a favored long range, hard hitting round in the future.

.50 Caliber Sniper Rifle Ammunition

.50 Caliber sniper weapons have limited use as dictated by the mission. Marine snipers, Navy SEALS, and Army Special Ops units have .50s as part of their weapons issue.

Table 7.3. Magnum Cartridge Specifications

	Bullet Wt	BC	MV fps	Mzl Energy	Rim Dia	Cs Length	OAL Ctg L
.300 Win Mag	190	0.533	3000	3796	0.532	2.62	3.34
.338 Win Mag	250	0.565	2700	4048	0.532	2.50	3.34
.338 Lapua	250	0.662	2950	4830	0.590	2.72	3.60
.340 Weatherby	250	0.565	2800	4353	0.532	2.82	3.56
.378 Weatherby	300	0.475	2900	5604	0.582	2.91	3.68

Air Force Explosives Ordnance Disposal (EOD) also use .50s for setting off explosive they don't want to approach. Because of the inaccuracy of most military Cal. .50 ammunition, the rifles are really more for destroying equipment than for engaging enemy personnel. Military snipers using the .50 caliber, want the most destructive ammunition available for the most effective results on enemy equipment. Although many exotic caliber .50 projectiles have been produced over the years, only a handful of these have been adopted by the military. Two of these are the Saboted Light Armor Penetrating (SLAP), and the other is the Multi-Purpose High Explosive-Incendiary-Armor Piercing Cal. .50, also known as MP or HEIAP.

During the 1970's, the Norwegian firm of Raufoss, began experimenting with a high

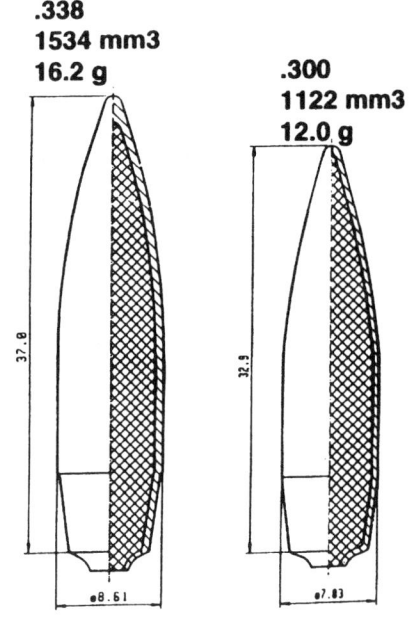

Fig. 7.1. Lapua's Lock Base design is also referred to as "rebated boattail". Bullet design allows for excellent retained velocity and accuracy at long range. Accuracy and power of .338 Magnum fills the gap between the .300 Winchester Magnum and Caliber .50 BMG cartridge for engaging personnel targets.

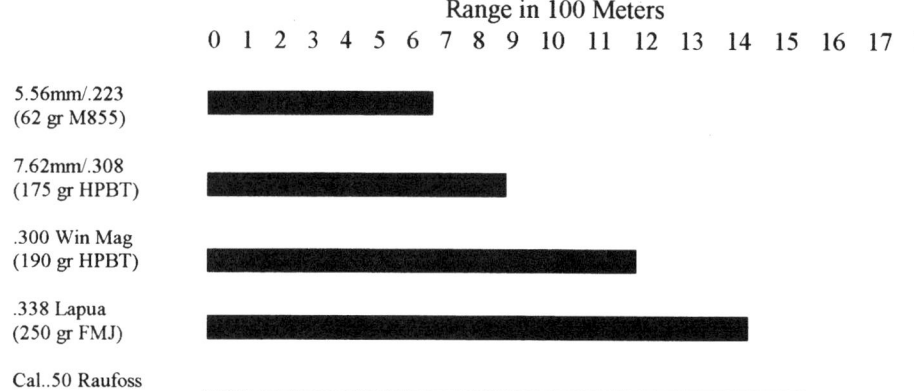

CHAPTER 7: The .308 Winchester, 5.56mm/.223, .338 Lapua, Caliber .50 BMG, and More

Malcolm Cooper (left), designer of Accuracy International's Arctic Warfare (AW) and Super Magnum Rifle, with Major John Plaster. Lapua .338 Magnum compared to the .300 Winchester Magnum is a non-belted case based on the necked down .416 Rigby. Norm Chandler is looking at camera from behind.
(Photo courtesy of Guy McCracken Jr.)

explosive projectile for the .50 Browning machine gun (BMG) cartridge. One of the first designs was successful and designated NM140. The explosive projectile is unfused so is detonated by a method known as Pyrotechnically Initiated Explosive (PIE) as designed by Raufoss. To detonate the explosive charge, a small amount of incendiary compound in the bullet tip is crushed during impact thus detonating the explosive charge. A hardened steel core increases penetration into thicker protective material and the zirconium provides a flash on impact so that the round can also be used for spotting. Winchester/Olin developed a similar PIE round during the 1980s known as the Winchester Anti-Light Armor Piercing (WALP). The projectile has a yellow tip and was not mass produced.

US Navy SEALs adopted a later version of Raufoss .50 BMG Multi-Purpose (MP) explosive round or High Explosive Incendiary Armor Piercing (HEIAP) with a more sensitive igniter. Known as the NM140A1, the Navy designation is the Mark 211. A primary concern about non-fused explosive projectiles is it's ability to be stable during rough handling. The Raufoss MP passes all military drop and rough handling tests. Projectile is identified at the tip with the standard USN color, green/silver. Specimen pictured is headstamped ".50 FNB 91". Cases were manufactured by the Fabrique Nationale plant in Belgium and loaded in Norway by Raufoss. Ballistics of the Mk 211 is identical to the US M8 Armor Piercing Incendiary (API) round which is what is normally used for practice by US snipers. Raufoss is also producing a Multi-Purpose Tracer (MPT) round designated the NM160. Bullet tip is colored green/red. Bullet construction is similar to the NM140 except for a small amount of tracer element in the core. Raufoss has developed a whole line of explosive projectiles up to 40mm.

Gunny Latimer (above), Scout/Sniper Instructor, circa 1984, fires an RAI (Research Armament Industries) caliber .50 rifle on Range 7, Quantico. This rifle was sold in quantity to the Government. The bolt seized up a lot and a rubber mallet was kept nearby to open the bolt. Scope is Weaver T-16. Weaver T Series scopes were very popular with AMUs and MTUs in the '70s and '80s. Ammunition is the M903 Saboted Light Armor Piercing (SLAP) round that fires a tungsten carbide projectile in an amber colored plastic sabot. Muzzle velocity is 3900 f/s and the sabot separates from the penetrator upon exiting the muzzle. The Barrett .50 Caliber M82A1A has a muzzle brake that interferes with the normal exiting of the sabot. Pieces of it shower down on the shooter after it explodes in the brake. The M903 was intended for machine gun use and has a significantly different trajectory than other caliber .50 ammunition types. Armor plate (below) is 1" rolled homogeneous steel. (Kent Gooch)

Silencer ammunition for sniper rifles.

Military snipers often support other special operations forces, like Special Forces or Rangers, in clandestine operations. If Marine and Army snipers are not actively engaged in the direct accomplishment of the mission, they may provide security to the main force. Sniper teams can provide security for a chopper landing zone (LZ), or provide countersniper and rear security support for the main unit to get in and out of a location. When snipers participate in the actual execution of the main mission and stealth and secrecy are to be maintained for as long as possible, the use of silenced weapons may be required.

Almost any caliber can be adapted to silencer use by loading the ammunition down to have a MV below 1100 f/s which is around the speed of sound. Silencers can only reduce the noise from the blast at the muzzle, but bullets traveling above sonic speed break the sound barrier and cause a loud "crack" that can be as loud as firing your weapon without a silencer.

An attempt to produce silenced M16A1s came about during the Vietnam War, but with little success. 5.56mm ammunition consisted of a variety of bullets from solid blunt lead to long heavy jacketed ones. Reduced propellant charges were used to keep the velocity down. Materials like oatmeal, cream of wheat, and paper were used inside the case to fill the air space, but most of these clogged up the gas ports of the M16A1s. The Navy SEALS tried a reduced velocity 5.56mm round loaded with a Sierra hollow point bullet. The intent was to kill guard dogs with silenced weapons, but the low velocity and light bullet did not do the job. .458 Winchester Magnum cases shortened to 1.5" and loaded to subsonic velocities with the 500 gr. FMJ bullet was a way to increase bullet energy and lethality. A few Winchester 70s in .458 Win Mag with silencers were sent to Vietnam and

CHAPTER 7: The .308 Winchester, 5.56mm/.223, .338 Lapua, Caliber .50 BMG, and More   143

*Fig. 7.2. Norwegian Raufoss Caliber .50 High Explosive Incendiary Armor Piercing (HEIAP) and U.S. Military M8-API projectiles. Raufoss bullet is also known as Multi-Purpose (MP). Upon hitting the target, the unfused incendiary composition ignites and burns. This in turn ignites the high explosive chemical which burns to a low order detonation . The zirconium element ignites and burns to increase the incendiary effect and duration . As the tungsten carbide penetrator pierces the target's steel or barrier, the mild steel cup which the penetrator is embedded into, holds the remaining components together for the 3 millisecond explosive detonation. Burning fragments damages the surface of the target as well as pass through the hole created by the penetrator. Raufoss claims that the complex damage created by the MP projectile defeats the self-sealing affects of aircraft fuel tanks. Armor penetration is also given as 11mm to a 45 deg. slope at 1000 meters. Both the MP and the US API rounds have similar trajectories. "From a drawing, courtesy of **Very High Power Magazine,** 1993 Issue #4, Fifty Caliber Shooter's Assn. Original source unknown."*

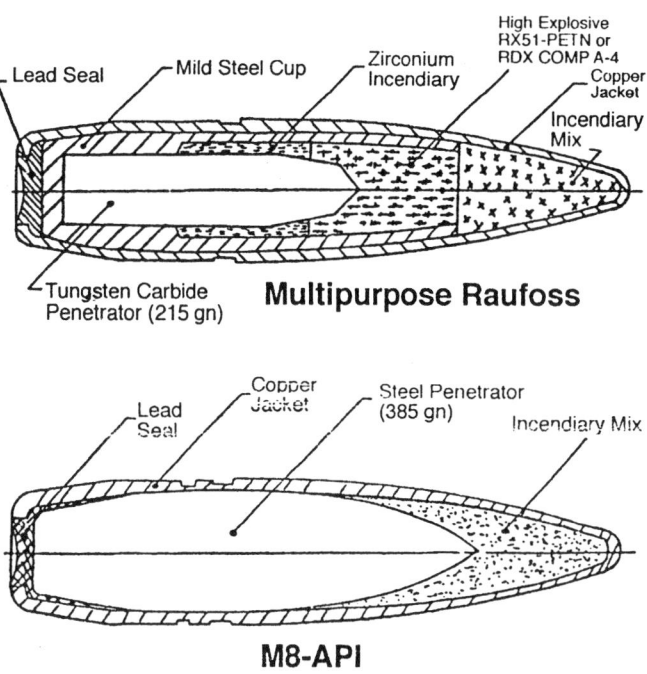

they proved a little more effective because of the higher energy produced by the heavier bullet. However, they lacked accuracy.

The idea of a silenced sniper round of greater lethality and accuracy than the 5.56mm, .458 Win Mag, or 9mm pistol round is a very sound idea. A law officer or military anti-terrorist sniper could take out individual terrorists isolated from a group without alerting the others. Much of the silenced rifle ammunition produced in the past were good attempts to make trade offs between weapons availability, accuracy, and lethality in trying to keep velocities below 1100 fps. To get better ballistics and lethality, a couple of new subsonic rounds are available today. Lapua lists a .308 Win subsonic loading of a 200 grain bullet with a muzzle velocity of 1066 fps.

The .300 Whisper

An effective silenced weapon round is the .300 Whisper. Designed by J.D. Jones, the cartridge case is a .221 Remington necked up to take a .308 bullet. In designing the cartridge, J.D. Jones wanted to use a bullet of superior BC even at the subsonic velocity. He also designed the round to fire from an AR-15. By using a smaller cartridge case than the .308 Win, cartridge efficiency goes up and powder charge remains small requiring a smaller silencer device. His original development loads used the 240 gr. Sierra MatchKing. MV was 1000 fps and at 300 yards the bullet was still traveling at 931 fps with 482 ft lbs. of energy. With a 100 yard zero, the bullet only rose 3.69" at around 50 yards and at 300 yards it drops 108". He recommends a 1-8" twist barrel and accuracy in his modified AR-15 was better than reduced loadings in 7.62/.308 Win loadings which usually runs around 4 moa. Both super-sonic and sub-sonic loadings are offered by Corbon. The tactical load uses a 220 gr. HPBT bullet that is longer than the cartridge case itself. At 1050 fps, energy is around 550 ft-lb. and at 100 yards, energy is still 500 ft-lb.. Corbon also offers a 125 gr. JHP and a 150 gr JHP at MVs of 2100 fps and 2000 fps respectively. The ballistics are simi-

Marine Scout/Sniper Brian Gauthier, 3/1, manning an observation post in Somalia with a Barrett. Unertl scope has special cam for caliber .50 ammunition. (Brian Gauthier)

L to R: M8-API has silver tip, Canadian MP round made by IVI with light blue tip, and Raufoss HEIAP round with silver/green tip. Raufoss cases are made in Belgium by Fabrique Nationale. (Mike R. Lau)

CHAPTER 7: The .308 Winchester, 5.56mm/.223, .338 Lapua, Caliber .50 BMG, and More 145

Sub-sonic silenced rifle ammunition. L to R: 5.56mm assembled by Ft. Benning MTU in 1970. Shortened .458 Winchester Magnum with 500 grain FMJ used in silencer equipped M70 in Vietnam/Cambodia, circa 1971-72. J.D. Jones/Corbon .300 Whisper shown with 220 and 180 grain bullet. (Mike R. Lau)

lar to the 7.62x39 round and Corbon claims accuracy is better with the .300 Whisper because of the better bullets. The cartridge's OAL was designed to function through a AR-15 or M16 variant by just changing upper receivers with a .300 Whisper barrel. Bolt rifles are also being produced for this cartridge and it would be a better choice for the law officer or military sniper than a semi-auto rifle. There is less noise produced by the operation of the action or the fired cartridge case bouncing around on a concrete floor to give the sniper's position away. Silenced sniper ammunition is not by any means as accurate, or as effective regarding terminal ballistics, as standard ammunition and weapons. But when the situations calls for use of silencer equipped rifles and sub-sonic ammunition, it's out there.

By having the most accurate ammunition available for sniper use, you eliminate one more variable that causes inaccuracy. As you will see in the following chapters, however, even having the most accurate rifle, ammunition, and ability to shoot 1/4" groups at 100 yards all day long, is not a guarantee that you will hit a target at long range.

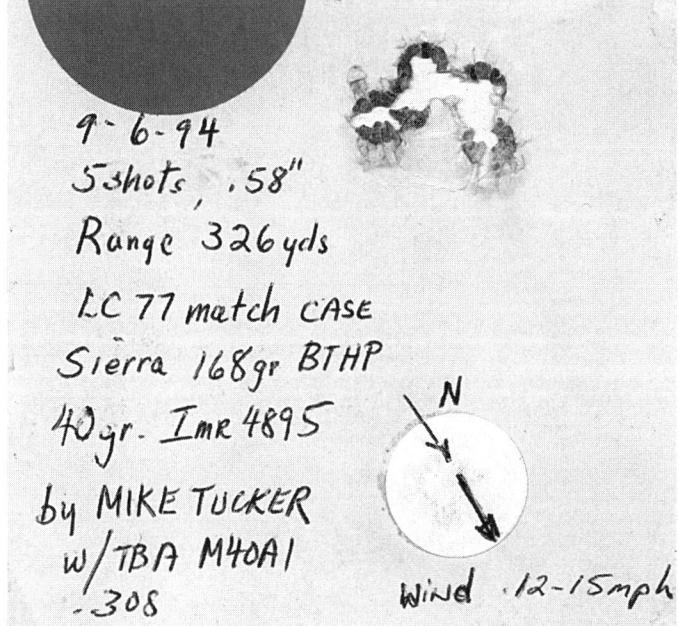

326 yd. target fired by Mike Tucker with one of his 26" barrel TBA M40A1s and .308 test handload on Jim Gannon's range. Note how wind moved group away from target. Below, Mike Tucker is firing Federal Match .308 Win ammunition through chronograph with TBA's velocity test rifle and 17" barrel. (Mike R. Lau)

CHAPTER 7: The .308 Winchester, 5.56mm/.223, .338 Lapua, Caliber .50 BMG, and More 147

Keith Scullin with his Savage 112 Varmint Stainless Steel in .223 caliber. Rifle has a medium weight 26" barrel and is also made in a left hand model. A unique feature of the Savage bolt rifle is the barrel assembly and the swivel bolt head design. During assembly, instead of finding a bolt that fits the action after the barrel and receiver are tightened, Savage assembles the action with a unique method called "swinging." This method is used because the barrel has no shoulder on which to tighten against the receiver face. The bolt is inserted and locked into the receiver. The threaded, shoulderless barrel, has a locknut screwed onto it and the recoil lug slipped over the threaded portion. With a minimum "go" headspace gauge in the chamber, the barrel is screwed into the receiver until the headspace gauge contacts the bolt. The locknut, with the recoil lug between it and the receiver, is then screwed tightly against the receiver and acts as the barrel's shoulder. The other unique feature of the Savage rifle is the pivoting or floating bolt head which allows both bolt lugs to make contact inside the receiver during firing. The theory is that during the firing vibration, the cartridge and the barrel flex or bend at angles not square to the bolt and receiver. This is due to tolerances of the mechanical fit of the components as well as the elasticity of the metals. During this flexing, the two bolt lugs alternately makes contact with the receiver lug areas as the bolt face tries to maintain it's square alignment with the cartridge and the barrel. The floating bolt head allows this alignment to be maintained. This is the same theory behind the PPC bench rest cartridge case, where it's semi-balloon head construction actually does the flexing to act like the pivot behind the bolt head. (Mike R. Lau)

CHAPTER 8: EXTERNAL BALLISTICS, INFLUENCES ON ACCURACY, AND BALLISTICS TABLES

If your rifle/scope/ammo will only shoot 3" groups at 100 yards (3 MOA) don't expect it to shoot 3 MOA (15" groups) at 500 yards, it will be worse. If you miss the target you might be able to say it was the rifle's or the ammunition's fault. I'll accept that. If you and your rifle are capable of 1/2 MOA accuracy or better, and you miss the 600 yard target, you screwed up. In the previous chapters we discussed the importance of a precision rifle and ammunition for sniper work as a way to decrease problems of accuracy and precision concerning yourself and your equipment. By doing so, you can work harder on controlling the **external influences** in your surrounding environment that affect your accuracy and the rifle's precision. In this chapter we will look at **(1) the external ballistics components, (2) the factors that define the ballistics, and (3)** some of the **external influences** that affect accuracy. In the next several chapters we will look at the more difficult **external influences**: wind, target distance, and target complexities.

External Ballistics Components

External ballistics concerns the flight of the projectile from the muzzle to the target. It deals with your rifle's zero, bullet velocity, energy, trajectory, maximum ordinate, time of flight (TOF) of bullet in air, bullet path, bullet drop, and wind deflection. **External ballistics components** apply to all rifle and ammunition systems and consists of the following:

Trajectory is the actual path a projectile takes after it exits the muzzle until gravity pulls it down. This path is computed mathematically and is defined by the bullet's weight and shape, velocity, air temperature, air pressure, angle of the rifle's barrel to the earth's surface, and several other factors as defined below.

Mid-range trajectory refers to height of the bullet path at the mid-way point of the trajectory.

Maximum ordinate is the highest point the bullet achieves in its trajectory above the line of departure. This point is usually a little further than the mid-range trajectory. For

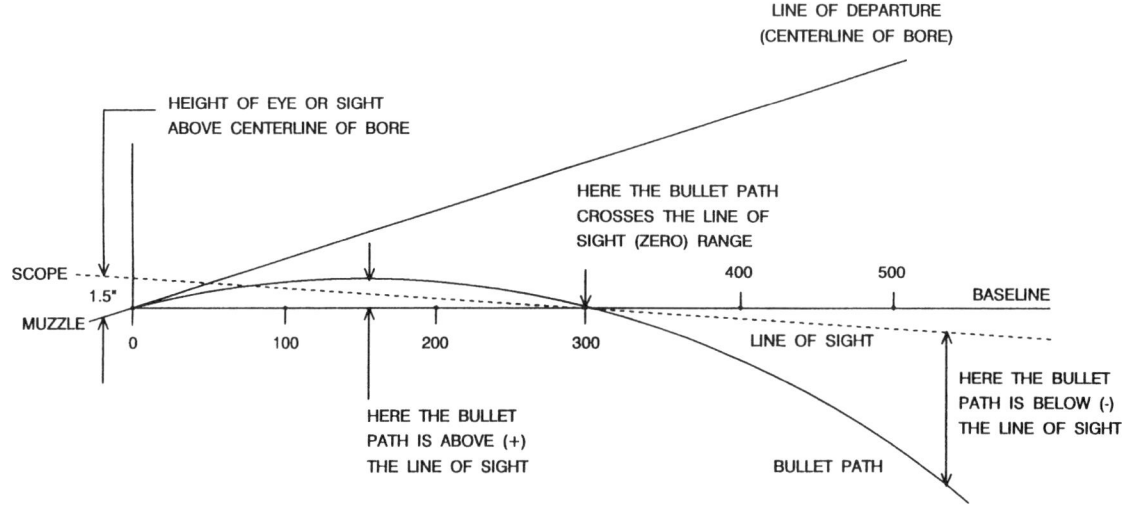

Fig.8.1. Trajectory Terms

example, with the .308 Win 168 gr. HPBT and an MV of 2600 fps, the maximum ordinate for a 400 yard zero trajectory is 13.6" at 215 yards. When the zero is changed to 800 yards, the highest point of the trajectory is 85.2" at 450 yards. This information can become important if you must fire under an overhead obstacle located between you and the target.

Zero Distance determines the length and height of your trajectory. It affects your bullet path and come ups.

Time of Flight (TOF) is the amount of time it takes the bullet to reach a specific distance.

For the sniper, it is used in the calculation for the amount of lead required to engage a target.

Line of sight (LoS) is an imaginary straight line from your eye, through the intersection of the cross hairs in the scope, to the target, and beyond.

Line of departure (LoD) is an imaginary straight line that represents the path that the projectile would take from the muzzle (follows the bore) if the velocity stayed the same and if there was no gravity present.

Come ups refer to the actual elevation adjustments in moa required to bring the *bullet path* back to your scopes line of sight or cross hairs. It is usually expressed in whole or fractions of minutes of angle (MOA). It will be discussed in the chapter on range estimation, but is mentioned here because you will see it in the ballistics tables produced in this chapter.

Bullet Path is the actual path of the bullet in flight and is measured in inches above or below the **LoS** at any given point along there. The *path* starts at the muzzle and is below the **LoS** because the sight is mounted above the rifle's bore. It may or may not cross the **LoS** before crossing the intersection of the **LoS** and the **LoD** which is your zero. From this point on the bullet falls below the **LoS** and is expressed as - (minus) inches. The **path** is the ballistics data you use to figure out how much elevation or *come ups* you need to bring the bullet back up to the cross hairs for a zero at another distance.

Bullet Drop is the number of inches the bullet drops from the muzzle toward the earth if the rifle was held parallel with earth's surface. The number of seconds it takes the bullet to fall to the earth's surface is the same time it takes for that bullet, fired from the rifle, to hit the earth's surface at a distance. The **drop** always starts at 0.0" at the muzzle and goes down giving -(minus) inches only. **Drop** is used to calculate the time of flight at any distance, but don't worry, ballistics programs will do these numbers for you now. Do not confuse bullet path with bullet drop. They are different and some references may refer to the **path** as the **bullet drop**. Do not be concerned with **bullet drop.** For us it has no other use than to calculate TOF.

Muzzle velocity is the speed of the bullet as it exits the muzzle. It is significant because it affects TOF, remaining velocity, accuracy, ballistic coefficient (BC), and terminal ballistics results. Sometimes measured 10 to 15 feet from the muzzle with a chronograph so velocity may be 15 to 25 feet less that actual from muzzle. Instrumental velocity for military ammunition may sometimes be taken at 78 feet. This is usually a two chronograph setup with the first screen at 3 feet from the muzzle and the second at 150 feet away. The midpoint of the setup is 75 ft. plus 3 ft. which is 78 ft..

Retained or remaining velocity is the bullet's velocity after initial muzzle velocity at a specific distance or when impacting the target.

Minute of angle (MOA) is a proportional unit of measure equal to 1/60 of a degree. Although the width of the angle is actually an arc, as it applies to shooting, 1 MOA means approximately 1 inch wide at 100 yards *or* 1.13" at 100 meters of range. 1 MOA at 1000 yards refers to approximately 10" wide and at 1000 meters is approximately 11.3 inches. (1 MOA = 1.047" at 100 yds)

Angle of Jump is the angle formed by the axis of the bore prior to firing and the bore at the time of bullet departure. This change in bore angle is caused by recoil, muzzle whip, and barrel vibrations.

Defining Your Ballistics and Trajectory

Factors That Define Your External Ballistics/Trajectory are:
 (I) Gravity.
 (II) Drag.
  A. Air Density.
   1. Altitude.
   2. Air Pressure
   3. Temperature of air.
   4. Humidity.
  B. Bullet Efficiency
   1. Ballistic Coefficient.
   2. Muzzle Velocity.

Gravity affects trajectory by applying a constant pulling force to the projectile causing the bullet to drop to the earth's surface. The shooter has no control over the force itself, but can alter the angle of the rifle's bore to change the position of the projectile above the earth's surface at a specified distance.

Drag is the other force on the bullet in flight that slows it down and affects the trajectory. Two factors create drag: ***Air density and Bullet efficiency.***

Air Density can be referred to as the "stopping power of air" because it is a medium consisting of moisture and gases through which the bullet must push through. Air density is measured as pounds per cubic feet and at sea level it is .076 lb/cubic feet. When you stick your arm out the window of a car that is moving you can feel the air's "stopping power". This resistance or "stopping power" changes with different air ***temperature, barometric pressure, humidity,*** **and** ***altitude***. Thus your trajectory and zero changes when these four factors change. There is no chart or formula for air density to correct your rifle's zero. You correct your rifle's zero and thus your trajectory for the effects of temperature and altitude only. Air barometric pressure and humidity effects will be corrected for when you adjust for the temperature and altitude.

Different *altitudes* have different air densities. Your rifle zeroed at a specific *altitude* has been zeroed at a specific *air density*. If you were zeroed at sea level (ASL), air density decreases if (1) *elevation or altitude* increases, (2) *barometric pressure* drops below sea level standard: 29.73 in Hg, (3) air *temperature* rises above standard temperature: 59 deg F, (4) or a combination of high temperature and high *humidity*. If you are zeroed at a higher altitude, when you return to sea level, the denser air creates greater drag on the bullet so it will impact lower. Use an *altitude chart* to correct your rifle's zero. If you don't have a chart for your particular ammunition, use a ballistic's program and create one.

Table 8.1 Point of Impact Rise at New Elevation For M118 SB:

With the rifle zeroed at sea level, come *down* in MOA from the above chart for the specific range at the corresponding elevation.

Range	1500 ft. moa	5000 ft. moa	10,000 ft. moa
100	.05	.08	.13
200	.1	.2	.34
300	.2	.4	.6
400	.4	.5	.9
500	.5	.9	1.4
600	.6	1.0	1.8
700	1.0	1.6	2.4
800	1.3	1.9	3.3
900	.6	2.8	4.8
1000	1.8	3.7	6.0

Temperature affects air density. Since the air is made up of gases and moisture, we know from physics that when the temperature is cooler, the molecules get closer together creating more weight per cubic foot volume. The cooler the air, the *higher* the air density, thus greater resistance. The warmer the air, the *lower* the air density, so there will be less resistance and your bullet will impact high. For 7.62mm and 308 Win ammunition, use the general rule: First you need to know at what temperature you are zeroed for. Lower your scope's elevation by 1 MOA for every +20 deg F change. Raise the sight's elevation or bullet impact 1 MOA for every -20 deg F change.

Temperatures affects ammunition. Not only is air density affected by temperature, but it also affects the cartridge propellant. Using ammunition in hot climates, or if just left in the sunlight, may cause an increase in the burning rate of the powder causing higher velocities. Higher velocities causes bullets to impact higher. Inconsistent velocities increase vertical dispersion of bullets on a target. Store your ammunition out of direct sunlight. Police snipers have this problem when they store their rifle and ammunition in the trunk of their police car during the summer.

An *offsetting factor* may occur when the lower temperatures at higher altitudes decreases your muzzle velocity, but the thinner air density only brings it back up to normal velocity. You may not have to make full sight adjustments or any at all.

Relative humidity (**RH**) tells us how much moisture is in the air and is measured by percent. 100% RH is when the air is saturated by water vapor. At 0% relative humidity the air is dry or unsaturated. Air density is *lower* in humid air and *higher* in dry air. However, the effect of RH on ballistics is minimal and, for all practical purposes, no correction data for your rifle is needed even at ranges beyond 800 yards. Since humidity is related to *temperature* and *pressure,* your sight adjustments should be made from those changes. The Marine Corps sniper schools teaches students that the affect on point of impact by humidity is minimum so no sight correction data is given. The Army's manual on the M14 rifle and marksmanship, FM 23-8, confirms the relationships between air density, air temperature, and relative humidity. Figure 151, Chart 3, of the manual:

CHAPTER 8: External Ballistics, Influence on Accuracy, and Ballistics Tables

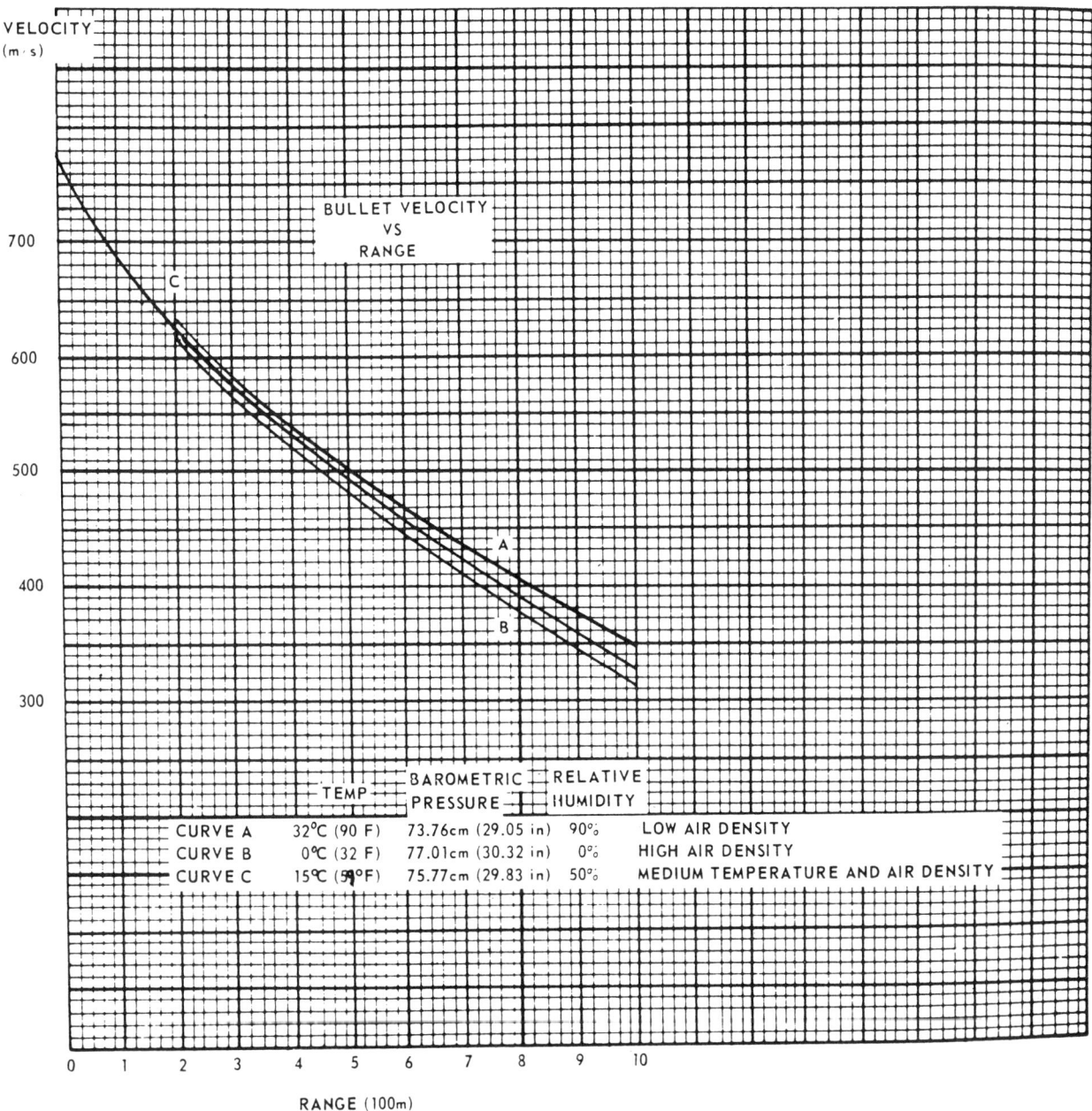

*Fig. 8.2. Velocity versus air density graph from **FM 23-8, M14 and M14A1 Rifles and Rifle Marksmanship**. Graph confirms that as relative humidity rises, air density is lower.*

Sierra Ballistics program also shows major changes in humidity has very little affect on ballistics.

Effect of Humidity on M118 SB
std temp 59 deg F, pressure 29.73 in

Relative Humidity	Vel (fps) @1000 yd	Path (in) @ 1000 yd
0%	1161	-424.5
50%	1164	-423.4
78%	1165	-422.8
100%	1167	-422.4

Bullet Efficiency: BC, Velocity, and the Trans-Sonic Region

Ballistic Coefficient (**BC**) refers to the aerodynamic efficiency of the bullet. It is a decimal number less than 1.000 that is mathematically derived by taking into account the weight, cross sectional area, and shape of the bullet. The standard projectile of perfect form has a BC of 1.000. There are different models used for calculating different BCs for the same bullet. According to the latest Sierra reloading manual, their preferred way to create the BC table is to measure BC with firing tests and use the G1 drag model or Ingall's table. This takes into account the rifle, cartridge, and bullet. Sierra believes their method is the best way for calculating BC for boattail projectiles. The results follow the Ingall's tables closely so BCs from this table can also be used to compare different bullets.

So what does all this BC mean?

A high **BC** means a flatter trajectory, decreased drag, better wind bucking ability, and higher retained velocity which could all result in greater accuracy at long distance. **BC** is used to calculate the retained velocity and time of flight from a given muzzle velocity at any specific distance. ***It defines the bullet's trajectory.*** The BC for a bullet changes as the bullet's initial velocity slows down at the longer ranges. To create a more accurate velocity chart for a particular load, a high, middle, and low velocity BC should be used at the specified *cut-off velocity* for each major velocity change.

For different muzzle velocities, the BCs and the cut-off velocities are given in the Sierra manual in the "Exterior Ballistics" chapter. Different muzzle velocities give different trajectories and changes in barrel lengths usually reflect this. Most tactical .300 Win Magnum rifles have 26" barrels. 5.56/.223 ballistics are usually for 24" barrels but M16A1s and A2s have 20" barrels. Savage 112 Series rifles in .223 Rem, .308 Win, and .300 Win Mag calibers have 26" barrels and so do the new Remington Police Sniper rifles.

For those of you who are interested in ballistics, there is Sierra's excellent Ballistics Program. It calculates all trajectory data including wind drift affects, time of flight, maximum ordinate, and much more. It can also calculate a BC if all you know is velocity between any two points and the distance between the points. A come-up table can be created for any loading once you have the trajectory table. Muzzle velocities change with different barrel lengths and this should be used if known. The best thing about the Sierra Ballistics Program is that it is very user friendly, quick, and you don't have to be a mathematician. (Sierra Bullets L.P., 1400 West Henry, Sedalia, MO 65301).

If you are issued a rifle and ammunition, you don't really have any control of the BC. Switching from Federal 168gr. match ammo to the new 175 gr. HPBT match ammo will help. Likewise, going from M118 SB or M852 to M118 LR will improve BC.

Speed of Sound is included here because drag is affected as sonic speed changes. Standard speed of sound at sea level is 1121 fps. Speed of sound decreases as elevation increases.

Table 8.3 Speed of Sound Vs. Altitude

Altitude	Speed of Sound
-200 ft	1124 fps
(ASL) 0	1120
1000	1107
2000	1092
3000	1078
4000	1063
5000	1048
6000	1033
7000	1018
8000	1002
9000	987
10000	971

CHAPTER 8: External Ballistics, Influence on Accuracy, and Ballistics Tables

Table 8.2 Ballistic Coefficients of Bullets (see note 1)

Ammunition Type	Bullet Wt. and Type	High BC	Boundary Velocity f/s	Mid BC	Boundary Velocity f/s	Low BC	
7.62 M80 Ball	148 gr FMJBT	0.404	not used	0.404	not used	0.404	note 2
7.62 M118 SB	173 gr FMJBT	0.494	1800	0.485	1800	0.463	
7.62 M852 Match	168 gr HPBT	0.462	2600	0.447	2100	0.424	
.308 Fed Match	168 gr HPBT	0.462	2600	0.447	2100	0.424	
7.62 M118 LR	175gr HPBT	0.502	2800	0.496	1800	0.485	
.30-06 Fed Match	168 gr HPBT	0.462	2600	0.447	2100	0.424	
.308 Fed Match	175 gr HPBT	0.462	2800	0.496	1800	0.485	
5.56mm M193	55 gr FMJBT	0.272	3000	0.245	2400	0.235	
5.56mm M855	62 gr SS109	0.324	2481	0.321	1974	0.311	
.223 Match	69 gr HPBT	0.301	2800	0.305	2200	0.317	note 3
.300 Win Mag	190 gr HPBT	0.533	2100	0.525	1600	0.515	
.338 Lapua	250 gr BT	0.662	not used	0.662	not used	0.662	
Cal. .50 MP	665gr HEIAP	0.600	not used	0.600	not used	0.600	note 4
5.45x39 Soviet	53 gr FMJBT	0.319	2387	0.318	1889	0.301	note 5
7.62x39 Soviet	122 gr FMJBT	0.148	not used	0.148	not used	0.148	note 6
7.62x54 Soviet	150 gr FMJ	0.408	2800	0.397	1800	0.387	note 7

Note 1: BC's boundary velocities for 168 gr. HPBT, 175 gr. HPBT, 5.56/.223 69 and 55 grain bullets, and .300 Win Mag 190 gr. are from the Sierra Reloading Manual, 50th Anniversary Edition. All other BC's and boundary velocities are derived from from existing velocity tables from other sources and Sierra Ballistics Program.
Note 2: Bullet weights of M80 ball run between 147.5 to 149.5 grs.
Note 3: Cal. .223 69 gr. HPBT MatchKing BC increases as velocity decreases.
Note 4: BC estimated from given bullet weight and M8 API Muzzle Velocity
Note 5: Soviet bullet is steel core 53 gr. FMJBT fired from AK-74 with 7.67" twist. MV is 2953
Note 6: MV of standard ball round is 2329 fps.
Note 7: Light Ball has no color tip. MV of SVD Sniper is 2723 fps.

Trans-sonic region is a term used frequently when snipers talk about how their bullet was affected by the wind at 900 yards or keyholed on the 1000 yard target. This region is normally reached when the bullet is between 1200-900 fps. Sierra Bullets conducted firing tests with different handgun loadings and different bullets being measured within this velocity region. They all exhibited the same thing. BC's for the bullets were fairly constant until it went into the transonic region at around 1160 fps, then the bullets went wacko. BC's went drastically high and low and then repeated this pattern with every 20 to 40 fps until at about 1050 fps when the BC settled down again. As a result Sierra created the three BC concept for specific velocities and bullets. For handguns these are the BCs before the trans-sonic region, in the region, and after the region. The BC calculated for the transonic region is an average of the many BCs occurring in that region. For rifle velocities we get three BC's also, but they represent the (1) high muzzle velocity, (2) range of velocity just below muzzle to after 500-600 fps falls of, and (3) below the middle range which would also include the transonic speeds. One theory points out that as the bullet's sound wave catches up to the bullet, the force of the sound wave creates a lot of air turbulence around the bullet causing it to be unstable. Nevertheless, what we are concerned about here is that our bullet's velocity stay out of the trans-sonic region for as long as possible, for as far as possible. We achieve this by using loadings that will give us higher velocities with bullets having better BCs.

Barrel Length vs. Velocity. A problem noted with ballistics tables in many articles and books on firearms and ammunition is the lack of information on the air temperature and barrel length used to get the data. The importance of this becomes apparent when you try to engage targets beyond 800 to 900 yards. At this range, many .308 Win/7.62mm bullets approaches sonic speed (about 1100 fps at sea level) and begins to lose stability. Because of this instability, the wind appears to also have a greater affect on the projectile. This is why the 26" barrel and the 175 gr. HPBT bullet for sniper rifles are recommended in order to stay out of the trans-sonic region out to 1000 yards at normal temperatures. For many LE snipers, this is not a concern unless you have to take a shot in a real tactical situation that is beyond say 600 yards. You could stay with your 168 gr. HPBT or 165 gr. SPBT in your .308 rifles.

While testing the new M118 LR ammunition with commercial powders for Paul Riggs, ARDEC, I checked out some Federal 168gr. HPBT match ammunition in my velocity test rifle. Using different barrel lengths, I confirmed what most already know: velocity decreases with shorter barrel lengths for most cartridges (some .22 rimfires are one exception). At 67 deg F, the following velocities at 15 feet from muzzle were recorded:

Table 8.4 Barrel Length vs. Muzzle Velocity

For Federal Match 168 gr. HPBT, 60 deg F.

Barrel Length	Vel (fps)	Vel loss (fps)	Cumulative (fps)
26"	2694		
25"	2654	40	40
24"	2616	38	78
23"	2592	24	102
22"	2572	20	122
21"	2566	6	128
20"	2524	42	170
19"	2509	15	185
18"	2477	32	217
17"	2426	51	268
16"	2414	12	280

Loss of velocity averaged around 28 fps per inch of barrel. Keep in mind that differences in loadings and other variables, such as throat length, type and amount of propellant, different barrels, etc., cause differences in velocity loss when barrel lengths and calibers change. Heavier bullets in the same caliber with same propellant has an effect like that of using a "faster" burning powder. Therefore, velocity changes are not as great when barrel lengths are changed.

Other External Factors that Influence Accuracy and Precision

The *most critical set of data* the sniper must learn to correct for are **wind, target distance, and target movement**. These directly affect the ballistics and projectile's accuracy once you have corrected your zero for the local conditions. These are also much more difficult to correct for than temperature and altitude. As new long range shooters soon find

Jim Gannon's range goes to 400 yards. This is the 326 yard firing point to bank of targets in top center of Oehler 35P chronograph. Always make sure you are aiming several inches higher than the housing for the electronic eyes in the bottom of the triangle by boresighting. One of us put three rounds through the top part of that lower housing before we figured out why the bullets didn't hit the target. Oehler's chono is very accurate and gives statistical data. It is easy to assemble and light weight. The only time readings are not accurate is when the wind causes the screens to move about. (Mike R. Lau)

CHAPTER 8: External Ballistics, Influence on Accuracy, and Ballistics Tables 157

Texas Brigade Armory's velocity measuring rifle consists of a Remington 700 short action and 9 barrels that measure from 16.25" to 26". Each barrel is a Remington take off and in near new condition. All are headspaced identically. Velocity tests of different types of propellant fired in different length barrels were conducted to study possible substitute propellants during testing of M118 LR ammunition. Test was designed to see at what barrel lengths does commercial ball and stick propellants show greatest rate of loss in muzzle velocity. All propellants tested showed major velocity loss when going from 26" to 20" barrels. (Mike R. Lau)

out, it's a whole new ball game when shooting out beyond 300 yards. Some kind of chart, table, or formula, are usually required for some of these unless you are so experienced that you can do without help. Each caliber and bullet type may require a different set of tables or charts so settle on one type of caliber and ammunition to keep it simple. Here is an outline of the factors to keep you organized as we go through this discussion. Parts II: A (wind affects), B (range affects), and C (target complexities), will be studied in separate chapters following this one. For this chapter we will only cover Part IV, light and rain affects and bullet drift.

Weather and Target Factors Influencing Accuracy Requiring Adjustments

    I. Wind
        A. Speed.
        B. Direction.
        C. Wind effects at different distances.
   II. Distance to Target
        A. Estimating Range.
        B. Firing up or down slopes.
        C. Adjusting sights for range.
  III. Target Engagement Complexities
        A. Moving target leads.
        B. Short Exposure Target.
  IV. Other factors affecting accuracy
        A. Light and Rain.
        B. Mirage.
        C. Bullet drift.

Accuracy as Affected by Light Conditions

Lighting conditions can affect shooting accuracy by making the center of mass on the target appear elsewhere on the target. Some common conditions of light and how it affects the sight image are:

(1) *Bright light* with no clouds, haze, or fog, can reflect or glare off a target causing it to appear larger. This is more apparent with iron sights and a scope's magnification may reduce this effect.

(2) *Bright light* hitting the target from the side, and the sun is low in the sky, will cause the target's center of mass to appear to shift toward the darker side. This is corrected by aiming a little further toward the bright side of the target. The telescope's magnification should reduce this effect.

(3) *Hazy light* is caused by fog, smog, dust, or humidity. It is not bright, but can be uncomfortable to the eyes. It can have the same effect on a bullseye target as bright light does in that it can make it appear smaller.

(4) *Light overcast* when there is a thin layer of clouds and no blue sky visible. Light overcast is comfortable to the eyes with no glare present. It is probably the best light condition for the shooter.

(5) *Dark heavy overcast* occurs when the sky is overcast and most of the light is blocked. The eyes will adjust to the dim light.

(6) *Scattered clouds* are broken up into small patches with the sun appearing at times. The sniper will have trouble adjusting to a target that is brightly lit sometimes and dark at other times.

(7) *Moving clouds* exist when the sky contains scattered clouds which move across the sky rapidly, making the sun appear only now and then. The sniper's eyes can fatigue rapidly due to changes from bright light to shadows.

The experienced sniper will overcome this problem by picking one of the two conditions in which to fire his shots. He will not try to fire under both conditions. This condition is the most difficult for the sniper because of the rapid light changes.

(8) *Rain* affects the shooter's concentration, visibility, and may also make weapon handling difficult. Rain also gets on the scope lenses. Rifle, scope, and ammunition should be protected from the rain until the sniper is ready to aim and fire. Light rain will not affect the accuracy of the bullet enough to be concerned with. Heavy rain and sleet can be a problem and destroy accuracy.

Accuracy Affected by Bullet Drift

When a bullet leaves the muzzle it immediately is pulled to the earth by gravity. Because it is falling, the air pressure under the spinning bullet is stronger than the air pressure on top. Like a tire on the ground, the bullet nose will turn slightly in the direction of the rotation imparted to it with the twist. For a right hand twist, the bullets turns slightly toward the right. As the bullet point turns slightly, its side is turned toward the frontal air resistance causing it to be pushed or drift slightly toward the direction the point is facing. This happens regardless of any wind drift affects. For a 150 gr. Caliber .30 M2 bullet this drift is about 6.7 inches at 1000 yards according to **Hatcher's Notebook**. The old 1903 Springfield rear sight slide had a correction for this drift. For the 174 gr. Cal..30 M1 bullet, drift is .95 inches as calculated by Wayne Ash in ***Precision Shooting***, page 23, April 1991 issue. Bullet drift gets excessive only when you get way out beyond 1000 yards. Don't worry about drift. Other factors, such as wind and range, that affect accuracy and precision are our greatest concern and we should concentrate on these.

CHAPTER 8: External Ballistics, Influence on Accuracy, and Ballistics Tables

KEEP IT SIMPLE, BUT BE ACCURATE

It's good to know how weather affects ballistics and the many things you need to do to correct them. However, Kent Gooch, Sam Chesnut, and I, decided that we all spend too much time in the class room and in the field worrying about the effects of weather on your trajectory. The best way to learn about weather affects on your shooting is to shoot in bad weather. Go to the range on the extreme cold and hot days. Go when its windy, foggy, or raining. Don't worry about people thinking you are stupid, because you will be the only one there. Compare your point of impact with those on normal days. Most of us don't have barometers or ways to measure humidity and you really don't need them. Take your chronograph and thermometer if you have one. Record your data on a chart for making zero adjustments at different temperatures. This will be for you only, your rifle, and ammunition.

About Ballistics Tables

When you look through other manuals and books on external ballistics, there seems to be a lot of data and information that you have to wade through. This is because each ballistics and data table is for a specific loading, with a specific muzzle velocity, and is in either yards or meters. In addition, there is also the separate data used for correcting external environmental affects on the projectile's flight, such as temperature and target move-

SFC Kent Gooch spent the last 20 years in both the Army and the Marine Corps. His life has been devoted to firearms and the teaching of weapons to others. In 1980 he graduated from the 3d Marine Division Scout/Sniper School and later from the 1st Marine Division Surveillance and Target Acquisition (STA) Platoon School. In 1981 he completed the Scout Sniper Instructor School at Quantico and went on to become an instructor with the Marine Corps MTU for 6 years which included the Scout/Sniper School, Close Quarters Battle (CQB) Course, and Small Arms Weapons Instructor School. Other assignments included Team Captain of the Parris Island Shooting Team, Senior Instructor of the USMC Security Force (MCSF) School in Valejo, California, Team Captain of the MCSF Shooting Team, and advisor for the Designated Marksman Program for Fleet Anti-Terrorism Security Team. After transferring to the Marine Corps Reserves, Kent enlisted into the Army National Guard. Army training included Special Forces Special Operations and one of the top graduates in numerous Marksmanship and Range Instructor/Safety courses. In 1996, he graduated from the U.S. Army Sniper Course given at Ft. Benning, Georgia. Today, SFC Kent Gooch is the Senior NCO of the U.S. Army National Guard Scout-Sniper School (NGSSS) at Camp Robinson, Arkansas.

ment, and each of these effects are explained in different chapters. Most of the important shooting data can eventually be combined into only a few ballistics/data tables for your one rifle/cartridge system. For easier referencing I have put the ballistics and firing correction data in Appendices I, II, and III.

Some of the ballistics tables and data charts in the Appendices and other data throughout this book were calculated from data generated by the Sierra Computer Ballistics Program. All trajectory data was calculated using standard altitude and atmospheric conditions which were used for many years by the U.S. Army Ballistics Research Laboratories at the Aberdeen Proving Ground in Maryland. These conditions also determine the air density and speed of sound.

Altitude:	Sea Level (ASL)
Barometric Pressure:	29.53 in Hg
Temperature:	59 deg F
Relative Humidity :	78%
Air density:	.0751 lb/cubic foot
Speed of Sound:	1120.27 fps

Data for rifles and ammunition at different locations may or may not coincide with the ballistics tables at the end of this chapter. You should verify your own weapon systems and modify your data accordingly.

The external ballistics in my tables for the M118 Special Ball fired from the 24" barrel, reflect nearly the same as that used by the Marines at Camp Pendleton. The Marine Corps Scout-Sniper School Lesson Plan gives the BC as .496 for M118 lot number LC-88/042 and .494 BC for other lots. These BCs are actually more in line with what the M118 will actually do, but they will not produce the same velocities in the Lesson Plan. To derive the velocities in my tables which came closer to the Marine's BC's, I changed the BCs as follows with the Sierra Ballistics Program: .486 for high and middle range, and .426 for lower. Cut-off velocity between middle and low was 1800 fps.

.50 caliber M82A1 Barrett rifles have 838mm or 33 inch barrels and factory listed velocity is 853m/s or 2799 f/s with M33 ball. The M82A2 and M90 bolt rifle have 736mm (29") barrels. .50 caliber Browning machine guns (BMG) have 36" barrels and muzzle velocity in Army manuals is given at approximately 3050 fps for these guns using ball, M2 or M33. M2 Ball and M2 AP have similar velocities while M8 API has a higher velocity because bullet is lighter. Most .50 Cal. ammunition follow almost the same trajectory out to 1000 yards with a time of flight difference of less than 1/10 second. This allows the different types of ammunition to be mixed in belts when fired from machine guns.

Ballistics Tables for the common sniper ammunition is found in Appendix I. Leads for moving targets are found in Appendix II. Wind adjustments are found in Appendix III.

CHAPTER 8: External Ballistics, Influence on Accuracy, and Ballistics Tables

CHAPTER 9:
RANGE ESTIMATION and SIGHT ADJUSTMENT

CPL Hengst spotting for CPL Crawford. (Mike R. Lau)

As a military or LE sniper, your "target range" is the battlefield or some wide downtown street. Distance to target is not a given and not in precise 100 yard or 100 meter increments. It is usually farther than or closer than anything you practice at on the local range. There are no wind flags and the terrain and foliage gives you no mirage to judge wind speed. The target is never the same size, always moving, and never travels the same speed twice. You just stirred up a fire ant mound and they are already assaulting you. Your head is hurting because you haven't slept in 24 hours and your trigger finger hurts because you smashed it on a rock crawling up here. You cannot get a natural point of aim because a bush is blocking your view and a large rock is under your hip. Your data sheets won't cooperate and give you the information you need. The calculator doesn't work. Must have been caused by the 3 inches of water in the depression you are laying in. When you finally wrestle the data from your log book, the target has moved and the data is no good. Luckily for you there are a few things you don't have to make decisions on: (1) you only get one shot at the target, (2) your shot has to be executed with the same precision as a high master competitive shooter, (3) you need to shoot quickly.

Unknown Distance Targets

In the field, the sniper's target distance begins as an unknown and is usually a distance he doesn't normally zero in practice for. This distance determines data for sight elevation, wind drift adjustment, and lead on a moving target. When the distance changes the data changes.

Before you can engage targets at unknown distances, you need three things: (1) a method for estimating the range, (2) data that tells how much elevation adjustment your scope requires to zero at that range, and (3) actual adjustment on the scope.

1. Ways to Estimate the Range

Laser range finders are a very accurate means and are becoming very popular as more manufacturers bring out new models and costs go down. A laser beam is aimed at the target and the time it takes for the light wave to return is calculated into distance. The Marines have fielded a high tech system where the Leica range finding binocular will send target distance to a GPS which itself is plugged into communication equipment. The GPS will give you the map grid coordinate of the target which is sent, via the radio, to the FDC (fire direction control) personnel who can arrange for an artillery or air strike on the target.

Maps are also accurate in determining distance if you use a small scale map and know the exact location of you and your target. Small scale is like 1:25000 while large scale is 1:100,000.

Simple methods, such as estimating 100 yard increments to the target, work good for shorter ranges and where a surgical or incapacitating shot is not required. However, unaided visual methods of range estimating can be confounded by light, terrain, shape of object, and background of target.

Another simple method is that of holding out your hand with thumb sticking up and comparing the size of your thumb nail to an object a distance away. Maj. John Plaster describes this technique in his book, **The Ultimate Sniper**, on page 281. To use this method, hold your thumb nail just below the person who is some distance away. Imagine the number of paces that a human target, 50 to 200 yards away, would appear to be walking across the top of your thumb nail. 4 paces across your thumb nail and the target is 200 yards away. 2 paces and the target is 100 yards away. The technique is feasible and actually works.

Always hold the thumb a specific distance from your face. Find the relationship that fits your own thumb size.

Fig.9.1. Ranging with the thumb.

CHAPTER 9: Range Estimation and Sight Adjustment

2. Range Finding with the Rifle Telescope's Reticle.

You may have thought John Plaster's example was amusing and appears to be impractical, but guess what? That same principle is used in most range finding rifle telescopes being made today. Hold up your thumb, or any object in your hand, next to the image of a distant target. The relationship between the viewed size of the two objects specifies an exact ground distance between them. That view where the two objects are compared is the image in your telescope. One of the objects is your scope's aiming reticle (your thumb nail) and the other object is the target.

Ranging with Variable Power Telescopes

A simple range finding system is found in most hunting and tactical scopes that have adjustable power. As the image of the target is increased or decreased in size with the power ring, it is compared or fitted to the fixed measuring device on the reticle. When they match in size, the distance to target is found by a yardage scale located on the power ring or on the scope's image itself. Some examples of these are the ART, Redfield's Accu-Trac and Accu-Range, and Burris' RAC. Many of these systems describe the use of the measuring reticle as fixed for one size target only. They are excellent for hunting deer and elk, etc. You

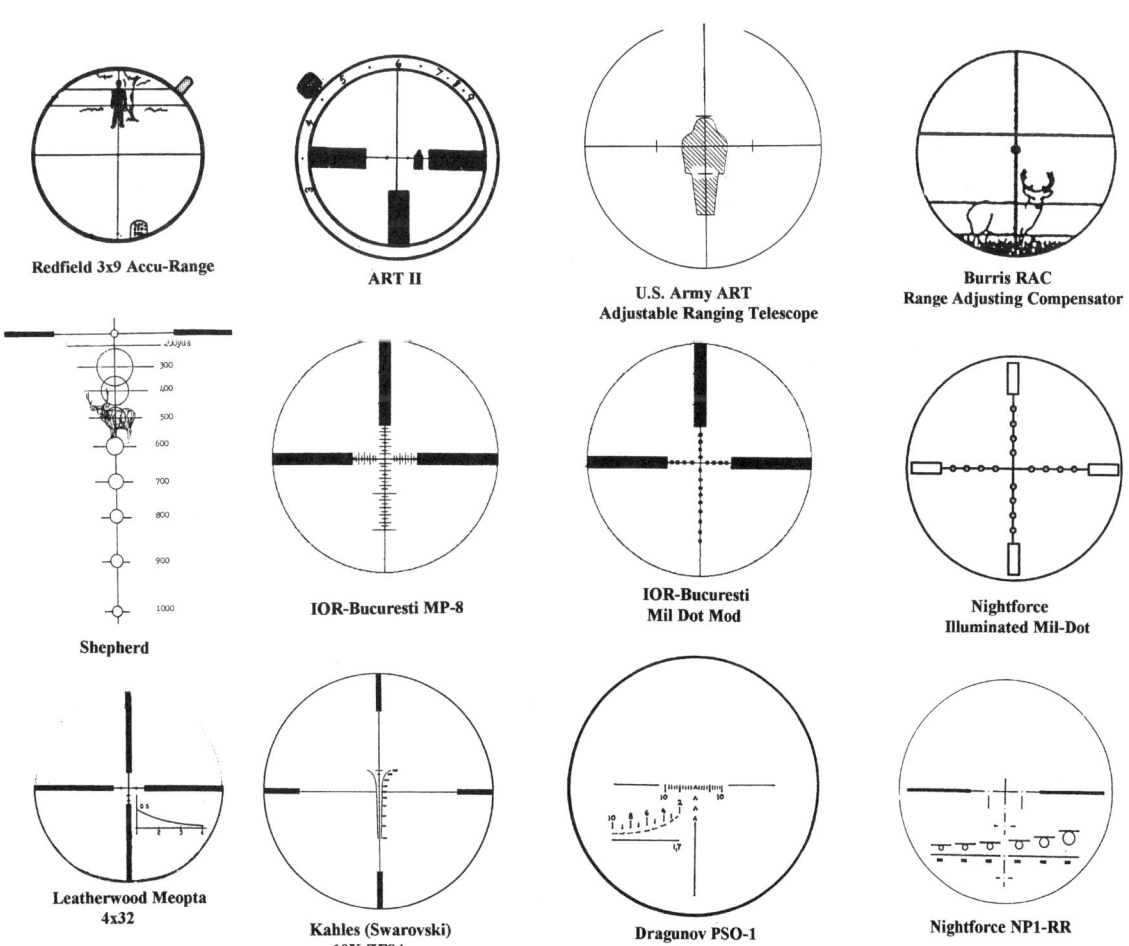

Fig. 9.2. Telescope Range Finding Reticles in use Today

can change the "value" of the reticle's measuring portion and also the "value" of the targets size, but then the range scale would be different. Some of these reticles are tied back into a cam operated by the power ring. When you fit the target to the comparing reticle, a calibrated cam located around the power ring will raise or lower the back end of the scope mounted on a moveable spring levered mount. A drawback to the use of cams in scopes is that the cam is usually calibrated for a single cartridge and load.

Ranging with Fixed Power Scopes (calibrating your thumb nail)

The image of the target is the same height as my thumb nail at one distance and half the height of my thumb nail at another distance. Instead of changing the size of the target image to fit my scope's reticle, I can have several reticles in the same scope so that each one represents a specific target distance. Some of these reticles are found on the Dragunov system, Meopta, Swarovski's "wedge" reticle, Nightforce, Shepherd, and the earlier Springfield Armory ART scopes. The nice thing about some of these scopes is that when you fit the target between the measuring lines you have automatically raised or lowered the rifle system for the hold off on the target. All you have to do is to fire. These range finding reticles are usually calibrated for a single type of ammunition.

These reticles may also be found in variable powered scopes, but can only be used at a specific fixed power, usually 10X or whatever power the manufacturer has scaled the reticle for. Sometimes the optics don't cooperate and you won't get the precise measuring of heights and widths as the manufacturer stated. For instance, instead of 10X you might get more accurate readings at 9.8X. You can check this out for yourself by measuring known height targets at known distances with your reticle.

All of these previous telescope reticles show a relation to the specific target height and target distance when compared to a measuring portion of your reticle. These are great to use if you get a clear view of the target height to superimpose the reticle on. In sniping applications this is usually not the case. What if your target is not the size in length or height as your reticle requires? Many times the "bad guy" will not show himself completely or his exposure to you is only fleeting. What if he is sideways to you, sitting, kneeling, or prone and facing you? What if your target is not animal or human?

Ranging with the MIL-Dot Reticle (calibrating with your thumb nail)

At a specific distance, I need four times the height of my thumb nail to cover the target. At another distance, the same target appears to be 1 1/4 times the height of my thumb nail.

Developed by the Marines in the late 1970's, the USMC Mil-Dot reticle in the Unertl 10X scope aids the scout/sniper in estimating distances.

To understand how this reticle works imagine a very small angle radiating away

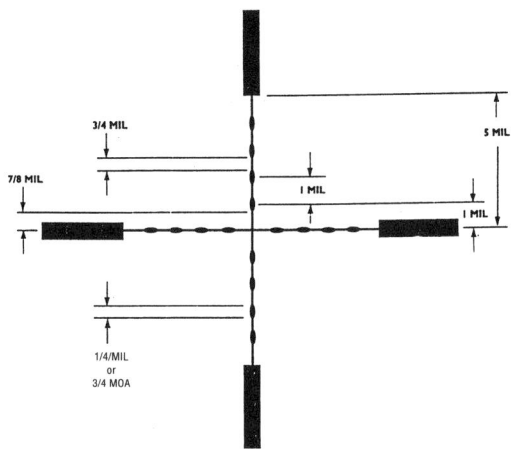

Fig. 9.3. USMC MIL-Dot reticle

CHAPTER 9: Range Estimation and Sight Adjustment

from you so that the farther away from you the wider the measurement between the two legs of the angle. We call this specific angle the MIL angle (for Milradian) and it has a specific height (or width) of 36" at 1000 yards. That same 36" will span 2 MIL angles at 500 yards and 5 MIL angles at 200 yards. Do not confuse MIL angle with the minute of angle (MOA).

MOA is also the width of the legs of an angle radiating away from you, but this angle is smaller, being only a fraction of a MIL. 1 MOA equals 1.0472" at 100 yds and 10.47" at 1000 yards. 1 MIL equals 3.6" at 100 yards and 36" at 1000 yards.

The two angles are not related to each other so they don't convert from one to another easily. Just keep in mind that each represent so many inches at different distances. We will use inches for all target heights, shot group size, bullet drop, wind drift, etc., regardless of whether we give distance in yards or meters.

Table 9.1 Relationship Between MIL, MOA, and Inches at Distances.

The value of MOA and the MIL in inches is incrementally proportional at the increased distance.

1 MIL ≅ 3.6" @ 100 yards	1 MIL ≅ 3.9" @ 100 meters
1 MIL ≅ 18" @ 500 yards	1 MIL ≅ 19.5" @ 500 meters
1 MIL ≅ 36" @ 1000 yards	1 MIL ≅ 39" @ 1000 meters

1 MOA ≅ .28 Mil at any distance
1 MIL ≅ 3.6 MOA at any distance

1 MOA ≅ 1" @ 100 yards	1 MOA ≅ 1.13" @ 100 meters
1 MOA ≅ 5" @ 500 yards	1 MOA ≅ 5.65" @ 500 meters
1 MOA ≅ 10" @ 1000 yards	1 MOA ≅ 11.3" @ 1000 meters

Having selected a method for estimating range, let's see how the Marine Corp MIL-Dot reticle is used to estimate distance.

Step 1: Obtain number of MIL spacings covering target.

On the reticle crosswires are a series of "football" dots (round dots etched on glass for the older Leupold Ultra scope) that are spaced 1 MIL apart. The football shape supposedly allows for a more precise measuring with their pointy ends. The dots themselves are slightly longer and slightly narrower than 3/4 MOA, thus the term "standard 3/4 MOA MIL-Dot reticle." When you superimpose the MIL-Dot crosswires of the scope on to a target, you count the number of MIL spacings covering that target.

1/23 Marine Scout/Snipers estimate the number of MILs covering a target to within 1/10 of a MIL. This requires holding the scope reticle on the target as steady as possible to insure that the most accurate MIL height estimation is made. For closer, larger targets, like full and half height silhouettes under 300 yards, 1/4 mil estimations will not create a significant error in estimating range.

Step 2: Height of your target

Most military and law enforcement sniper's data or log book have charts depicting heights and widths of human targets and common silhouette targets. Tables are also useful showing heights in inches from 9 to 72 inches and their relationship to yards or meters distance corresponding to mils. (See Fig. 9.5)

Besides human targets, there are numerous other objects that can be used to determine target distance. When using the mil-dot reticle, the object's height must be known and it must be small enough to be measured in mils in your scope's image. A table of larger MIL-heights and heights or widths of common objects is given in the chapter on log book management.

Step 3: Calculation to get distance.

Once you know the height of the target, for example: 72" or 2 yards for 6 foot person, you can calculate the distance to the target. Distance can be figured using a chart or formula that tells you how far a target is if you know the target height and number of MILs. (See Fig. 9.5)

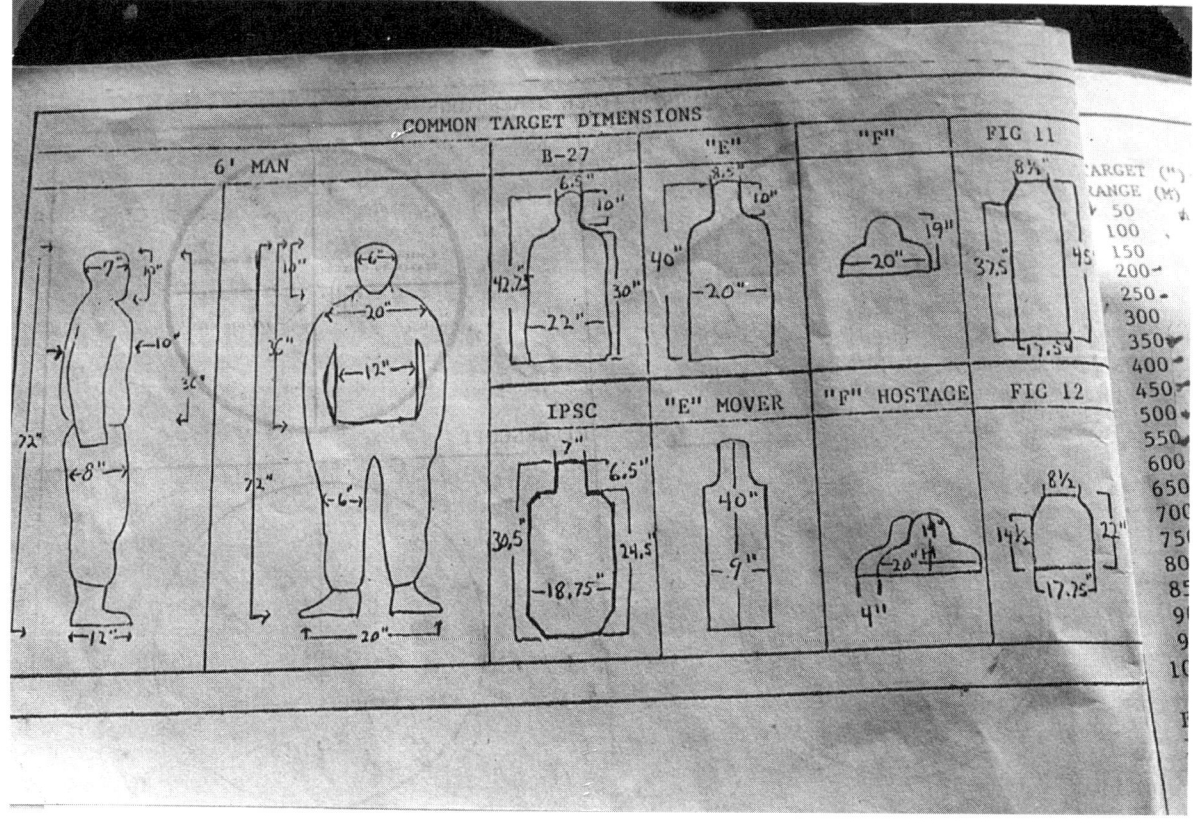

This is a page from the data book of SFC Pete Carpentier, Texas National Guard. The page is a wealth of information on target nomenclature and dimensions. (Courtesy Epitacio Carpentier)

Another way to get target distance is to use the following formulas:

$$\frac{\text{(Height or width of target in yards)} \times 1000}{\text{Number of MILs covering target ht. (or width)}} = \text{Distance (YDS)}$$

$$\frac{\text{(Height or width of target in inches)} \times 24.5}{\text{Number of MILs covering target ht. (or width)}} = \text{Distance (Meters)}$$

25.4 is the factor for converting inches to millimeters.

Conversion Factors tables are handy when using formulas for calculating data. With a little use, most of these conversions will be stuck in your memory. There is a complete conversion table in the chapter on logbooks.
(1) Meters to Yards: Add 10% of the meters to come up with yards

Example: 550 meters + 55 = 605 yards

(2) Yards to meters: Subtract 10% of yards to get meters.

Example: 600 yards -60 = 540 meters

Example: Target Height in Yards = 6 ft = 2 yards

Target Height in Mils = 4 mils

$$\frac{2 \times 1000}{4} = 500 \text{ yards}$$

Importance of estimating range accurately

Both the Army and the Marines emphasize the importance of estimating target distance as accurate as possible at the longer ranges. As just mentioned, target height is estimated to the nearest 1/10th of a mil when using the Unertl or M3A day-optic. This is especially more critical when the targets are small in size such as the 40" silhouette. With 7.62mm ammunition, there is almost 6 MOA

CHAPTER 9: Range Estimation and Sight Adjustment

mil-RELATION RANGING TABLE 1.016

TARGET (") →	6	10	18	20	22	24	30	36	40	42	45	72
RANGE (M) ↓												
50	3.0	5.0	9.1	10.0	11.2	12.2	15.2	18.3	20.0	21.3	22.9	36.6
100	1.5	2.5	4.6	5.0	5.6	6.1	7.6	9.15	10.0	10.6	11.4	18.3
150	1.0	1.7	3.0	3.4	3.7	4.0	5.0	6.1	6.8	7.1	7.6	12.1
200	0.75	1.25	2.3	2.5	2.8	3.0	3.8	4.6	5.0	5.3	5.7	9.15
250	0.6	1.0	1.8	2.0	2.2	2.4	3.0	3.7	4.0	4.25	4.5	7.3
300	0.5	0.8	1.5	1.7	1.9	2.0	2.5	3.0	3.4	3.6	3.8	6.1
350	0.44	0.7	1.3	1.5	1.6	1.75	2.2	2.6	2.9	3.0	3.3	5.2
400	0.38	0.63	1.1	1.25	1.4	1.5	1.9	2.25	2.5	2.6	2.9	4.6
450	0.34	0.55	1.0	1.1	1.25	1.4	1.7	2.0	2.25	2.3	2.5	4.0
500	0.3	0.5	0.9	1.0	1.1	1.2	1.5	1.8	2.0	2.1	2.3	3.7
550	0.28	0.46	0.83	0.9	1.0	1.1	1.4	1.7	1.8	1.9	2.1	3.3
600	0.25	0.42	0.76	0.84	0.9	1.0	1.25	1.5	1.7	1.8	1.9	3.0
650	0.24	0.39	0.7	0.78	0.85	0.93	1.2	1.4	1.55	1.65	1.75	2.8
700	0.21	0.36	0.65	0.7	0.8	0.87	1.1	1.3	1.5	1.52	1.6	2.6
750	0.2	0.34	0.6	0.68	0.75	0.8	1.0	1.2	1.4	1.44	1.5	2.4
800	0.19	0.32	0.57	0.64	0.7	0.75	0.95	1.1	1.3	1.33	1.4	2.25
850	0.18	0.3	0.54	0.6	0.65	0.72	0.9	1.05	1.2	1.25	1.35	2.15
900	0.17	0.28	0.5	0.57	0.6	0.67	0.84	1.0	1.1	1.2	1.3	2.0
950	0.16	0.26	0.48	0.53	0.59	0.64	0.8	0.96	1.06	1.12	1.2	1.9
1000	0.15	0.25	0.46	0.5	0.56	0.6	0.76	0.91	1.0	1.1	1.15	1.8

FIND TARGET SIZE IN INCHES ON TOP LINE. FOLLOW COLUMN DOWN TO TARGET MEASUREMENT IN mils. FOLLOW LINE LEFT TO RANGE IN METERS.

MIL height and distance chart from SFC Carpentier's data book. Target size in inches is given at the top and MIL height in the main body. Range is derived from the left column. This is a good chart because it is for 50 meter increments. (Courtesy Epitacio Carpentier)

Inches =		9	12	16	18	20	22	24	28	32	36	60	66	69	72
Yards =		.250	.333	.444	.500	.556	.611	.667	.778	.889	1.000	1.667	1.833	1.917	2.000
MILS	3/4	333	444	593	666	741	815	889	1037	1185	1333	2223	2444	2556	2667
	1	250	333	445	500	556	611	667	778	889	1000	1667	1833	1917	2000
	1 1/4	200	266	355	400	445	489	534	622	711	800	1334	1466	1537	1600
	1 1/2	167	222	296	333	371	407	445	519	593	667	1111	1222	1278	1333
	1 3/4	143	190	254	285	318	349	381	445	508	571	953	1047	1095	1143
	2	125	167	222	250	278	306	334	389	445	500	834	917	959	1000
	2 1/4	111	148	197	222	247	272	296	446	395	444	741	815	852	889
	2 1/2	100	133	178	200	222	244	267	311	356	400	667	733	767	800
	2 3/4	91	121	161	182	202	222	243	283	323	364	606	667	697	727
	3	83	111	148	167	185	204	222	259	296	333	556	611	639	667
	3 1/4	77	102	137	154	171	188	205	239	273	308	513	564	590	615
	3 1/2	71	95	127	143	159	175	191	222	254	286	476	524	548	571
	3 3/4	67	89	118	133	148	163	178	207	237	267	445	489	511	533
MILS	4	63	83	111	125	139	153	167	195	222	250	417	458	479	500
	4 1/4	59	78	104	118	131	144	157	183	209	235	392	431	451	471
	4 1/2	56	74	99	111	124	136	148	173	197	222	370	407	426	445
	4 3/4	53	70	93	105	117	128	140	164	187	210	351	386	404	421
	5	50	67	89	100	111	122	133	156	178	200	333	367	383	400
	5 1/4	48	63	85	95	105	116	127	148	169	190	318	349	365	381
	5 1/2	45	61	81	91	101	111	121	141	162	182	303	333	349	364
	5 3/4	43	58	77	87	97	106	116	135	155	174	290	319	333	348
	6	42	56	74	83	93	102	111	130	148	167	278	306	320	333
	6 1/4	40	53	71	80	89	98	107	124	142	160	267	293	307	320
	6 1/2	38	51	68	77	86	94	103	120	137	154	256	282	295	308
MILS	6 3/4	37	49	66	74	82	91	99	115	132	148	247	272	284	296
	7	36	48	63	71	79	87	95	111	127	143	238	262	274	286
	8	31	42	56	63	70	76	83	97	111	125	208	229	240	250
	9	28	37	49	56	62	68	74	86	99	111	185	203	213	222
	10	25	33	44	50	56	61	67	79	89	100	167	183	192	200

Fig. 9.5 MIL height and distance chart in yards from Premier Reticles.

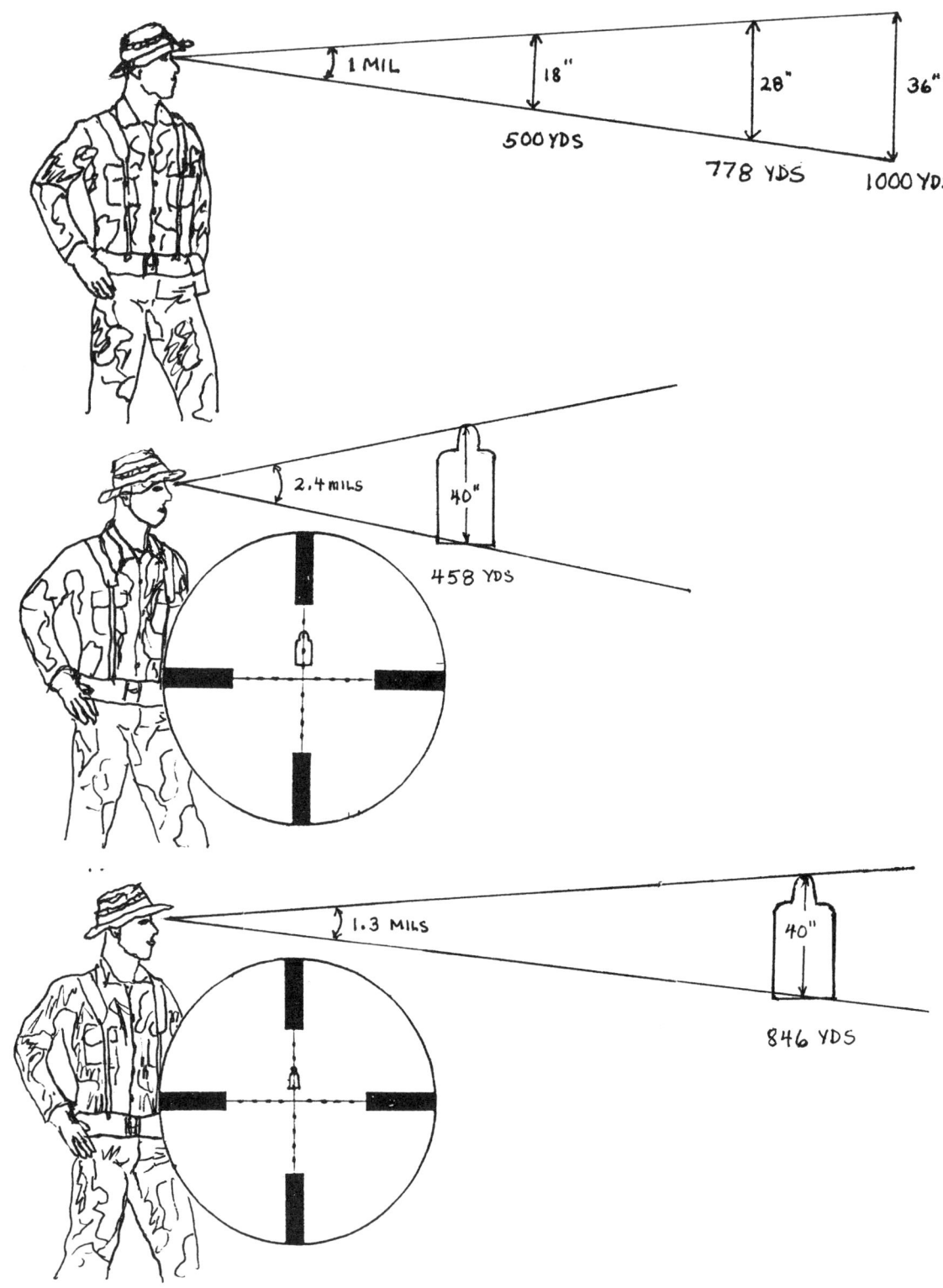

Fig. 9.4. Ranging with the MIL-dot Reticle. Target distance is obtained by measuring target with mil scale in scope reticle. Target height in yards or inches must be known to determine the distance using a chart or mathematical formula. MIL angle or MIL height, Target height, and Distance are all related. Any one of these can be found if the other two are known using the range formulas for yards or meters.

of elevation when going from 700 meters to 800 meters. This breaks down to around one MOA elevation for every 16 meters of ground distance. If you incorrectly estimated the ground distance to target by 50 meters at a distance beyond 700 meters, you could be off the center of target by about 20 to 24 inches.

SFC Gooch believes that it is necessary to break the mil dots down to 1/10th of a MIL when estimating range to a small (around 40" or so) target. When estimating range on a larger target, say a full height 6 foot silhouette, the criticality of estimating to 1/10th of a mil is unnecessary, For example, a target is at 700 meters, Milling out only a 40" section of the full size target, he got a reading of 1.5 MIL height or 677m distance. If instead, 1.4 MIL was used for target height, the estimated distance is now 725m. This is a difference of 48 meters for the 1/10th MIL difference. On the other hand, if the 72" target is measured the full height, 2.6 MILs gives 703 meters distance. Measuring the target to the nearest 1/4 MIL or 2.5 mils gives a range of 732 meters. This is a 29 meter difference for the 1/10th of a MIL change in target height. Kent Gooch also makes note to remember to focus the parallax before trying to estimate MIL height or your reading will be off. It gets worse as your target is further out. From 800 to 1000 yards, the M118 bullet drop between 50 yard increments goes like 53", 61", 68", and 78".

3. The Come Up Table

Now that we estimated range accurately we need to adjust our scope for that range. When we adjust our scope, we are adjusting the trajectory to impact on the line of sight at the new figured range. Since we are normally zeroed at 100 yards. we will *come up* in elevation from there. Come ups are simply the amount of elevation, or depression, you need to adjust for to zero at the new target distance.

Come ups for each caliber at standard temperature and pressure at sea level (STP) are found in the ballistics tables located in Appendix I. This data tells you how much elevation in MOA or inches to adjust for each range in 100 meter or yard increments. There is also cumulative comeups in case you go from your 100 yard zero directly to your new range.

Using the Come Up Tables and Extrapolation

Come ups are calculated directly from the bullet path in inches at each successive 100 yard increment. For example look at the M118 LR with the 175 gr Sierra HPBT fired from the 24" barrel:

Range (yards)	Velocity (fps)	Path (in.)	Come ups to range (moa)	Come ups cumulative (moa)
600	1648	-102.4	4.5	17
700	1515	-154.4	5	22
800	1394	-221.7	5.75	27.75

1 MOA elevation for 700 yards moves the bullet up 7 inches. If the 700 yard bullet drop is 154.4 inches from the 100 yard zero, then 154.4 divided by 7 equals 22 MOA. If you crank up 22 moa adjustment on your elevation knob, you bring the point of impact back up 154.4 inches to rezero at 700 yards. The *"come ups* to range" column shows you the moa change from the previous range. To engage a target at 800 yards after shooting at 600 yards, you would crank up 5 moa (for the 700 yard zero) and then 5.75 moa (800 yard zero), or a total of 10.75 moa from the 600 yard zero.

Battlefield range for sniper's are usually never in even increments of 100 yards or meters. To create a table of comeups for every 5 yards can be done, but is not necessary. Snipers in the field extrapolate the come ups for odd distances. Suppose I estimated target distance to be 758 yards. Using the table for the M118 LR/175 gr HPBT Fed Match with 24" barrel again, I see that between 700 and

172 *The MILITARY and POLICE SNIPER*

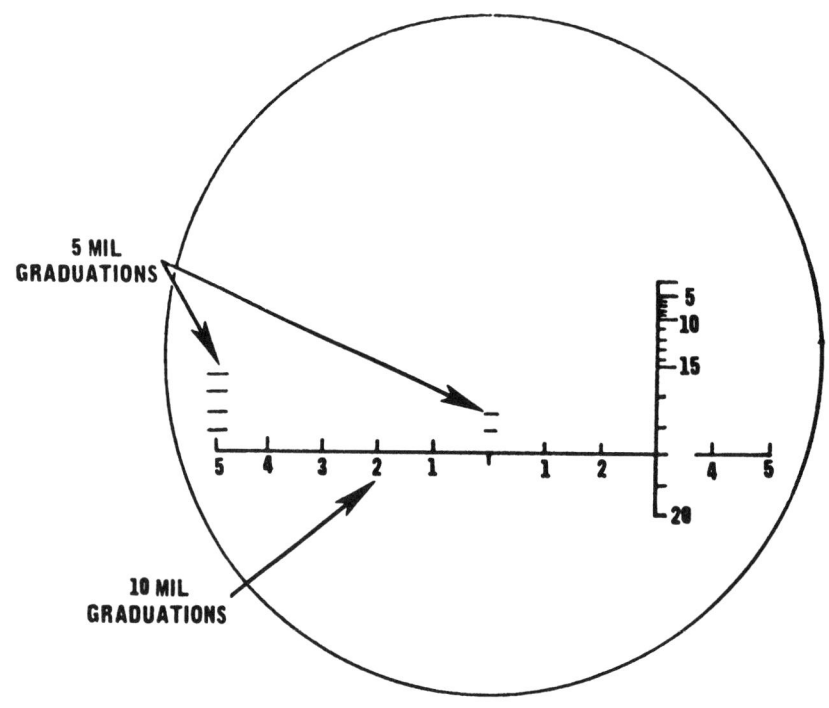

MIL scale in M17A1 binoculars are in 5 mil increments. Many military scopes and binoculars have similar MIL scales in them that can be used to measure MIL height. (Mike R. Lau)

CHAPTER 9: Range Estimation and Sight Adjustment

800 yards, the come up is 5.75 moa. Since 758 is approximately halfway between 700 and 800, I can add to my 700 yards come up half of the 5.75 moa which is rounded to 2.75 or 3.0 moa. I'll go the higher since the range was actually 8 yards beyond halfway. My come up at 758 yards is now 22 + 3.0 = 25 moa. Checking this with my Sierra Ballistics Program, my bullet drop at 755 yards (5 yard increments minimum) is 186.2 inches. Dividing 186.2 by 7.5 (1 moa = 7.5 inches @ 750 yards) I get 24.9 moa. My estimated 25 moa come up is close enough for precision government work!

With the Mark 4 M1, crank the elevation knob up 25 moa. With the Unertl 10X, turn the main elevation knob to the 700 yard tick mark and add 3 moa with the fine tune knob below. Alternatively you could come up to the 800 yard tick mark on the big knob and come down 3 moa on the lower dial. For the M3A, the complete calculation should be done in meters. Take 10% of 758 from 758 yards to get 682 meters for target distance. Work the come up in the same way using the meters table. The M3A's elevation only adjusts in 1 moa increments so you can round off the come up to nearest whole moa. If round up, hold low on center of mass. If you rounded the come up to a lower moa, then hold slightly above center of mass.

A couple of things you should know about come ups. First, there *will be differences between the come up table data and what you actually shoot* because of your own area's environmental conditions, elevation, and the way you shoot. The come up data from any book or manual will usually get you on target. You must get your own come up data by actually shooting the distance. Or create a table using a ballistics program and verify by shooting. Second, the come-ups you obtain by actual firing may not work for another person using the identical load and similar rifle as

M40A1   Ser#6737705

RANGE	CLICKS	MOA	BULLET PATH (inches)	HOLDOFFS (MILS)
25	-5	-1 ¼		
50	-3	-3/4		
100		------		
125	+2	+1/2	½	
150	+3	+3/4	1	½
175	+5	+1 ¼	2	½
200	+7	+1 ¾	3 ½	½
225	+10	+2 2/4	5 ½	¾
250	+12	+3	7 ½	1
275	+15	+3 ¾	10 ¼	1
300	+16	+4	12	1 ¼
325	+20	+5	16 ¼	1 ½
350	+23	+5 ¾	20	1 ½
375	+26	+6 2/4	24 ¼	1 ¾
400	+26	+6 2/4	26	1 ¾
450	+36	+9	40 ½	2 ¼
500	+43	+10 ¾	53 ¾	3
550	+49	+12 ¼	67 ½	3 ½
600	+53	+13 ¼	79 ½	3 ½
650	+66	+16 2/4	107 ¼	4 ¾
700	+74	+18 ¼	127 ¾	5 ¼
750	+88	+22	165	6 ¼
800	+93	+23 ¼	186	6 ¾
850	+107	+26 ¾	227 ½	7 ½
900	+120	+30	270	8 ½
950	+140	+35	332 ½	9 ½
1000	+154	+38 2/4	385	10 ½

Fig.9.6. Sam Chesnut's personal come up table for 26" barrel Texas Brigade Armory M40A1 rifle, SN 673705, with 168 gr. HPBT .308 Win Federal Match.

yours. My come ups are slightly different from Sam Chesnut's for the same rifle and ammunition because Sam is a much larger person and he holds his rifle much tighter so he gets less muzzle whip. His come ups are about what is shown in the charts for both M118 and .308 Federal Match with 168gr HPBT with a muzzle velocity of 2700 fps. Check your own comeups at these ranges for yourself. However, the data in the ballistics tables will put you near the target's center of mass if you do all of the other adjustments correctly.

Sniper tip: Target is missed and the spotter cannot see the bullet strike. Most likely the bullet went high or you would have seen the impact out front depending, of course, on the terrain. A couple of ways to get back on target if you think your zero was correct: (1) If you have trouble estimating range accurately, and you think the range is between two distances, adjust for the closer distance. You may get a chance to observe the bullet strike in front of the target and can adjust from there. (2) If you think the bullet went over the target, have the spotter get directly behind and above the shooter and spot along the rifle's line of sight. It is easier to spot the bullet trace if there is some kind of background such as a number board, burm, hill, building, etc. You should have a feel for where you think the range and elevation should be about and then drop several more minutes off your elevation so your bullet will impact low on the next shot. Or check to see if your scope got loose. You know your rifle better than anyone else. Crank down maybe 2 or 3 MOA and get the bullet to impact low. You may get lucky and hit the target, if not your spotter will see the bullet hit the ground and can adjust from there.

Notes on Laser Range Finders

With all the laser range finders on the market today, it is a wonder why the military does not issue one to every sniper team. Many devices are not that expensive, and good useable ones, like the Bushnell, cost as little as a few hundred dollars. The more expensive ones, like the Leica 7x GEOVID, may run about $3000. Think about it, a sniper with a 1 moa rifle and a laser range finder could engage more targets accurately at long range than a shooter with a 1/4 minute rifle and no laser range finder. If the military can spend big bucks on sniper weapons and training, think how much cost in those areas could be saved by increasing the high tech weaponry of the sniper or ordinary soldier just by issuing him a laser range finder. This is not to mean that the laser range finder replace good weapons and training, but to supplement it in order to increase probability of hit. Varmint hunters are already way ahead in this aspect as they are the one's demanding the devices right now.

A laser beam is a light that is intensified and sent in a particular direction. Military laser range finder systems use Yttrium Argon Gas (YAG) and are of such high intensity that the light can actually damage the eye if it is directed toward a person. Because of this power, military lasers are sometimes classified as weapons and can be issued for training only if controlled or for use on ranges. Military laser devices are rated to a mil-spec number that classifies them as to eye safety. Commercial laser devices use less intense light and are safer to use. Cost of a device is dependent on the technology and ability to pick up targets.

When selecting a commercial device, compare one laser range finder to another with the following factors: (1) Lasers source of energy. (2) Beam divergence. (3) Receiving sensor. (4) Type of target it will be used for. (5) Accuracy. (6) Cost.

The lasers's *light source* is an LED (light emitting diode) sort of like that which lights up your VCR or calculator when you turn the power on. It uses low amounts of battery power to light up and the beam of light is almost in the infra-red spectrum so cannot be seen without a IR device. The light energy

CHAPTER 9: Range Estimation and Sight Adjustment 175

does not get weaker the further out it goes. As the light beam goes further it gets wider.

Beam divergence is measured by its height and width in mils. The Bushnell Yardage Pro 400's beam divergence is 4 mils high and 2 mils wide. At 1000 yards the beam covers an area 4 yards high by 2 yards wide. Although the light wave is as strong at 1000 yards as it was when it left the device, it is not as effective because it is scattered and is hitting all sorts of other objects, such as the ground in front of the target. Ranging is made even more difficult when you are close to the ground and so is your target. The beam is sensing the ground way before it senses your target. This is why you sometimes get a better reading if you put the crosshair of the device slightly above the object. The Leica's beam is .3 MIL high and 1 MIL wide. This makes it more useable at long range even when you and your target are low to the ground. The narrower the beam, the greater the danger to the eye, and the higher the cost.

Your ranging device has a ***sensor*** (another LED) that reads back the reflection of the light with the same wavelength. Unfortunately the sun emits some of the same wavelength so objects around the target can reflect this also. To get around this problem, Bushnell uses "first return measurement." The light waves hit the target first and then hits the background objects next. The first return light will be that of the target so it is the one giving you the reading. However, there still is the problem with the ground and the wide scattering of the beam which will limit the accuracy and range of the device.

Target characteristics is the fourth factor affecting the accuracy of your ranging device. Large targets are of course easier to range. Shiny objects reflect light better and dark colored, rough textured objects absorb a lot of the light. The angle to the target determines how much and when the beam sees the ground first. Objects behind the target interfere a lot, especially when you cannot hold the device very steady. With the lower cost devices, trying to range on a human target standing in front of a building in the background may be difficult if you are at the limits of the device.

Accuracy and Cost

The higher the accuracy of the device the higher the cost. Bushnell's device is actually an excellent device for the price of around $250. It can sense silhouette size targets out to 500 yards. Bushnell has brought out a new model, the Yardage Pro 800, that ranges out to 800 yards and cost around $400. The Laser Impulse 2K, made by Laser Technology, Inc., ranges silhouette size targets out to 900 yards, larger objects to 1500 yards, and buildings out to 2000 yards. It costs around $1500, looks to be the size of a small video camera, and can be operated with one hand or from a tripod. A similar device is made by North American Integrated Technologies. Simmons is bringing out their Laser Mag 600 in 1998. The 7X device weighs only 19.5 oz. and will range to 600 yards. Tasco also has a device, but I could not find info on it. Swarovski markets their RF-1 at $3000. It ranges out to 1100 yards.

Leica's GEOVID is a 7x 42 range finding binocular that weighs around 3.3 lb.. It is eye safe and is the choice of many military forces. It puts out a beam of .3 mil high by 1 mil wide and will range to 1500 meters with a + or - accuracy of 2 meters. Included in the device is an electronic compass that is adjustable for declination so it can be used anywhere. Leica's sensor is called "pulse gating." It senses all light being reflected back from 25 yards to 1500 yards in 1 yard increments. "Bins" or storage devices, collect the sensor's electronic signals and the "bin" that collects the greatest number of readings is the range displayed. Leica's narrow beam can range a dark silhouette in front of a large object at 1100 meters. Although Leica may say their device can only range to 1500 meters, on clear bright days, the GEOVID has been known to range much farther than that. According to one user, it detected a paper cup on a highway beyond 600 yards. The optics resolution is excellent and it will also read while it's raining or snowing.

The Leica, Bushnell, Simmons, and Swarovski models are all available from the manufacturer or from the southwest US distributor, S.W.F.A., Inc., P.O. Box 69, Desoto, TX 75123.

Finally

A quick and practical way to get distance is with the range finding reticle found in most rifle telescopes. You already have the scope on the rifle and it is already pointed at the target. All you have to do is use the reticle on a known target dimension, make a calculation or read a scale, adjust the elevation or use Kentucky windage, and shoot. Find the system that gives you the level of precision you are striving for, is easy to use, and fast. If you have poor equipment, change it. If you don't have a choice on equipment then figure out a way to make it work and master that.

Sniping is practical shooting for the precision rifle shooter. Factors such as size of target, target distance, wind velocity and direction, temperature, humidity, movement of target, equipment and ammo, sniper's position, and firing technique, can all work against the sniper in hitting the target if not done right. To be successful is to understand each one, practice, and make them work for you. In competitive shooting, if you master the basics and increase your skills, they work for you because all your competitors on the firing line have to deal with them also. In real world scenarios involving the military or police sniper, determining target distance is extremely important because for most situations it is not a given. Range estimation has to be precise the first time. You will not get sighting shots and the first, and usually the only shot, is for keeps.

Strive for precision, but keep things simple.

CHAPTER 9: Range Estimation and Sight Adjustment

CPL Benett Thomas and SGT Sam Chesnut. (Mike R. Lau)

CHAPTER 10:
"This Ground Fire Was A Godsend To Us", UPHILL AND DOWNHILL SHOOTING

To many, he was an "all American boy." He was an Eagle Scout, a scoutmaster, an alter boy, and served in the Marine Corps. Friends knew him as a happy, friendly person, who never said bad things or got angry at others. Raised in a middle class family, he worked a part-time job, took 14 credit hours at the University during the summer, and was on the honor roll several times. Underneath all of this, he was an antisocial psychopath. He hated his father, had guilt about his wife, and was angry at his mother for leaving his father. He was a heavy gambler, wrote bad checks, and was heavy into pornography. He was court martialed while in the Marines for illegally carrying a pistol, loaning money at high interest rates to fellow Marines, and threatening to beat up another Marine. He beat his wife and as one doctor stated, "He was mean as hell."

A little before noon on August 1, 1966, after killing his wife and mother, Charles Whitman hauled a foot locker to the observation deck of the tower on the campus of the University of Texas in Austin. From the trunk, he took out a 12 Ga. shotgun, three pistols, a .35 Cal. rifle, a Cal..30 Carbine, and a 6mm Remington with 4X Leupold. He opened fire on the people below. Firing from all sides of the tower he hit students, school staff, visitors, and even onlookers a few blocks away. Over a hundred Austin Police and Dept. of Public Safety (DPS) Officers responded. In addition, members of the Secret Service, the Texas Rangers, State Capitol Police, and Campus Police assisted. The Austin PD had only twenty .35 caliber rifles at the time, and these were kept at the police building. They were mainly for riot control and had open sights. Among the law enforcement officers present there were only a few rifles being used. Most of the return rifle fire came from civilians.

Robert A. MacQuigg already graduated from UTA and was working in the Biology Building Lab near the tower. As he left the building for lunch, he heard a shot and saw someone firing from the tower. He realized what was going on outside and helped a police officer get people into the building. He told the officer he had a .30-06 rifle with a 4X scope at home. They went out the back of the building and the officer drove MacQuigg home in the police car. He had no ammunition so they stopped at a hardware store and bought some 150 gr. soft point ammunition. On return to the campus, another officer joined them and they headed up to the roof of the English Building. The officers had binoculars and spotted for MacQuigg as he fired at Whitman. His rifle fire forced Whitman to use the drain holes as firing ports so MacQuigg aimed for those, hoping to ricochet one through. MacQuigg fired 31 rounds at the observation deck, but because of the steep angle of fire and his own poor cover, he was not able to hit Whitman. From below, law officers returned Whitman's fire with handguns and a few had hunting rifles, but all were ineffective. Bullet holes peppered the sides of the building from halfway on up.

James M. Damon, a graduate student, saw Whitman shooting and also three injured students lying in the open area around the tower. He raced back to his car and drove 2 1/2 miles home to get his .30 Carbine and ammunition. He drove back and was allowed to pass on foot through a police road block. While running toward the Academic Center, he stopped and fired two rounds at Whitman. When he got to the building he fired 15 rounds from the first floor. Knowing he couldn't get a good shot from this position, he decided to go to the rooftop. When he got to the fourth floor, some students laughed at him for carrying the rifle. The librarian locked the door to the roof and refused to let him up there. She berated him for bringing a gun on campus. When he tried to explain to her the need for suppressive fire, she threatened to revoke his library permit. After ten minutes of arguing with her, a custodian finally let him onto the roof. Damon fired 35 more rounds at Whitman and ran out of ammunition. Shortly afterwards, a flurry of shots were heard coming from near the top of the tower and it was all over.

Charles Whitman was finally killed by Austin police officers Ramiro Martinez and Houston McCoy, who were assisted by Officer Jerry Day and a civilian, Allen Crum. The incident lasted around 1 hour and 30 minutes. Eighteen innocent persons were killed and 30 wounded. Most of this occurred within the first 20 to 30 minutes until about the time MacQuigg, Damon, and a few other civilians and police, returned Whitman's fire. One police officer had a .243 hunting rifle with a scope on it. The combined efforts of the civilians and police with rifles from below, kept Whitman pinned to one side of the tower, allowing the 4 men to approach and kill Whitman on the observation deck of the Austin tower.

Austin's Police Chief, Robert Miles, praised the civilian riflemen in helping to bring the incident to an end. "I don't want to condemn their action...", he said. "This ground fire was a Godsend to us, inasmuch as it pinned Whitman down in one spot on the west wall."

Uphill and downhill shooting.

The concept of aiming low when shooting uphill or down *assumes* that *you have adjusted your sights for the slope distance*. If you don't bother to figure distance to a target on a slope, and you are zeroed for a shorter range, you just might hit the target if you aim at the target. When you zero your rifle horizontally to the earth, you are adjusting your elevation to correct for the gravity and weather effects on the bullet for a specific distance. Since gravity is a 90 deg. force to the earth,

CHAPTER 10: "This Ground Fire Was a Godsend To Us", Uphill and Downhill Shooting   181

The large square building in the middle is the Academic Center where Damon fired at Whitman at the top of the University of Texas tower. In the lower right corner is the rectangular English building where MacQuigg made his way to the rooftop to return Whitman's fire. (Photo courtesy of The National Rifle Association of America.)

the bullet being fired uphill or downhill, only gets affected the horizontal ground distance to the target. If you zero for the slope distance, you over elevated your sights to the aiming point. This is why hunters aim low when shooting targets uphill or downhill. It assumes you estimated the distance correctly and corrected elevation for the slope distance.

The greater the angle, the more you have to lower your elevation (depression) until you get to about 60 deg. A 45 deg. angle would give you the farthest horizontal ground distance, geometrically. Beyond 45 deg., you are actually decreasing the equivalent horizontal ground distance. However, beyond 45 deg., you may still experience further need for more lowered elevation until you get to around 60 deg where it should max out.

John Plaster noted in his excellent book, **The Ultimate Sniper,** about a method of correcting uphill or downhill angle shooting. He calls it the "quick fix", after he learned it from the FBI and modified it. (See page 199 in John's book.) At 45 deg. angles, the longer slope distance, is reduced by the decimal constant factor of .70, to give you the horizontal ground distance. For a 30 deg. angle the slope distance is reduced by .90. The Marines have been teaching and using this method long ago. They call it the Angle Firing Chart. Although labeled for use with M118 Special Ball, it can be used for any ammuni-

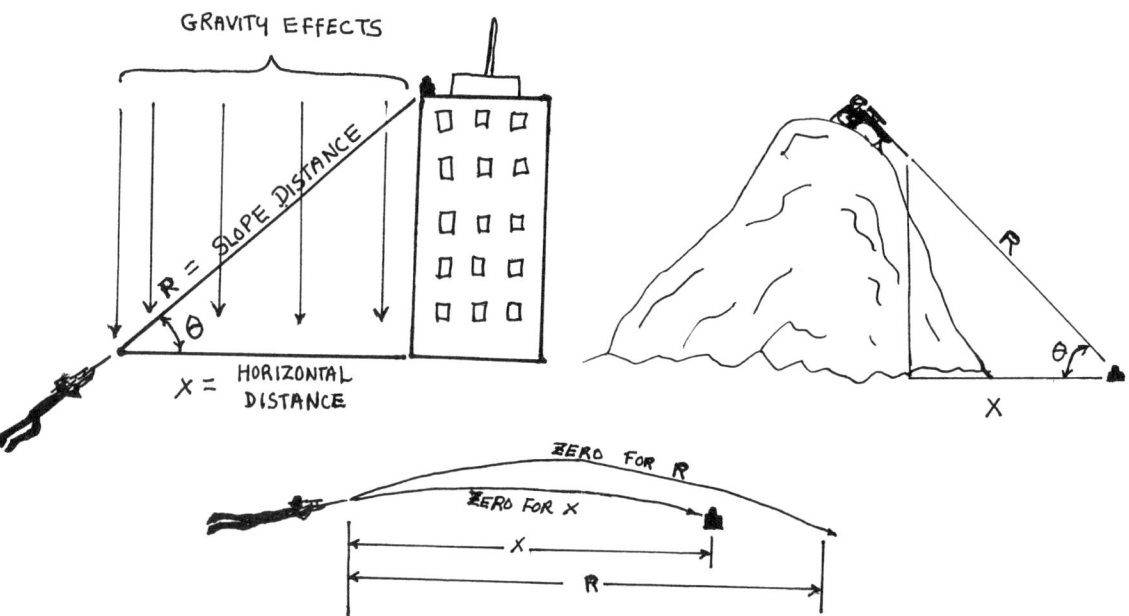

Fig. 10.1. Gravity forces only act perpendicular to the surface of the earth. It doesn't give a hoot if you are shooting uphill or downhill. It will only provide the same amount of force as if you were shooting horizontal to the earth's surface. If you are zeroed for the slope distance, R, you are zeroed for a longer distance. This is why you aim low or rezero for the horizontal distance.

tion. The chart gives you the factors for nearly all angles to 90 degrees. All you do is multiply the target distance by the factor and you get back the precise horizontal ground distance (if your distance to target was accurate).

The Angle Firing Chart is based on a simple trigonometric function: using the cosine angle to determine the legs of a right triangle. (And you thought the trig class in high school was a waste of time.)

The relationship of the length of the legs "x" and "R" are defined by the cosine of the angle θ (theta).

$$\cos \theta = \frac{x}{R} = \frac{\text{horizontal distance}}{\text{target distance along slope}}$$

manipulate the equation to get:

horizontal distance = cos θ x target distance along slope

or

$$\text{target distance along slope} = \frac{\text{horizontal distance}}{\cos \theta}$$

Angle Firing Chart
(Any caliber and velocity, UP or Down)

Slope Angle	Cosine	Slope Angle	Cosine	Slope Angle	Cosine
0	1.00	30	.87	60	.50
2.5	1.00	32.5	.85	62.5	.46
5	.99	35	.82	65	.42
7.5	.99	37.5	.80	67.5	.38
10	.98	40	.77	70	.34
12.5	.97	42.5	.74	72.5	.30
15	.96	45	.70	75	.26
17.5	.95	47.5	.67	77.5	.22
20	.94	50	.64	80	.17
22.5	.93	52.5	.61	82.5	.13
25	.91	55	.57	85	.09
27.5	.89	57.5	.54	87.5	.05
				90	.00

Formula:

Slope Range X Cosine of Slope Angle = Horizontal Range

ZERO for HORIZONTAL RANGE

Fig 10.2. Marine Corps Angle Chart gives slope angles and corresponding cosine factors to calculate for the horizontal distance.

The numbers in the chart are the cosine constants for each angle. John's use of .90 for the 30 degree angle is rounded off from .87, which is the cosine angle. The hardest part is for you to estimate the angle. The Marines have a very simple method to measure the angle of the slope using the map reading protractor. Tie a small weight to a piece of string and hold the other end to the center of your map reading protractor. Hold the protractor vertically and let the weight hang. Sight along the 90 degree line on the protractor to the same angle as the slope you are firing. The hanging string will give you the angle of the slope as read on the protractor.

The main points are:

(1) If you use this Angle Firing Chart to figure your horizontal zero, adjust your scope for the horizontal zero and aim on target.

(2) If you adjust your scope for the up or down target distance and don't adjust for the horizontal distance, AIM LOW!! How much? You can get the difference in bullet path (minus inches) from the horizontal distance and slope distance. Make yourself a "quick fix" chart for your rifle and ammunition. The greater the angle, the more compensation you need to make.

(3) With the Sierra 168 gr. Match, for shooting within 10 deg. up or down from horizontal, there is less than 1/2" point of impact difference out to 300 yards. For a 30 deg. slope you will stay within 2.0". The longer the range the higher the bullet impact for the zero at the slope distance. For most shooting, slight angles won't make a lot of difference unless it is to be a precision surgical shot.

(4) Most of us can't tell the difference between a 45 degree and a 43 degree angle unless you had a surveyor's instrument or calibrated eye balls like robo-cop. There are a couple of ways, besides using the cosine

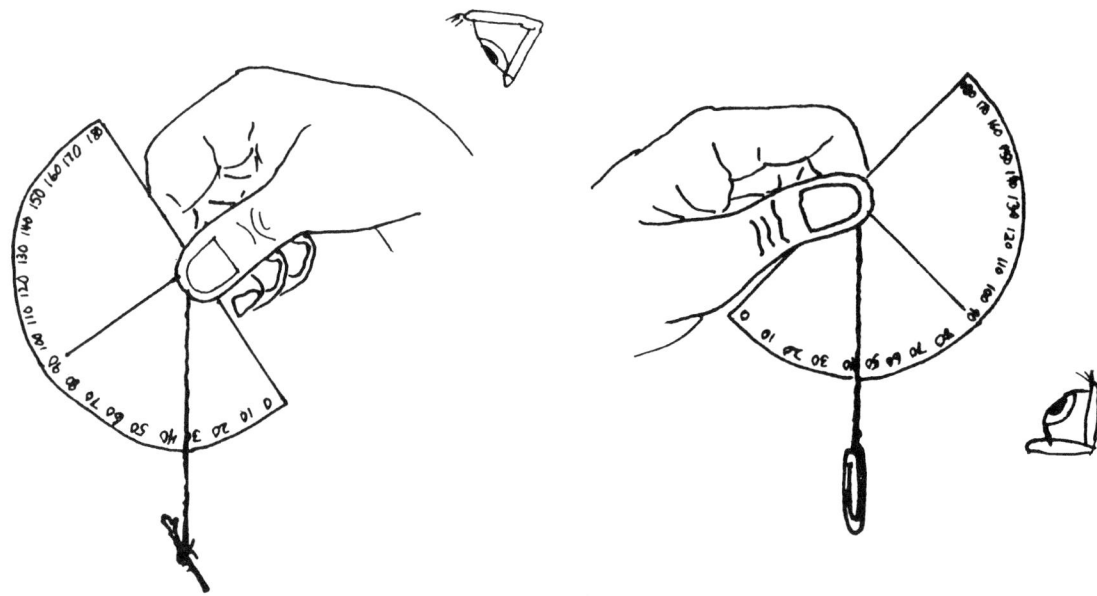

Fig.10.3. Marines measure slope angle with a protractor, string, and a small piece of weight.

table, you can determine the horizontal distance to a target on a slope. One is with a map. Another is to mil a known height object at ground level, in line and directly below the target for your horizontal distance.

5) Another way to estimate horizontal ground distance is to use the Pythagorean Theorem. If you know the height of the building or feature the target is standing on, and also the distance to the target on the slope, use this relationship to get the ground distance: (See Figure 10.4)

(6) Looking at the cosine chart, you can see that the cosines starting at zero degrees get smaller until you get to 90 deg. This means that as the angle gets steeper up to 90 deg., the corresponding ground distance gets shorter.

(7) A flatter trajectory cartridge like the 7mm Remington Magnum, such as the Secret Service counter-sniper uses, will shoot closer to the line of sight than high trajectory cartridges, when shooting at angles.

(8) When measuring heights of targets with scopes that have "comparing reticles", like mil dots, remember that the target height is shortened because you are looking at it from an angle. A more accurate mil reading may be made by measuring the width of the target if it is facing directly toward or away from you or is presenting a side view.

(9) Use John's "quick fix" method, or aim low, for easy to identify angles like 30, 45, 60 degrees with their corresponding cosine reduction factors of .87, .70, and .50. You may not get a chance to adjust your scope for the angle shot in an urban situation, unless you determine you must make a precision surgical shot and you have the time. Military and police snipers making long range shots should attempt to estimate the angle as close as possible and adjust their scopes.

CHAPTER 10: "This Ground Fire Was a Godsend To Us", Uphill and Downhill Shooting

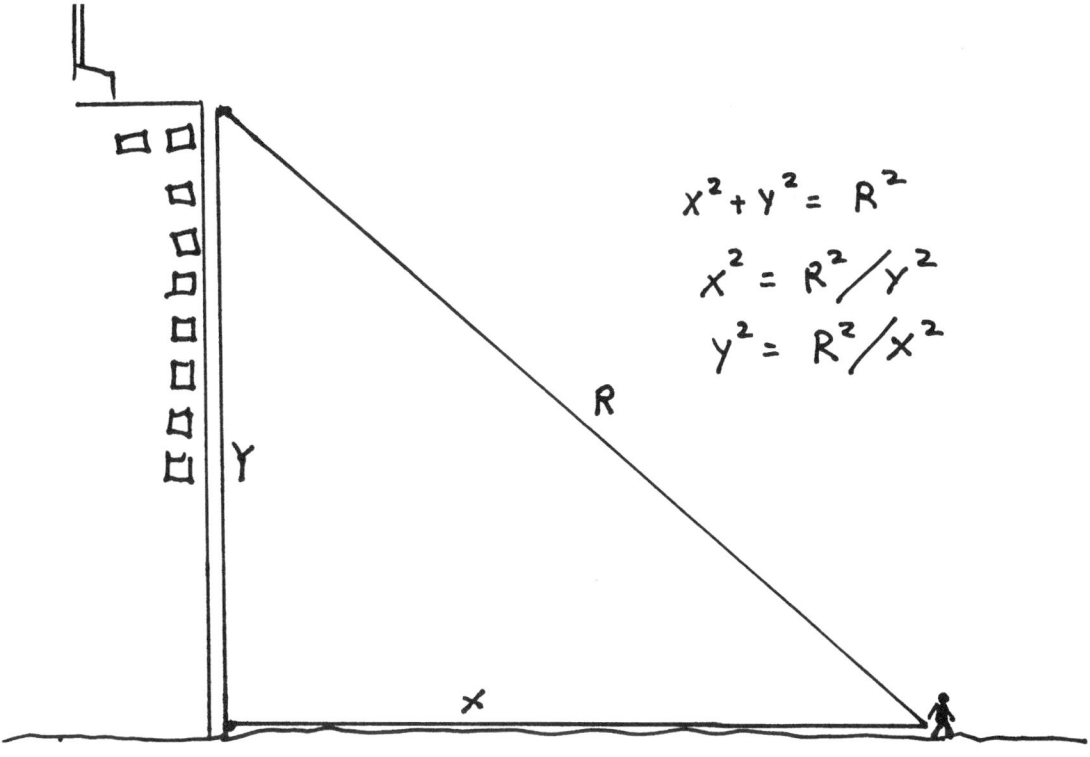

$$x^2 + y^2 = R^2$$
$$x^2 = R^2/y^2$$
$$y^2 = R^2/x^2$$

Fig.10.4. The Pythagorean Theorem can be manipulated like the cosine angle formula to get other distances.

Finally, keep things simple. You don't have to memorize the cosine chart or even understand why you use it or why you shoot low if you don't use it. The more you know about it the better decisions you can make when taking the shot. Just make sure you make an accurate adjustment when shooting uphill or downhill, especially if you have to make a precision surgical shot.

SFC Mike McClellan and Maj. Ron Wigger, U.S. Army National Guard Marksmanship Training Unit, Camp Robinson, Arkansas.

CHAPTER 11: "Shoot, Damn it, Carlos!" WIND READING AND SIGHT ADJUSTMENT

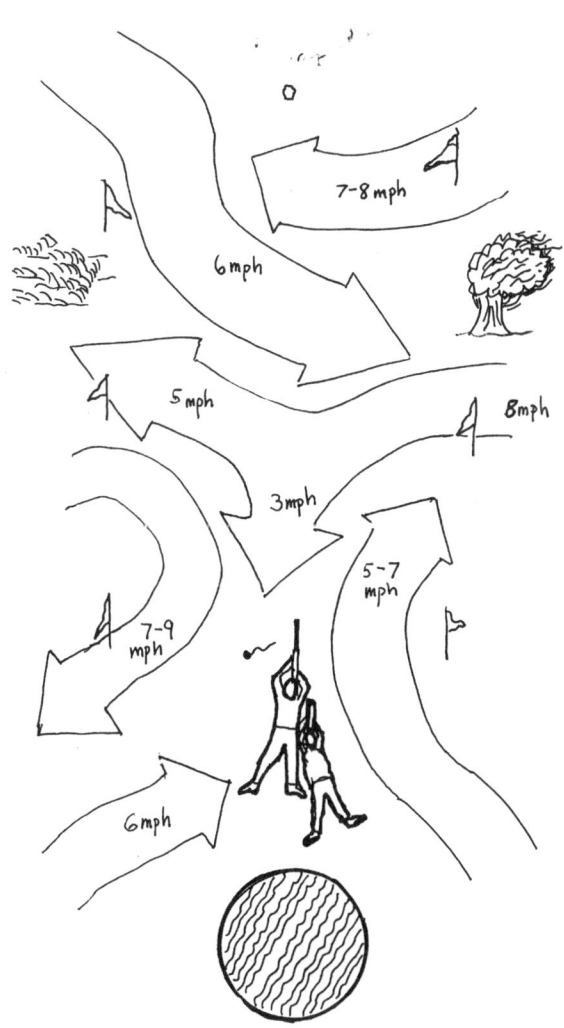

Fig.11.1. Don't let squirelley wind conditions scare you. On the range or in the field, wait for normal conditions or use the friendly mirage.

What is wind?

Heat from the sun and rotation of the earth causes slow rotation of the air masses in the atmosphere. We like to shoot early in the morning because the wind is less. As the temperature goes up, the wind speed picks up until it slows down again late in the afternoon. As these large air masses move about the earth, they drag on buildings, trees, hills, and other features, creating turbulence, especially close to the ground. Because of this turbulence, the speed of the wind is always changing. Wind direction is always changing too, because the large air mass rotates. This is why you see the wind direction change dramatically during the course of the day. The ground provides resistance to air movement so winds speeds are different as you move further off the ground. For example, if the wind speed at the surface is 10 mph, it may be 12 mph at 2 feet above the ground and 12 1/2 mph at 5 feet. At 10 to 20 feet above the ground, it may be 15 mph.

Why shooters need to know wind speed and direction.

We all know what wind does to a bullet. It pushes it off it's straight line path to the target. For the inexperienced shooter, and even some experienced, wind is scary. For one thing, you can't really see it. You can only see what it is doing to something else. "I have to trust a tree to tell me how fast the wind is moving?" For the sniper, estimating wind speed and direction may be harder than estimating range. There is no device like mil-dots to help you determine wind speed halfway to the target, there are no wind flags, you can use an anemometer, but it only tells you the wind speed around you. Your observer has a spotting scope that could be looking for mirage, but he is more concerned with watching the bullet's impact so he can correct your shot. He is usually focused on the target, so you won't turn on him and say, "What? You didn't see where the bullet went?"

When you make an estimation of the wind speed and direction, and then blow the shot, you lose confidence in your ability. You try it again and it gets worse. You thought the wind was blowing from the right so you corrected for it, but the bullet missed the target to the left. Hmmm, too much windage. So you take off some right windage. Up comes your target back up from the pits and there is the spotter, way off to the right. It's embarrassing because everyone can see your target and you are the only guy with the spotter off the silhouette. So you turn to your buddy shooting next to you and ask him what he is using for windage. He gives it to you. You look in your scope at the mirage and then at the flags to make like you know what you are doing. You put the windage on your scope and fire. When your target comes up you smile, because the spotter is perfectly centered on the upper chest of the silhouette. However, you had no idea what you were looking at in your scope, and the flags really didn't correspond to the windage you put on your scope. For the next shot the wind changes, and we are right back where we started from.

The Simple System

To make wind shooting easier, competitive shooters develop a system for estimating wind speed, direction, and how much windage is needed for sight adjustment. A system that is simple, yet efficient for the kind of shooting you do, insures the data is complete and as accurate as possible. You will eliminate confusion as to what data or observation is important and what is not. If you follow your system, you can make better judgment calls. You will know you are making the same correct call for the same wind conditions and your confidence in wind shooting will increase. The same system that a competitive shooter uses can be applied to sniping. Competitive shooters also like simple, quick, and effective systems. They don't like lugging a computer, a slide rule, a calculator, hygrometer, thermometer, anemometer, and a meteorologist, to the range or field anymore than you do. In many situations, you, the sniper, have more time to think about the wind factors and make accurate calculations from your hide, than the competitive shooter does on the range.

Our example of a simple system is similar to many that competitive shooters use and also the military. However, I hope to present it to you in a more organized manner so it does not seem like too much information is being thrown at you. A lot of police snipers get wind reading classes and some are lucky enough to have a place to practice wind shooting. Many of those that do, don't practice enough which means you forget what you are supposed to do. This can also happen with military snipers. In addition to the basics there will be some advanced information for your knowledge and/or use.

A simple system is:

(1) *Estimate range to target and adjust scope for range*

(2) *Determine wind speed*

(3) *Determine wind direction*

(4) *Assign wind value for effective wind*

(5) *Determine windage adjustment from table or formula*

(6) *Adjust for windage on your scope*

(7) *Determine wind constant or normal condition*

(8) *Learn how the wind around you changes*

Steps 7 and 8, determining wind constant conditions and learning wind changes, are important for a competitive shooter because he is going to shoot a number of rounds, over a longer period of time, from the same location. As a sniper, this may not be as important if you have to move quickly into a position, fire only one round, and leave the area. See it's already getting easier. This may become important if you have to stay in an area a long period of time, fire one or several shots, and then move or continue to stay in the general area and shoot again. But then you might have more time to do so. Learn steps 7 and 8 because they will give you more knowledge about understanding steps 2 and 3. Sometimes you won't wind shoot following this system's sequence, so use what is best for you. Step 1, estimating the target distance, is covered in detail in another chapter.

So you won't get lost:

(1) **Wind velocity or speed** is in miles per hour and does not consider wind direction.

(2) **Wind direction** is given in "clock" direction (more on this coming up)

(3) Wind **velocity AND direction**, when accounted for together, is **EFFECTIVE WIND.** (Explanation coming up)

(4) **Wind value** is what amount we reduce a 90 degree cross wind when that same wind speed comes toward us at a different angle. The resulting wind after being reduced is the *effective wind*.

Step 4. Value of the wind.

Let's do Step 4 first, *wind value*, so that we will have a better understanding of *wind value* when we look at Steps 2 and 3, *wind direction and speed*. You will see why we did this as you go along. In actual use of your system, wind value will be determined after you get the wind's speed and direction.

Wind direction is observed using the clock method. Position yourself in the **center** of the clock so that the wind direction is imagined blowing toward you from all sides. An inexperienced shooter will imagine the wind blowing across the center of the range as blowing from 3 to 9 o'clock with his firing position at 6 o'clock and the outer edges of the field or range as the clocks outer edge.

If you read wind direction with you at 6 o'clock, this manner you will get incorrect wind direction observations. When the wind changes to blow from a direction other than from directly left to right and you use the center of the range as its direction, you could misread the true affect that wind has on the bullet. The reason why competitive shooter's tie a streamer to their scope stand, is not to give

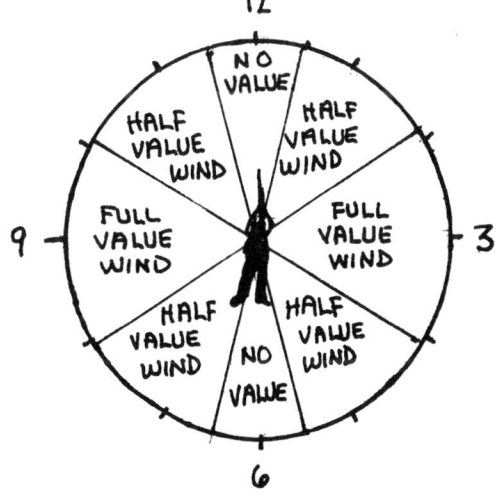

Fig. 11.2. Reading wind correctly puts shooter in the center of the clock. 6 o'clock shooter position will cause problems with wind values when estimating wind direction.

```
   20°      35°      45°      60°      90°
  5 mph    9 mph   11 mph   15 mph   22 mph
```

Use the formula: $\frac{\text{Angle of Flag}}{4} \approx \text{wind speed}$

Fig. 11.3. Wind flags can be used to determine wind speed.

them wind speed, it is their to tell them what direction the wind is blowing toward them. They put themselves in the center of the clock face.

A 5 mph wind blowing from 3 o'clock deflects a bullet 10 inches to the right of the aiming point on a target at 400 yards. The same 10 mph wind now coming toward the shooter from 2 o'clock, only moves the bullet 5 inches of center. The wind coming from behind at 6 o'clock or from the front has no affect on the bullet being pushed off course. Wind blowing from 3 and 9 o'clock are *full value* winds because it has the greatest effect on the bullet. The same speed wind blowing from 2 o'clock has less effect and is given a *1/2 value*. The 12 and 6 o'clock wind has *no value*.

Sniper Note: 6 and 12 o'clock wind directions will slightly increase or decrease the point of impact of the bullet at long range. A similar effect can be seen when an aircraft in flight has a tail wind or nose wind. The effect is that the wind will shorten or increase the time of arrival at the airport. In most cases this effect on the bullet is minimal if the wind speeds are less than 10 mph and under 800 yards. This minimal effect is shown in the following data and note the minimal bullet path change across each row:

Table 11.1 Bullet Path Change with Head or Tail Wind

Wind speed 10 mph, direction:

@Yards	6 o'clock	No Wind	12 o'clock
500	-62.9 in.	-63.2 in.	-63.5
800	-219.9	-221.7	-223.5
900	-408.5	-413.1	-417.7

Steps 2 and 3: Estimating wind speed and direction

Wind flags show *wind speed* on the practice range and most high power rifle ranges have them.

This is the military method that was determined using a large artillery flag of a specific size and weight and was dry. Many ranges and some flag manufacturers know what material and what size to make these flags. Smaller, lighter flags give greater angles for the same wind speed, just as wet flags are heavier and show lesser angles. For the sniper in the field, there will be no standard wind flags put out by the bad guys so you can get a better shot at them.

Most military training manuals says that wind, half way down range, is the one to look for. This is correct, but M/SGT Jim Owens, High Master High Power Rifle competitor makes a point that a *strong close-in wind* causes greater bullet drift than a weak wind down range. A strong close-in wind will start to veer your bullet off of it's course earlier causing the bullet's flight to be further off-course all the way to the target. A strong wind, down range, doesn't start pushing the bullet until it is closer to the target so the bullet's wind shift is less for the same strong wind closer to you. Some might argue the point, but Jim bases this on experience. My own experience says Jim is right. Go out and try this yourself. You should learn by experience which wind is going to effect your bullet the most and adjust for it.

Wind direction is very critical to shooting and should be determined accurately because it will affect the "value" or amount of affect the wind has on bullet drift.

Wind flags also show wind direction and have a second set of angles showing you the wind direction at the flag's location. It is a usable indicator, but remember the flag is high off the ground and it is the wind direction down range, not near you.

If you have a lot of wind flags on the range, sometimes you will see some moving in one direction and others moving in other directions. Jim Owens says to pick the flag that is closest to you and in your line of sight. If it's the one at 800 meters then that is the one you use. Do not combine two or more flags' readings and use an average. It may not work.

Anemometers or **wind gauges** are accurate for telling you what the wind speed is around you. There are some accurate low

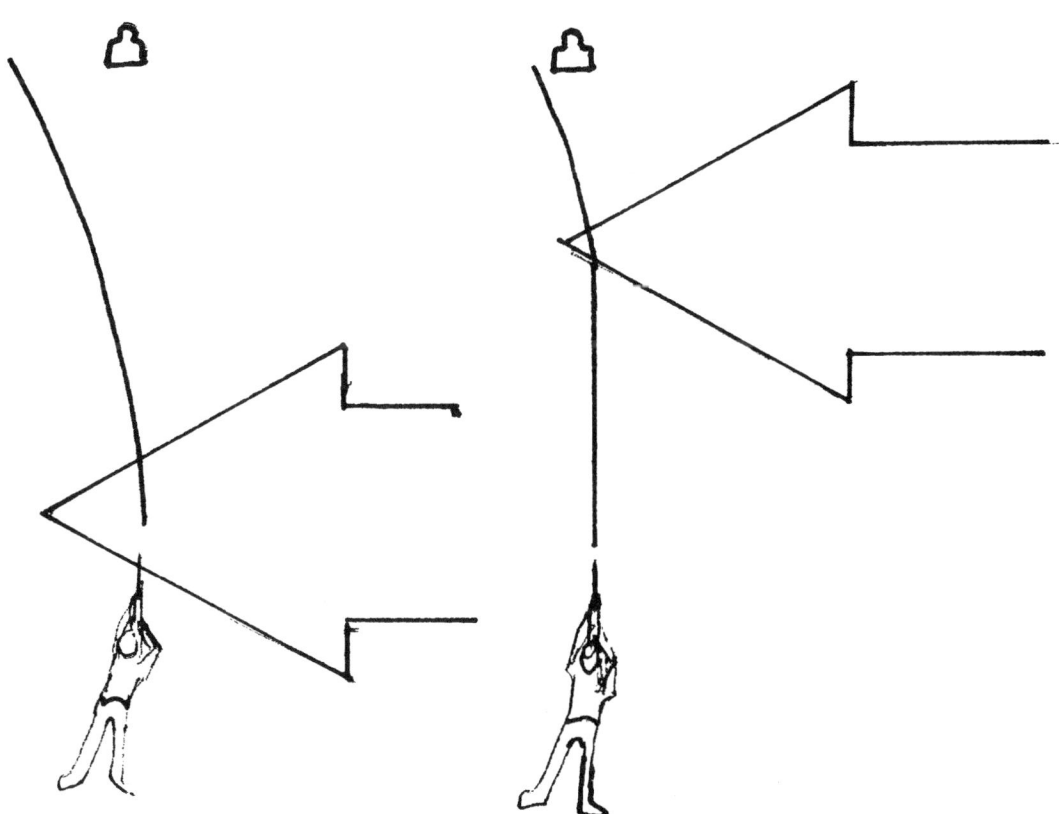

Fig. 11.4. A close wind will effect your bullet longer and push it farther off course than a far wind of the same value.

Fig. 11.5 Wind flags are also used to show wind direction where flag is.

cost wind gauges that can be purchased for less than $20 from shooter's supply companies. Some gauges are directional and will show the highest reading when facing into the wind direction. Someone might tell you that it's only good for the wind around you and doesn't tell you anything about down range, but it has its uses. You can use it to get good wind speed averages. You can see if the speed at your location usually means a wind speed or direction change down range. Or you can measure the wind hitting your face so you know what 3 mph or 7 mph wind feels like. Heck, you might just come out to the range one day when the grass is being cut and put up some wind flags. Then walk to the middle of the range and watch the flags. See how the cut grass blows around and the uncut grass sway. Observe what different size and type of trees on all sides of the range do. Then walk to each side of the range and make the same observations with your wind gauge in hand. Wait until later in the day and do it again. I bet you will learn a lot about the wind, your range flags, and the trees on your own range, by using your anemometer. The next time you attend a match or practice at that range it won't seem so scary, because after that long talk, you and the wind are now friends.

Observations are the sniper's main wind speed gauge. We did some when we went out to observe the trees and grass on the range with our wind gauge. Army and Marine snipers, and competitive shooters, all use basically the same observations to determine wind speeds. Some of these may also give you an idea of which way the wind is blowing, such as smoke and loose objects blowing around. To keep things simple, for when we calculate windage for our scope, we are going to use a single speed from each group of speeds, i.e., for 8-12 mph wind use 10 mph.

- 0– mph — cigarette smoke rises straight up
- 1–3 mph — can hardly be felt on face, but smoke will drift
- 3–5 mph — will be felt lightly on face, leaves rustling slightly
- 5–8 mph — leaves and twigs in trees move constantly
- 8–12 mph — will raise dust and loose paper
- 12–15 mph — will cause small trees to sway
- 20–25 mph — will cause large trees to sway

Check everything between you and the target. Look for the wind that will have the greatest effect, but don't ignore minor wind. See if it is going to have a greater effect because of where it is at or if it has more "value". A slower wind coming from 3 or 9 o'clock may have more "effectiveness" than a stronger wind blowing at you from an angle. For wind direction where you are at, dropping dust and wetting your finger may be the only means you have when in the field and they work a lot of times so use it.

Mirage will be the most accurate indicator of effective wind along your bullet's path if it is read correctly. It will allow you to see very small changes in wind movement if you are attempting to make a precision shot at long range.

What is mirage?

Mirage is caused by the rising of heat from the ground and is easily seen on hot sunny

CHAPTER 11: "Shoot, Damn it, Carlos" Wind Reading and Sight Adjustment

days across unbroken terrain. As the warm air rises, it mixes in with the cooler air above it. The warm and cool air have different densities. Sun light going through this air is bent and you see the one going through the warm air differently than the light passing through the cold air. (This is sort of like holding a stick in the water. The stick looks bent where the two mediums, air and water, come together. The two images of the stick, the one in the water and the one in the air, look different because of the light going through the different mediums.) Mirage appears as wavy streaks going up (if no wind) as warm air rises. Unfortunately, the mirage gets weaker as the ground gets cooler. Also if the wind exceeds 12 to 15 mph the mirage will disappear as it gets blown away. On broken terrain, such as shooting across a draw or valley, you most likely will not be able to see a mirage because you are shooting high over the ground.

Why use mirage for wind estimations?

Mirage in wind reading should be used whenever possible because it is the most accurate means to judge wind speed and direction. It is a better indicator than the range flags or simple observation methods because it combines all of the small variances in wind speed and direction and takes into account both to give you the *effective* wind!!

The large truck appears to have just passed through water in a low area. This is not water, but mirage. (Sam Chesnut)

To observe mirage for wind shooting, back the focus off 1/8 to 1/4 turn counter clockwise. Your focus should be between 1/2 to 2/3 the distance to the target. (Sam Chesnut)

How to observe mirage for wind shooting

Observe mirage, for wind speed and direction, by focusing on the target *with your spotting scope* and then turning the focus knob or eyepiece 1/16 to 1/4 turn *counterclockwise* or however your spotting scope works. You want to observe mirage about 1/2 to 2/3rds of the distance and in front of the target. This may contradict the close vs. far wind affect theory, but it is a tried method so stay with it and check it out yourself. It is easier to observe the mirage's movement if you have a solid background that is out of focus behind the mirage. As the mirage gets weaker because of cooling, a good background will make viewing easier. An example of a good background is the out-of-focused target number board, a building, a large rock, vehicle, or other large object near your target. Remember not to focus on the background of the target or the target itself. Look for the mirage along the edge of the object on it's side or top or both. Mirage will be seen as moving wavy lines (sort of like a moving sinusoidal wave seen on an oscilloscope for all you electronic types out there). Focus on only one or two of the wavy lines, not the entire scope's view. It is easier to detect small changes in wind direction and speed using this method, but if there is no object in the view to the target then don't worry about it.

Fig.11.6. Mirage is easier to see when there is an unfocused object in the background. Competitive shooters look at only one or two lines of mirage. Small changes in effective wind can be detected this way allowing them to make 1/8 moa adjustments.

The waves can be seen to move rapidly, or slowly, at angles, straight up, or across from one side to the other, in your spotting scope's view. What you see is the speed of the <u>effective</u> wind. When there is no wind or the wind is blowing from directly behind or toward you, the heat waves can be seen going up or "*boiling.*" Remember: *mirage observed in your spotting scope with the target in focus or focused behind the target is incorrect!!* High quality spotting scopes, like the Kowa TSN-1, can pick up mirage better than cheap optics even as the ground gets cool. The direction of mirage movement is known as **CLASSIFICATION** of mirage.

When the mirage is "boiling," the wind has no effective value on windage so this is a good time to shoot with zero windage on your scope. The Marines make a simple rule: silhouette size targets from 300 to 1000 yards needs to have windage put on *anytime* you see lateral movement of the mirage. I would add, that if you need to make a precision or surgical shot at under 300 yards you also need to correct for windage.

Many times you will observe the wind change direction and go in the opposite direction. As the wind begins to change direction, the mirage will slow down and then shake or "*shiver*" right before you see it change direction. It sort of is like a boil except the wave is not a slow rising one, the peaks and valleys of the wave gets sharp and shakes very fast or "*shivers.*" The first time you see it you'll know it, because it will be obvious. It only takes a few seconds to happen so you might miss it.

When you see this happen, a lot of times the wind will suddenly change directions immediately after the shiver. You will also notice that when this happens, a lot of shooters on the line miss their targets completely because they are adjusted for the opposite, previous wind direction. Hold up on your firing until you see the actual change and then correct for the new wind or wait till it goes back to the "constant condition" (we'll discuss this shortly). Be careful, however, as sometimes the wind may not change directions after the short shiver and return to the same directions. Experienced target shooters spend most of their time looking at mirage and other wind factors. They spend the least amount of time on the rifle.

The more you use mirage the more confidence you will have in your own ability in interpreting it. For the experienced competitor, wind is not a handicap, it is a blessing because he or she knows how to read and adjust for it while everyone else is struggling to stay on paper. Because experienced competitive shooters rely heavily on the accuracy of mirage to give them wind data, they have turned it into a fine art that will give them an edge over other competitors. The competitor will look for small changes in mirage to get 1/8th value. He or she determines this by observing very small secondary mirage lines moving at a slightly different angle than the heavy lines.

Two Problems Associated with Mirage

(1) Mirage can distort and sometimes completely block your line of sight. This is

CHAPTER 11: "Shoot, Damn it, Carlos" Wind Reading and Sight Adjustment

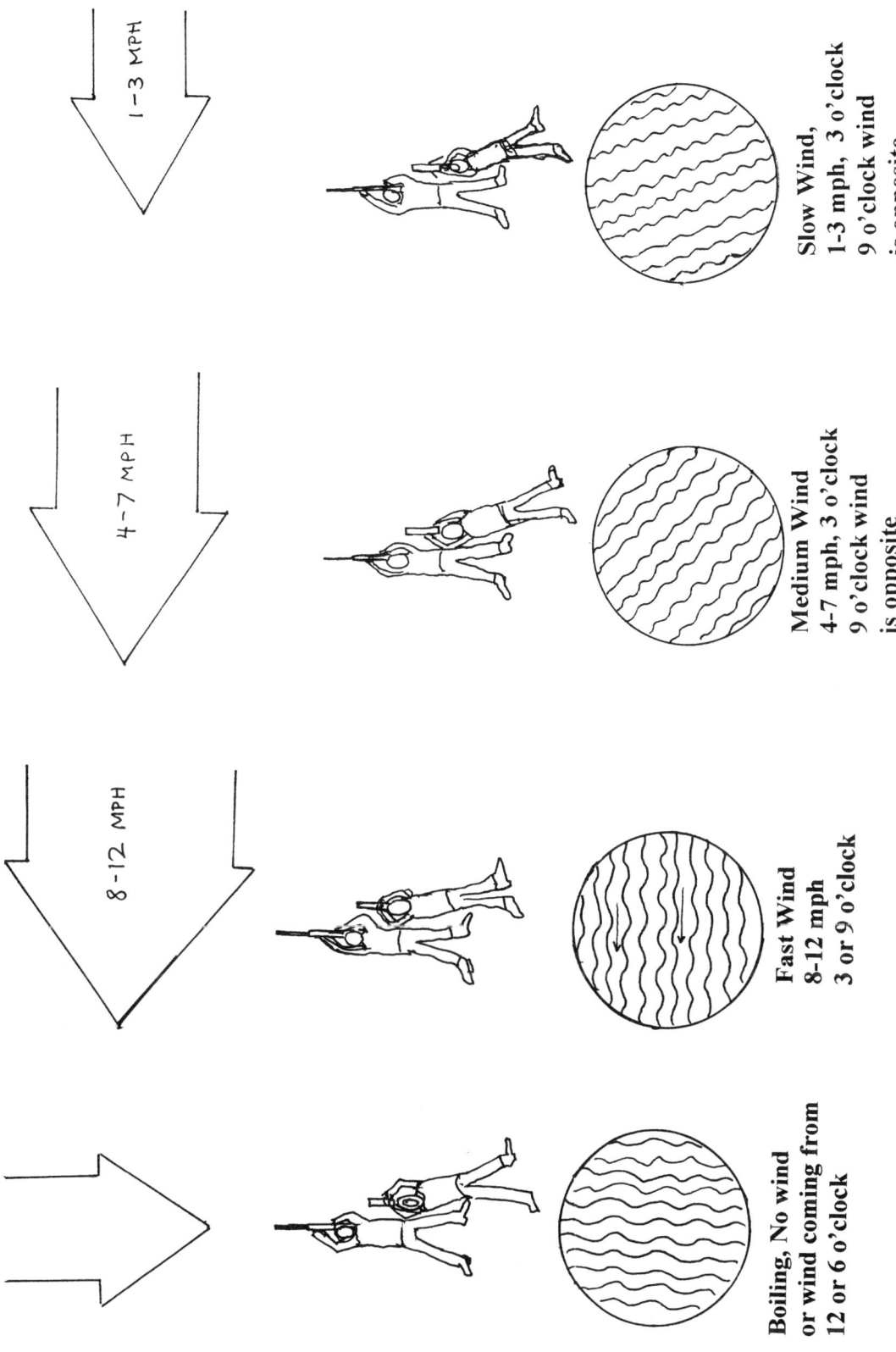

Fig.11.7. Mirage Classifications: Boiling, Fast, Medium, and Slow.

196 *The MILITARY and POLICE SNIPER*

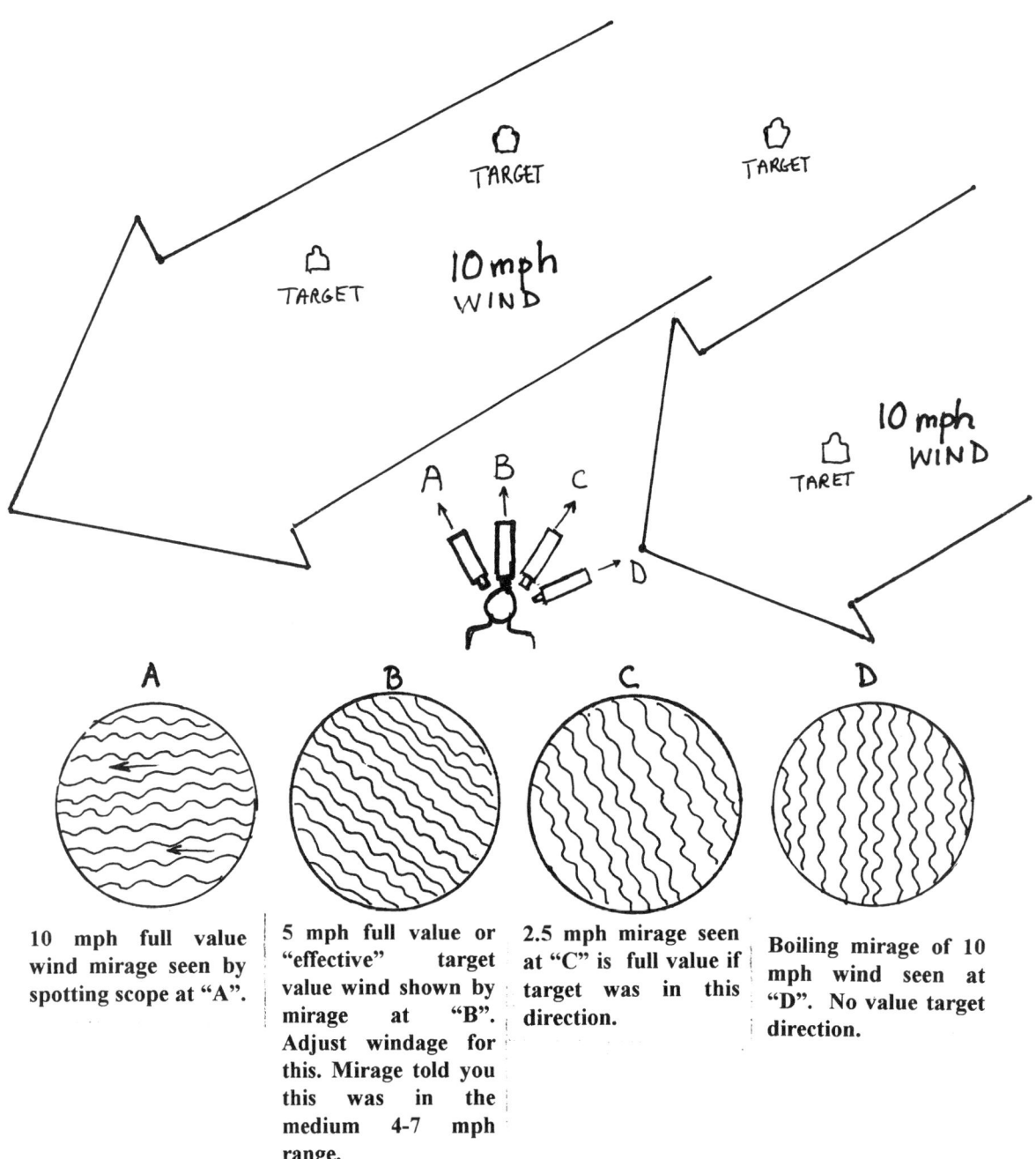

A	B	C	D
10 mph full value wind mirage seen by spotting scope at "A".	5 mph full value or "effective" target value wind shown by mirage at "B". Adjust windage for this. Mirage told you this was in the medium 4-7 mph range.	2.5 mph mirage seen at "C" is full value if target was in this direction.	Boiling mirage of 10 mph wind seen at "D". No value target direction.

Fig.11.8. This illustration shows you why mirage is usually the best method for determining wind effects. You don't really have to determine full wind speed and direction and then take half or quarter value. "Effective" wind mirage already takes into consideration the speed and direction together, along your line of sight to target. It tells you the full value wind adjustment.

CHAPTER 11: "Shoot, Damn it, Carlos" Wind Reading and Sight Adjustment

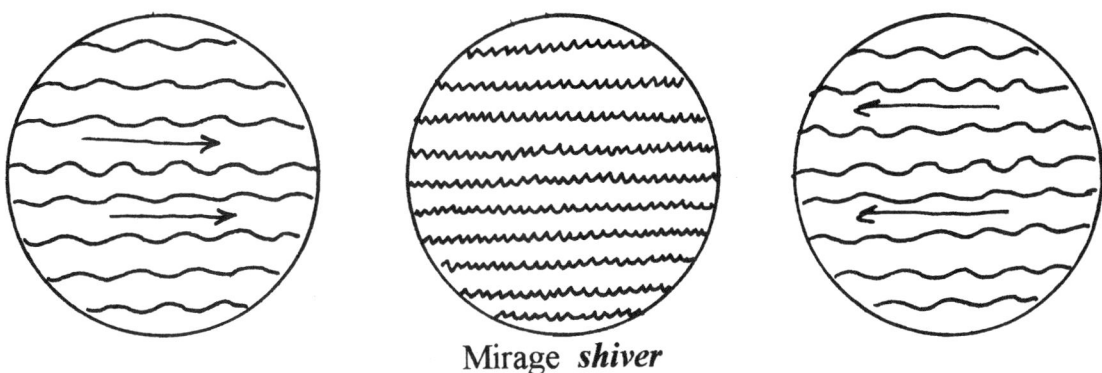

*Fig.11.9. Mirage **shiver** means most likely wind will change directions. Be careful because sometimes the wind will not change direction and continue in original direction before shiver.*

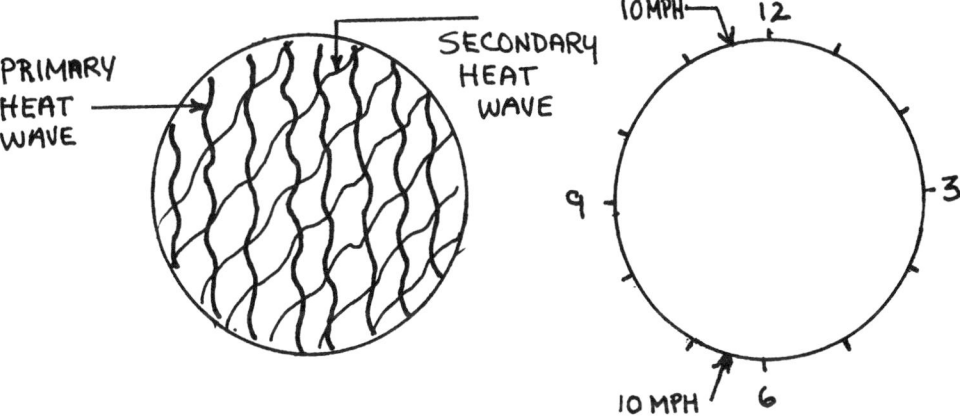

*Fig.11.10. Mirage with **secondary heat waves** is used by advanced competitive shooters for very fine adjustments. In tactical shooting we will not be concerned with this unless you want to use it.*

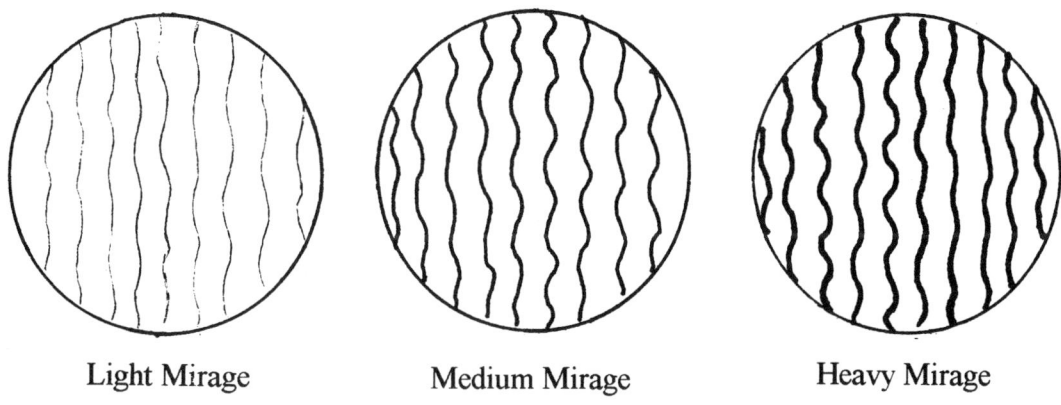

*Fig.11.11. **Categories** of mirage densities. Mirage seen through the rifle telescope will distort target image and makes accurate mil height or width sizing difficult.*

Sighting on a target through mirage will distort and displace target image. (Sam Chesnut)

especially a problem when shooting in hot weather or hot climate areas like Saudi Arabia. The heat waves rise high off the ground so it doesn't matter from what position you shoot or even if you try to take a little higher position off a vehicle. The target will also appear to be moving up or down or from side to side with the heat waves. Unless you have no choice, it is best not to shoot through heavy mirage because of this. When your rifle barrel gets very hot the heat waves coming off the barrel will add to the ground mirage. Increase in mirage *density* increases distortion and is **CATEGORIZED** as follows:

(2) Besides distortion and movement, there is target displacement of the image when seen through the mirage. The target image displaces in the direction of the mirage movement and is independent of wind speed and direction. The amount of displacement depends on the mirage category. If the mirage is boiling, the target displaces upward. If it is moving at an upward angle (3-7 mph wind) the target image is displaced up and over slightly. If the mirage is flat, indicating 9 o'clock or 3 o'clock cross wind, the target is displaced laterally only. When the shooter fires at a target distorted by mirage as seen through the rifle telescope, the bullet will impact slightly away from the target in the direction of the mirage movement. The shooter can compensate for the image displacement, but still must adjust for windage as seen by the mirage through the spotting scope. Target displacement may be as much as 1.5 moa, but the amount of displacement can vary a lot depending on where you are on earth. Again, if you are in a very hot climate, like Saudi Arabia or the Libyan desert, you need to determine what the mirage target displacement correction is by firing tests or your Intelligence Officer may be able to get that for you from other units. A simple test can be made to check displacement of the target by placing a rifle scope in a rest and aligning the cross hairs on the bottom and one side of the target. Do this when it is cool in the early morning. As the mirage increases and changes direction during the day, you can watch the amount of displacement that occurs.

CHAPTER 11: "Shoot, Damn it, Carlos" Wind Reading and Sight Adjustment 199

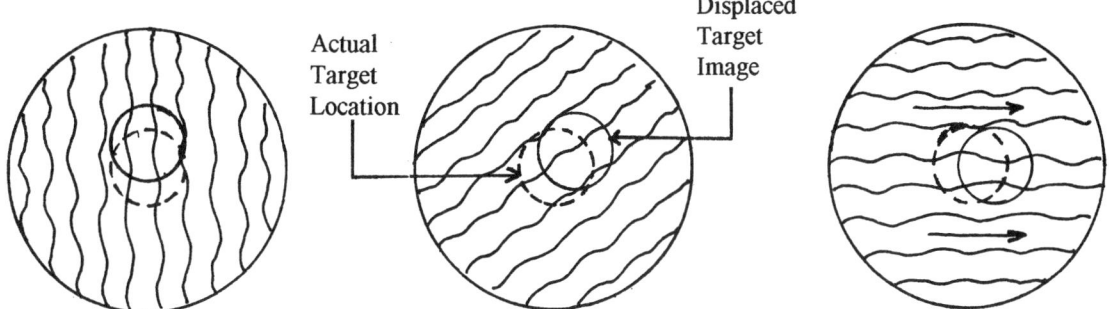

Fig.11.12. All categories of mirage as seen through the rifle telescope will displace the target image. Amount of displacement may be dependent on environmental conditions in your local area. With heavy mirage going left to right or boiling, the displacement could mean as much as 1.5 moa sight adjustment in the opposite direction of the mirage's movement. Light mirage may displace target image .25 to .5 moa. Effective mirage moving at slow or medium speeds may displace target anywhere from .25 to 1.5 moa laterally and .25 to 1.5 moa horizontally. Remember this is only if the target is being distorted by the mirage as seen through your focused rifle scope.

Keep mirage reading simple for tactical shooting. An experienced competitive shooter knows all the different combinations of categories, classifications, primary and secondary heat waves, target distortion, and displacement effects of mirage. This could run well over 25 combinations. The experienced competitor has seen and fired under most of these mirage conditions and has them recorded in his log book for record. He or she knows what sight adjustments are required for each. As a tactical shooter you will not have this kind of experience because many of your shots are not adjusted by mirage nor are you shooting under similar weather conditions and terrain each time. You also don't practice with this much detail. Keep it simple. (1) Attempt to use mirage when you can because it is the best effective wind indicator. (2) Use only the Classifications for identifying effective wind. (3) Know how mirage affects target distortion and displacement. (4) Shoot through heavy mirage only if necessary.

Step 6. Adjusting For Windage on Your Scope

Now that we know the wind's speed, direction, and value, we need to correct the windage knob on our scope. There are several ways you can do this.

(1) *Wind charts* are good for the competitive shooter because they convey windage corrections in number of clicks or moa to the shooter. They are also easy to use and don't require mathematical calculations. The charts can be drawn to show more precise wind speeds or direction. Some things to watch for:

(a) Make sure you are using the correct chart for the ammunition type and target distance. Wind charts usually are produced for competitive rifle target distances: 200, 300, 600, 800, 900, 1000 yards.

(b) Insure the number of clicks given, is the same as that on your scope. Most windage charts are for 1/2 minute click National Match M14's with iron sights. Or 1/4 minute match rifles using 168 gr. Federal Match. Redraw the "rosette" and correct the clicks if you need to. Some charts that are given in meters are sometimes converted from charts that were originally in yards without changing the values of the clicks. Compare charts by different makers.

(c) Jim Owens turns out very good wind charts for the NM M14/M1A with 1/2 moa adjustments and 168 gr. with 1/4 moa adjustments. In addition, the book shows the flag angles and mirage diagrams. Printed on laminated paper, the book is water re-

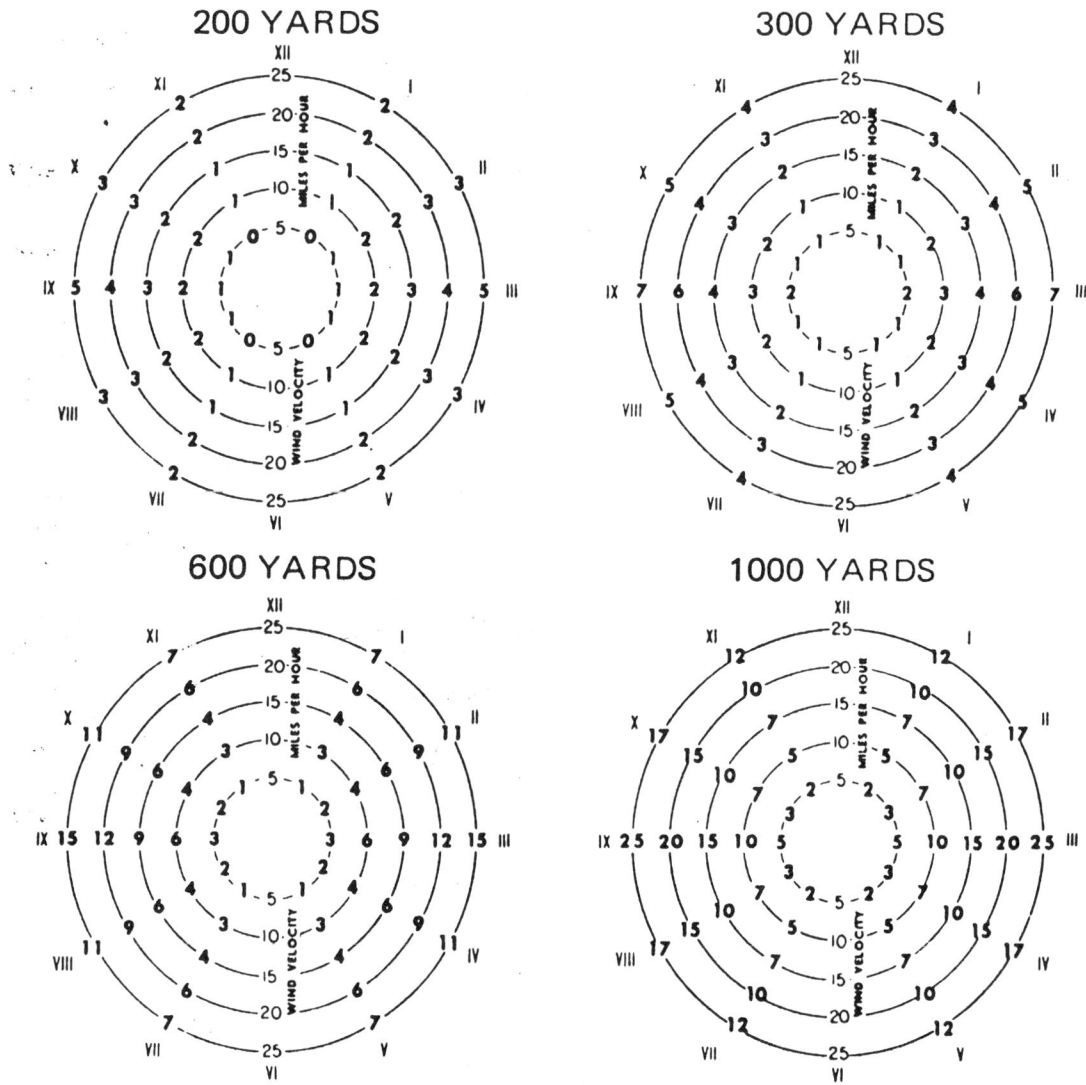

WINDAGE DIAGRAMS
Based on 7.62 N/M ammunition with 173 gr. B.T. bullet, M.V. 2610 f.s.

Circles represent wind velocity as indicated.
Roman numerals represent wind direction.
Dark numbers are number of 1/2 moa clicks.

Fig.11.13. Windage rosette chart from Marine Corps log book.

sistant and is not expensive. (M/SGT Jim R. Owens, 2621 W. Pierce St., #101, Milwaukee, WI 53204, all books and videos are between $10-$20).

(2) *Windage Adjustment Charts* are taught to the basic military sniper student because they are very accurate and easy to read. They are usually made up for 7.62mm M118 or Match ammunition for 24" barrel military sniper rifles, or for the 22" barrel M14. See the chapter on the Log/Data Book for the Marine Corps windage adjustment chart for M118 ammunition.

(3) *Wind Formula* is accurate and easy to use and remember. The Marines like this one for M118 ammunition. Express the range as a single digit, i.e., 700 meters = 7.

$$\frac{\text{Range} \times \text{Wind Velocity}}{\text{Constant}} = \text{MOA Adjustment}$$

Constants are: 0-500 (yds or meters) = 15
600 = 14
700 = 13
800 = 12
900 = 11
1000 = 10

Example: 10 mph wind from 9 o'clock, with target at 730 yards.

$$\frac{7.3 \times 10}{13} = 5.6 \text{ or } 5\ 1/2 \text{ moa windage (full value)}$$

Make sure you use the *"effective"* wind speed for this formula so you get the right windage. The constant for the divisor represents a figure that takes into account the bullet weight, density, velocity, air resistance, distance to target and sight movement. It is ammunition specific. The Marine Corps wind formula and windage adjustment chart follow closely to each other and has been verified with the Sierra Ballistics Program at STP.

Step 7. The Normal Condition or Constant

Normal or constant condition refers to the wind speed and direction that occurs the *majority of the time.* Wind direction and speed are taken together and the normal condition represents one specific wind condition only. If there is no wind most of the time, then your constant or normal condition is "no wind." If the wind is from 8 o'clock, and is 7 mph most of the time, then this is your normal condition. Competitors zero or fire their sighting shots during this condition. Whenever you see this condition you shoot. In rapid or slow fire target shooting, you may be able to shoot several rounds this way until the wind changes and then will have to "hold off" or use Kentucky windage. You can check mirage, look at the wind flags, or use the observation methods between each round fired. You will lose a shot if the wind changes and you continue to fire without adjusting for the change.

The target shooter will begin looking for the normal condition the moment he or she is on the firing range. If he is spotting and scoring for another shooter, or is allowed to observe the firing line while his relay is waiting to get on line, he will also use this time for getting the wind constant. He can observe the flags while setting up his equipment and then check mirage during prep time, if he is allowed. Sometimes the shooter will continue to observe the wind to make sure he has the right normal condition even after the "commence fire" command is given. Once he gets a good reading on the normal condition he can rapidly fire several shots, while quickly checking between each shot, to see that the normal condition is still there. The inexperienced shooters will be trying to adjust zero for each wind change and will be chasing the "spotter" around the target adjusting for it instead.

If the sniper is given ample time to determine the normal condition, he can correct his scope for it and engage the target or targets at this time. Remember what the normal condition windage is for the distance. Sometimes the target will appear only momentarily when the wind conditions are not normal. To prevent confusion during wind changes, come back to zero windage and then readjust. If the normal condition returns you know exactly what to put back on the scope for windage without having to do any calculations.

"In practice, if you only wait for the normal conditions to return, you will shoot a high score, but you will never learn how to correct windage for any other condition." M/SGT Jim R. Owens, USMC (Ret.), NRA High Master competitive rifle shooter.

Step 8. Changing wind conditions.

Observe fish tailing, gusts, and dropping off or dying wind conditions. Note how often and how long these lasts. Note what leads up to a normal condition and what follows. The sniper should observe changing wind from the hide, so he might be able to know when and how long he might be able to wait before he takes the first shot. Or, he may have to plan to take the shot before a difficult wind change occurs.

Another System: THE BRITISH SYSTEM

The British Army has used this method for a long time to calculate moa adjustments. You can memorize the entire chart. Some say this system is simple, but for me it isn't. There are too many things that have to be kept track of. It is also not as accurate as the Marine Corps windage chart beyond 500 meters for 7.62 M118/Fed Match and accuracy gets worse as wind speeds gets higher and distance gets farther. If you find this method easier to use and accurate for you and you like it, then by all means use it.

(a) Estimate range to nearest 100 yards.

(b) Start by determining "BASIC MOA" windage at target range. This corresponds to what a 15 mph, full value wind, would require for windage in moa at your actual target range. "BASIC MOA" for 600m = 6 moa, 800m = 8 moa, 900m = 9 moa. etc.

(c) Next determine "OBSERVED SPEED" of wind velocity by observation, flags. Do not account for wind direction at this time. Mirage is "EFFECTIVE" wind so is not part of this wind reading system.

(d) Next classify "OBSERVED SPEED" of the wind: as "*light*" (if 3 mph), "*medium*" (if 7 mph), or "*heavy*" (if 15 mph).

(e) If "OBSERVED SPEED" is determined to be "*light*" then use 1/4 of the "BASIC MOA" value. If "OBSERVED SPEED" of wind is determined to be "*medium*" then use 1/2 of the "BASIC MOA" value. If "OBSERVED SPEED" is determined to be "*heavy*" then use full "BASIC MOA" value. The resulting full or reduced "BASIC MOA" figure is your "FULL VALUE MOA" (3 or 9 o'clock) windage for the 3, 7, and 15 mph winds. It *does not* take into account the observed wind direction, unless the wind is actually blowing from 3 or 9 o'clock.

BRITISH WIND TABLE

BASIC TABLE IS BASED ON A 15 MPH FULL VALUE WIND

Range	100	200	300	400	500	600	700	800	900	1000
MOA	1	2	3	4	5	6	7	8	9	10

Light wind (3 mph) use 1/4 of basic value.
Medium wind (7 mph) use 1/2 of basic value.
Heavy wind (15 mph) use basic value.

Fig.11.14. Table for the British Wind System

CHAPTER 11: "Shoot, Damn it, Carlos" Wind Reading and Sight Adjustment

(f) Next we adjust for the true wind direction. For other than observed full value (3 or 9 o'clock) wind, we need to adjust the "FULL VALUE MOA" we got in Step (e) for the real wind direction. Determine the "OBSERVED DIRECTION" of the wind by observation methods or flag. From its clock direction, reduce the "FULL VALUE MOA" by 1/4 or 1/2. This is the "EFFECTIVE MOA ADJUSTMENT."

(g) For winds between the "*light*" (3 mph), "*medium*" (7 mph), and "*heavy*" (15 mph) we need to fine tune by extrapolation. Take a fourth or half of the "FULL VALUE MOA" and add it back. This will be the ADJUSTED EFFECTIVE MOA. Example: For a 10 mph wind at 1:30 o'clock, with target at 800:

Range = 800 meters

BASIC MOA = 8

OBSERVED SPEED = 11 mph

Classify Wind = 11 mph is classified as MEDIUM 7 mph wind so will be 1/2 Basic MOA

FULL VALUE MOA = 1/2 (medium 7 mph) of BASIC MOA (8) = 4

Wind direction = 1:30 o'clock so will reduce FULL VALUE MOA by 1/2

EFFECTIVE MOA ADJUSTMENT = 1/2 of FULL VALUE MOA = 2

Since 11 mph is between classifications of 'medium 7 mph' and 'heavy 15 mph' we need to add a little more moa . Lets add 1/2 of the EFFECTIVE MOA.

ADJUSTED EFFECTIVE MOA = 2 +1 = 3 moa adjustment.

Sniper Tip: Hold Offs for Wind. Army and Marine snipers learn to "hold off" or "Kentucky Windage" for situations where there are sudden wind changes or you don't have time to adjust sights because target exposure will be short. The MIL-Dot reticle can also be used to hold left or right of the target for windage or to lead a moving target. For instance, instead of cranking the windage knob to move the bullet impact left 2 MOA (20.94") to correct for left wind at 1000 yards, you could hold the crosshair 1/2 MIL (18") left of center of the target. A target at 600 yards, and moving at 2 mph, requires a lead of 21.6" or approximately 3/4 MIL with the 7.62mm M40A1.

If you don't have a wind deflection chart in inches you can convert moa to inches at the specific range. You can also convert inches to Mils and use your MIL-Dot. Keep the table simple, use large increments of wind speed and round off to 1/2 MILs. The next chapter will cover moving target leads in detail.

For small changes in wind requiring only a quarter moa at longer ranges or 1/2 minute at closer ranges the spotter can tell the shooter to hold on a shoulder or at the edge of target.

KEEP IT SIMPLE, BUT BE ACCURATE

Long range sniping is wind shooting. Develop your own system for wind reading and practice it. I believe wind effects and ranging are the sniper's two most difficult shooting problems. More targets are missed because of these two factors than any other. If you don't get these two right, your zero will be incorrect and it won't matter how well your shooting technique is or how well your rifle shoots. You will miss your target. You cannot became a true expert if you don't understand why you are doing something. Wind shooting requires a lot of personal judgment that relies heavily on experience as well as knowledge.

Captain Jim Land, USMC, watched from the stands as Corporal Carlos Hathcock prepared to fire his last shot from the 1000 yard line. There was only one minute left before cease fire would be called, ending the Wimbledon Cup match. Carlos bent over to look at the mirage in his spotting scope and quickly laid back onto his rifle. He noted the range flag as he focused on the cross hairs. Only twenty seconds remained. He glanced at his watch and five more seconds passed. He shifted his toe and put the cross hair at 7 o'clock on the bullseye. Still focused on the crosshair, Carlos noticed the range flag dip slightly. With only a few seconds left, Jim Land's patience got the best of him, "Shoot, damn it, Carlos," he yelled. As if CPL Hathcock could hear the shout, he fired. "Cease fire! Cease fire!" came over the loudspeaker. After the targets of the other shooters came up from the pits, the announcer spoke, "Ladies and Gentlemen, we will now disk the score of the 1965 National Champion, Marine Corporal Carlos N. Hathcock II, of New Bern, North Carolina". As the target came up, he saw through the spotting scope, the red disk covering the bull's eye. When the red disk was lowered he saw the white spotter near the inside edge of the black. Oblivious to the cheering crowd, he told himself the wait for the slight break in the wind was worth the gamble. He won by four inches.

Carlos Hathcock. (Mike R. Lau)

CHAPTER 12:
"We'll take 'em", MOVING TARGETS

One day in March, 1967, 24 year old Marine Sergeant Carlos Hathcock, along with his observer, Lance Corporal John Burke, set out from their base camp, Hill 55, to check out possible enemy activity in the Ca De Song river valley, Republic of Vietnam. Two years earlier, the river valley received it's nickname, "Elephant Valley", when other Marines, overlooking the valley, observed Viet Cong using eight elephants to transport heavy weapons. Two years earlier, another happening took place that will play an even greater significance in the outcome of the events that are about to happen. On August 26, 1965, Carlos Hathcock won the 1000 yard Wimbledon Cup Match at Camp Perry, Ohio.

Besides his .30-06 Winchester Model 70, Sgt. Hathcock carried his .45 Caliber M1911 pistol. LCPL Burke operated the radio and was armed with his M14 rifle. Later that evening, the two scout/snipers settled into their observation hide at the edge of the valley. From their position, they would be able to observe the Ca De Song river, a thousand yards away, and the surrounding rice fields. When morning came, there was low ground fog in the distance that only gave them a field of fire to about 800 yards. As the mist cleared, the two Marines observed a company of North Vietnamese infantry, about 150 strong, crossing the rice paddies on the near side of the river. Within a few moments, the column would be crossing within 500 yards of their front. Sgt. Hathcock and LCPL Burke thought about calling in artillery fire, but Hathcock made the final decision.

"We'll take 'em," said Hathcock, and told Burke to begin with the rear of the column and he would take out the leader at the front. With the crosshairs of the 8 power Unertl carefully leading it's target, the first round from the M70 dropped the lead soldier. LCPL Burke fired and the rear soldier fell. A second round from the Winchester dropped another soldier carrying a pistol. The NVA column broke and ran toward the cover of a nearby dike. The Marines opened fire again and several more enemy fell dead. Expecting the NVA to attack, the Marines carefully moved to a new position 50 feet away. During the lull in firing, one of the enemy leaders came up from behind the dike to see if the threat had left. Carlos dropped him with a single shot. Suddenly, four NVA from each end of the dike came charging out to assault the sniper's earlier position. Hathcock engaged the four on the right and Burke took the left four. Three enemy in each group were hit and the remaining two retreated. They never made it back to the dike. The short range AK-47s could not counter the effective long range fire from the two Marines.

An NVA officer got up and ran toward the river about 500 yards behind the dike. Before he could disappear in the fog, Hathcock dropped him with one shot from the

Model 70. Hathcock and Burke then moved to yet another position where they waited for sundown.

Hathcock and Burke observed the dike until they could no longer see. Radioing for artillery illumination, the parachute flare revealed the NVA trying to move toward some empty huts a thousand yards away. With careful aim, they opened fire at the head of the column and several were hit. The rest ran for cover back to the dike. Waiting in the darkness, Hathcock and Burke hoped the soldiers would try to make a run for it again. When he felt it was time, Hathcock requested another flare. A smaller group was running for the huts. Both men opened fire on the fleeting targets and they dropped to the ground to return fire. The men behind the dike returned fire on the Marine's position also, as Burke told Artillery to keep illumination rounds coming. Hathcock and Burke continued to fire at the soldiers lying in the open. Carlos told Burke to take well aimed shots. One after another, the Unertl's crosshairs settled on a target and each time the Model 70 spoke, an enemy soldier was killed. None of those that tried to flee came back to the dike. Then Burke took on the men at the dike with his M14. They were all coming at the two Marines. Continuing their deadly precision shooting, so many of the enemy were hit that they broke off the attack and ran back to their cover. Artillery kept the night illuminated and occasionally Hathcock and Burke fired a round at the dike to remind the enemy they were still out there. On the third day, the Marine snipers could have called in artillery fire on the soldiers, but decided to hold the NVA in their position until Marine reinforcements could sweep the area. The NVA took no action that day and night came. Calling for illumination all night, nothing happened. Hathcock and Burke took turns observing. The fourth day came and like the previous, nothing happened, that is, until dark. This time, a larger group of the enemy made a run from behind the dike. Afraid they would

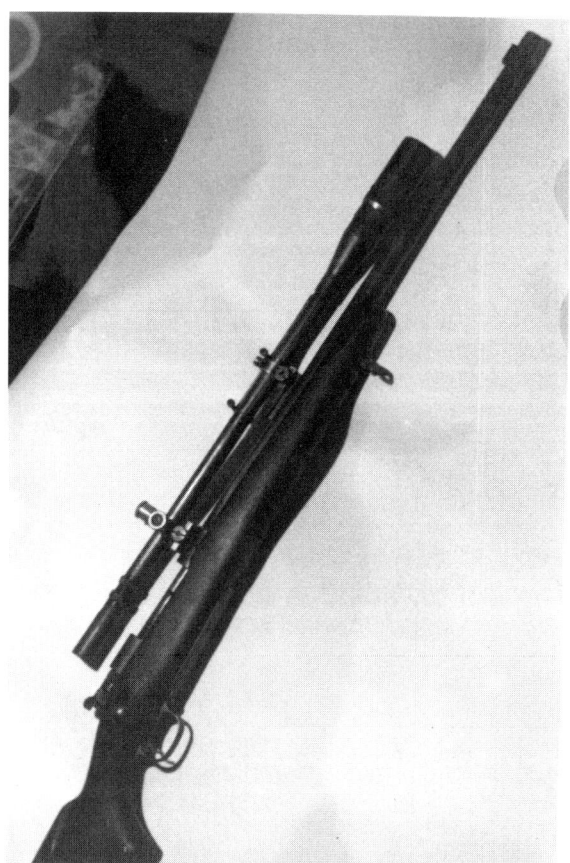

WWII/Vietnam Era USMC Winchester Model 70 sniper rifle with Unertl telescope. Recoil spring was removed from around the scope so when fired, the scope had to be returned back to its original position in the mount by hand. Sporter stock with the rounded narrow forend was preferred by Marines for field carry over the wide target style forend. (Photo courtesy John Halpin)

get around them, the two Marine snipers took careful aim, bringing down yet more of the enemy. The remainder of the group ran back to the cover of the dike. Desperately, the enemy formed another platoon and once again assaulted the Marines, while the remaining NVA company opened up to support the attack. The AK-47 projectiles fell harmlessly into the jungle as Hathcock and Burke had already moved to yet another position. Aided by illumination from artillery and with precision rifle fire, the snipers broke up yet another attack after inflicting numerous casualties. As another group prepared an attempt to break out, Hathcock decided it was time to move.

CHAPTER 12: "We'll take 'em", Moving Targets

Military match ammunition was issued only in 20 round boxes after the M1 Garand and 1903 Springfield became less popular for match use. During the Vietnam War, instead of fumbling around with loose cartridges and boxes in the field, Army and Marine snipers assembled the Cal. .30 M72 (shown) and 7.62mm M118 Match rounds into 7.62mm ball ammo charger clips. The 5 round clips were placed into discarded M80 ball cloth bandoleers which had 6 pockets holding 2 clips per pocket. Sniper ammunition is easier to find and carry this way when carrying a lot of equipment. (Mike R. Lau)

As he and Burke covered each other during the move, the assault came. Once again, the Marines' precision fire was lethal. Every time Hathcock fired, the caliber .30 boattail bullet found its mark. The assault up the hill broke off and the NVA ran back to their cover. Another attempt at midnight to reach the huts was made, but was turned back again by carefully aimed shots. During the early morning hours, the two scout/snipers worked their way to a new position to the ridge overlooking the end of the dike on the side away from the huts. That morning, Carlos, was ready to implement his final plan. He fired a shot at the near end of the mud wall. One of the enemy tried to see where the shot came from and Hathcock dropped him with the next shot. A few of the enemy got up and ran from the far end of the dike toward the huts. The Marines fired several more rounds at the near and middle end of the dike hoping more would come out running. A couple of them did. As the snipers quietly withdrew over the ridge, artillery rounds were already on the way down into the valley. A Marine unit was waiting nearby to sweep the area. It is unknown how many enemy casualties were caused by Hathcock and Burke because these were never confirmed.

When Sgt. Carlos Hathcock and LCPL John Burke set out to recon Elephant Valley, they had no idea what they were getting into. Five days ago, when he left the safety and com-

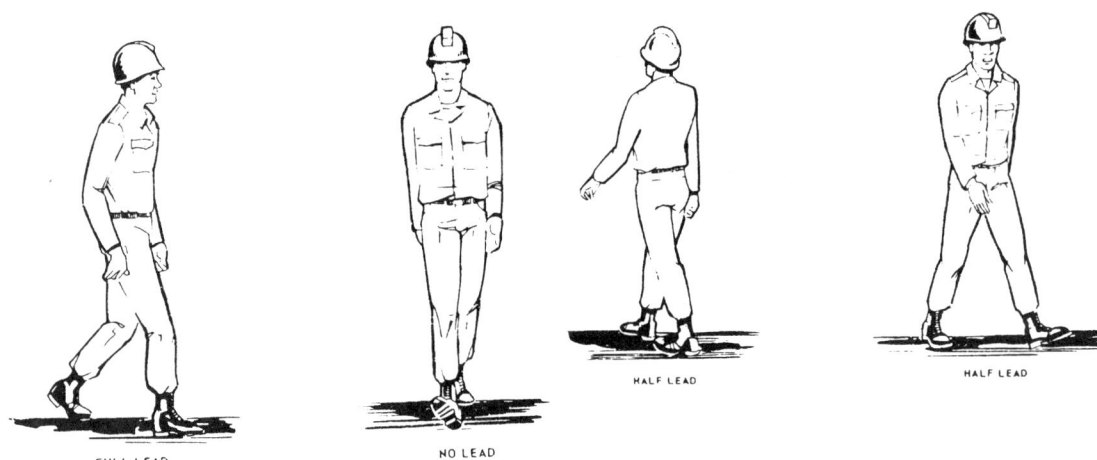

Fig. 12.1. There are three basic target movements : (1) laterally at 90 degrees to you, (2) oblique or around 45 degrees toward or away from you, (3) directly away or toward you. These are determined by the amount of the target's upper body that we can observe.

fort of Hill 55, Carlos tied his bandoleer of M72 match cartridges around his waist. The bandoleer held his entire caliber .30 ammunition load, 65 rounds.

ENGAGING MOVING TARGETS

To Estimate Target Lead You Will Need the Following:

1. Distance to target.
2. Bullet's time of flight.
3. Speed of target.
4. Angle of target movement.
5. Wind conditions.
6. Determining amount of lead.

Distance to target is determined exactly like distance to any target as explained in an earlier chapter.

Time of flight for bullet to reach target distance is given in the ballistics chart for the specific ammunition type. The M118 LR's time of flight for 24" barrel at STP, calculated with the Sierra Ballistics program is given here so that we can use it to do examples. TOF for other calibers are given in Appendix II.

Table 12.1 Time Of Flight (seconds) for M118 SB and LR

Range	Meters	Yards
100	.1	.1
200	.3	.2
300	.4	.4
400	.6	.5
500	.8	.7
600	1.0	.9
700	1.2	1.0
800	1.4	1.3
900	1.7	1.5
1000	2.0	1.8

Speed of the Target is the most difficult variable to obtain. A running target should not be engaged unless it is fairly close or you have no other choice. Your probability of a miss is great and there will be danger to hostages and innocent bystanders.

Table 12.2 Personnel Target Speeds in FPS and MPH

Movement	Speed of target
Slow patrol	1 fps/.8 mph
Slow stroll	2 fps/1.3 mph
Slow walk	4 fps/2.5 mph
Fast walk	6 fps/3.7 mph
Run (10 min mile)	10 fps/6 mph
Sprint (7 min mile)	13 fps/8.6 mph

CHAPTER 12: "We'll take 'em", Moving Targets

Angle of movement is divided into three directions. Only two require different amounts of lead.

(1) Full value lead is required when the target is moving 90° to your line of sight. Only one of the target's shoulder and side is seen. Target width is approximately 12" for a full size average man.

(2) Half value lead target presents only one upper arm and a part of the upper chest or back. The target may be moving away or toward you at this angle of about 45 degrees. The target width is about 16" for the average size man.

(3) No lead targets presents a full frontal or back view and is moving toward or away from you. Target width is about 20" for a full size man.

Wind speed corrections are explained in a previous chapter.

Determining amount of lead can be done with a table of predetermined leads or with formulas to calculate either lead in feet or lead in mils.

(a) Determine Lead (Feet):

TOF (sec) X Target Speed (fps) = Lead (feet)

(b) Convert Lead in Feet to Mils, using YARDS or METERS distance.
R (YARDS) = Range in hundreds of YARDS
R (METERS) x 1.1 = R (YARDS) = Range in hundreds of YARDS

$$\frac{\text{Lead (ft) x 12 in} - 6}{\text{R (hundreds of YARDS) x 3.6 in}} = \text{Lead in Mils from leading edge of target}$$

Example: Range is 575 YARDS, TOF at range is .7 sec.,
Target Speed is 4 fps

.7 sec x 4 fps = 2.8 ft. for lead

$$\frac{(2.8 \text{ ft x 12 in)} - 6}{5.75 \text{ (hundred yards) x 3.6}} = 1.5 \text{ mils lead}$$

The divisor "3.6" represents the number of inches in a mil at 100 yards. The Army rounds this down to 3.5 because it is easier to divide in multiples of 3.5, i.e., 3.5, 7, 10.5, 14, etc.. For meters conversion, the divisor should be 3.9" per mil per 100 meters. You can round this to 4.0 to make it simpler to use. TOF for yards is fairly close to TOF for meters so it won't hurt to learn and have one set of TOF for both.

(c) Table of Leads in MILs for M118 SB and LR

Table 12.3 TARGET SPEED (fps) LEAD in MILS CHART

For M118 SB and LR, 24" Barrel M40A1

	(slow patrol)	(slow stroll)	(walk)	(fast walk)
Range	1 fps	2 fps	4 fps	6 fps
100 yds	0	0	0	1/4
200 yds	0	0	1/2	1-1/4
300 yds	0	1/4	1-1/4	2-1/4
400 yds	0	1/2	1-1/4	2-1/4
500 yds	1/4	1/2	1-1/2	2-1/4
600 yds	1/4	3/4	1-3/4	2-3/4
700 yds	1/4	3/4	1-3/4	2-3/4
800 yds	1/4	1	2	3
900 yds	1/2	1	2	3-1/4
1000 yds	1/2	1	2-1/4	3-1/2

Make up your own table using the above formulas for other distances and speeds. This will give you some practice in using the formulas. Define aiming points with the mil dot and the width of target. For example: for 500 yard shot, 3 mph requires 19" lead from front edge of 12" wide laterally moving target. At 500 yards this is approximately 1 mil dot lead. 1 mil @ 500 yds = 5 x 3.6" = 18".

(d) *Common leads* require no calculation and are taught by the Army sniper schools.

Table 12.4 Hold Offs for Moving Targets Strolling at 2-3 fps For M118 SB and LR, with 24" Barrel M40A1

Under 200 yards	Aim center of mass.
300 yards	Crosshair at leading edge.
400 yards	A line of light between

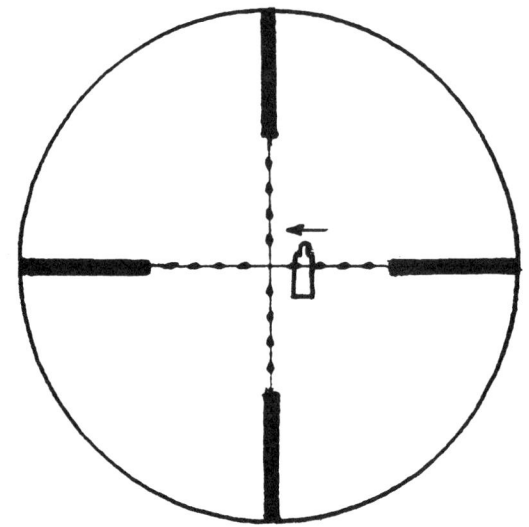

Fig. 12.2. 1 MIL lead on forward edge of target moving from right to left. Target is at 500 yards and is 40" high and 12" wide.

	crosshair and target leading edge.
500 yards	1st mil dot on trailing edge or 6" lead on leading edge.
600 yards	1st mil dot center of mass or 15" lead.

Methods of Engagement.

Moving targets are the most difficult types of targets to engage. A military sniper may have to engage moving targets at much longer distances and requires wind and range estimation corrections. Engaging moving targets is sort of like wind shooting. Instead of correcting for wind deflection, you will be correcting for target deflection. It requires you to lead the target. When wind and target movement is present it is even more difficult to shoot.

(1) *Tracking* requires the sniper to establish an aiming point ahead of the target's movement and maintain this lead until the weapon is fired. The sniper must follow through with his movement during firing to insure the correct lead. This is not a simple movement to do if you are in the prone position and have your elbows planted in the ground. You can also be hindered by the fixed position of the bipod. Using a backpack instead of a bipod will give you more room to move the rifle laterally. Longer range targets are not as difficult because of the smaller image in the scope. The tracking method works well with the sitting, kneeling, and standing position. Use the *tracking method*:

(a) When very close to the target and field of view is limited.

(b) If you are going to receive a "fire on command" during a hostage situation.

(c) When firing on a fast moving target.

(2) *The Ambush method* places the cross hairs well forward of the moving target and then held stationary. The target will walk into the location of the selected mil angle on the horizontal wire that represents the correct aiming point on the leading edge of the target. Use this method when:

(a) the target is seen as small enough in your scope's field of view to allow you enough time to find a fixed position ahead of him and make a precision shot.

(b) the target is moving at a constant pace and over a fairly well defined route.

(c) the sniper can select the place to engage the target.

(3) *Engaging a stop and go target* requires the sniper to track and hold until the target holds long enough for the sniper to make a good shot. The sniper must not fire while the target is moving because he is not using a lead on this type target.

(4) *Engaging the bobbing target* requires the sniper to look for a pattern to the multiple short exposure target. This pattern could last for a few seconds several times a minute or for a few seconds once every other day. If a pattern is found then the sniper can align his scope in the general vicinity and wait for the target to appear. Use the lowest power on your scope to give you the widest field of view. Fixed 10X scopes are difficult to use if the target is

CHAPTER 12: "We'll take 'em", Moving Targets

This photo shows a moving target being walked by a handler in the butts. The large white rock located on the burm in front of the target can be used as a reference for ambushing the target if the target moves across it as seen in your scope. Put intersection of the cross hairs about 8" to 10" above the top of the burm and align the vertical cross wire with the rock. Put a little amount of pressure on the trigger. When the target appears to be at the correct lead, in mils or feet for the speed and distance, take up the remaining trigger pressure evenly. At this distance of 100 yards, if the target is moving at a slow stroll, aim for center of mass and you will get a hit on the center of the target. (Mike R. Lau)

inside of 100 yards and constantly moves to another location before reappearing. The field of view of a scope that is set on 10X is only about 11 feet at 100 yards.

Some Helpful Notes when Shooting Moving Targets

(1) Sniper has a tendency to watch the target and not the crosshair. Train yourself to focus on the lead aiming point of your cross hair to maintain the correct lead when you fire.

(2) There is a tendency to jerk the trigger and flinch when firing at moving targets. This is especially apparent when using the ambush method or engaging fast moving and bobbing targets. Practice to eliminate this.

(3) Failure to adjust for wind. Put your windage on the scope so you can concentrate on the lead.

(4) For a right handed shooter, sometimes it is harder to engage a target moving from left to right than from the other direction. The lead may have to be increased by 2 times to overcome the natural hesitation in follow through because the rifle butt is cramping the shoulder and upper body. This hesitation is sometimes difficult to overcome even by experienced shooters. The ambush method may resolve this problem because it does not require follow through.

(5) If the target is walking or running towards a car, building, bunker, or other cover, you may be able to wait until he gets to that

location and has to make a turn along side a building or open a door. He may not be moving for a moment in time, long enough for you to get a good shot.

(6) When practicing, always keep your data book handy and record your shots so you can study them. Verify your leads and speeds so you can get familiar with them and don't have to do calculations the next time you recognize the target speed and distance.

(7) Keep it simple: TOF x fps = lead in feet. You can stop here and engage target with this distance by estimating number of feet forward of center of target. Or: 1 MIL = 3.5" (.3 feet) at each 100 yard *or* 100 meter increment. Mentally multiply .3 by the number of 100 yard increments and round to whole number. Mentally divide that number into the lead in feet using round numbers. Example: Target at 650 meters, 4 fps.

(a) 1 sec x 4 fps = 4 feet lead;

(b) .3 x 7 = 2 feet (1 mil @ 700 yds);

(c) 4 divided by 2 = 2 mils lead from center of target.

Carlos Hathcock examines reproduction Winchester M70 .30-06 sniper rifle with Unertl Scope at the 1996 SOF show in Las Vegas. Carlos' love and respect for firearms is reflected in his own subtle way when he turns and smiles at the author and says, " The bolt face needs cleaning." (Mike R. Lau)

CHAPTER 13: MANAGING THE LOG/DATA BOOK

Most of us that do some kind of military, law enforcement, or long range competitive shooting, keep some sort of shooting data in log books. These include data on the rifle, ammunition, types of targets with dimensions, come-ups, range and wind charts. I own several different military and law-enforcement log books and also have seen many examples of pages that were created by shooters who believe theirs is the best log book page ever created. What I notice about all of these books and pages is that none really tell you how to fill them out. It is assumed that you already know what to write down, but in most cases you can figure this out for yourself after you use the data collection sheet a couple of times. Data and interpretation of data for the sniper can become a tedious chore. Remember one thing, the purpose of using this data is to increase your probability of making a first round hit on the target. It is not there to boggle your mind or make you lose your shot because the target disappeared while you were calculating the distance. When you become a Master Sniper like Carlos Hathcock then you can get by without having to look up data or do calculations. Until then we need the help from data tables and calculations. The next chapter will give you examples on how the sniper uses the data.

From your own experience and priorities, organize your own log or data book that is easiest to reference and exclude charts and tables that do not apply to your rifle, ammunition, or the type of shooting you will not do. Some persons have separate books for data and firing records. Do whatever pleases you. Loose leaf binders like Sam Chesnut's or mine allow you to reorganize your book after a while. Index tab the sections of data you need if the data is not on a firing record sheet, even if there is only one page that has the tab. Write on the index tab clearly and in big letters. Put the most important formulas or data on the first page where the index tab is located so you don't have to thumb through the pages. Rewrite data and formulas so that it is easily read in poor light. Don't try to resize all the information onto one page. It only makes it hard to read and could cause you to slow down or make an error in a formula. **Keep it simple.**

DATA/LOG BOOK ESSENTIALS

(1) Ballistics Table for your particular rifle ammunition will have:
   (a) Velocity to 1000 yards or meters if .308 Caliber.
   (b) Maximum ordinate to range.
   (c) Bullet path (- inches)
   (d) Come Up table for each 100 yards or meters
   (f) Cumulative Come Up to range

(2) Zero Temperature Deviations Correction chart. (Also carry a small thermometer).

(3) Target or Object dimension chart.

(4) Mil Height to Distance conversion chart.

(5) Formula for converting Mils and Target Height to Range.

(6) Wind Correction chart with at least Full Values to 1000 Yards or Meters

(7) Formula for Converting Moving Target Leads into Mils.

(8) Cosine Angle Firing Chart.

(9) One type firing record data sheet. If you use more than one type of data sheet and it is in another location in your book, you may lose track of round counts, temperature affects on zero, mirage corrections, etc.. You don't want to be thumbing through your data sheets, while firing, trying to locate a note made a half hour ago.

(10) Range Card.

(11) Metric to Yards/Inches Conversion chart.

Essential if you travel to different locations to shoot.

(1) Zero Corrections for Altitude Changes.

(2) Know how to convert Celsius to Fahrenheit and vice versa. The rest of the world uses Celsius (Centigrade).

(3) Other types of Record Firing Sheets for different type targets.

Recording your data.

When you collect data on shooting conditions for your log book, **keep data simple**. Don't write paragraphs even if it is remarks. Develop a system of simple codes, acronyms, abbreviations, and drawings that records your information with one stroke of your pencil and is still understandable. Here are examples.

Ammunition. Type, Manufacturer, and, Lot number. 168 gr Federal Match is "168FM", Black Hills is "168BH". If you are using the same ammo as the last data sheet use ditto marks.

Humidity: High (H), Normal (N), Dry (D)

Light: Bright (B), Hazy (H), Overcast (O)

Mirage: Light (L), Medium (M), Heavy (H). A single wavy line through the circle representing effective mirage is all that is needed.

Wind Direction: Draw arrow on chart and write down speed.

Wind Velocity: Estimate 0-3 mph, 3-5 mph, 5-7 mph, 7-9 mph, and use 2, 4, 6, 8, etc.

We know it's mph so don't write "mph".

Light Direction: Lay open data/log book on ground and hold pencil straight up with it's point in the block where you would record this information. Draw an arrow in this space in direction of shadow made by your pencil.

Firing records should be neat and readable. You don't have to fill in every block or space. The main things to watch for are temperature, wind, and range.

You decide what is important and what is not through experience. It is your book.

CHAPTER 13: Managing The Log/Data Book

I. The Data/Log Book

Marine Scout/Snipers at Quantico recording hits on target. (Photo courtesy Kent Gooch)

Brian Gauthier, gave author his data/log book. Made out of OD Cordura canvas by Raine, Inc. (P.O. Box 585, Daleville, Indiana 47334). Outside of case is covered with OD duct tape for waterproofing. Plastic notebook with plastic binder rings won't rust. Map is laminated in plastic and clear plastic pocket is Velcro™ sealed. Has pocket for securing back cover of USMC data book. Additional pockets located behind map pocket and log book hold pencils, calculator, thermometer, ruler, and protractor. You don't need a military type data book like this USMC one. It has a lot of pages of blank firing records that you will never use. (Mike R. Lau)

216 *The MILITARY and POLICE SNIPER*

Navy Seal Log Book made by Trigon Technologies has waterproof data and record sheets. High quality Cordura case includes Silva type 10 compass and thermometer. Naval Special Warfare (NSW) members can request through your commander the version with classified data directly from Trigon Technologies. Request "NSWC Crane Stock Number 7610LLLT90404". Civilian version without classified data is available from Iron Brigade Armory or Texas Brigade Armory. Cost is around $85.

Log book with calculator belonging to National Guard shooter. This one is simple and practical. You will need blank scratch paper for calculations and a calculator. Keep only essential data pages in your book. Be brief and write so it is easy to read. (Mike R. Lau)

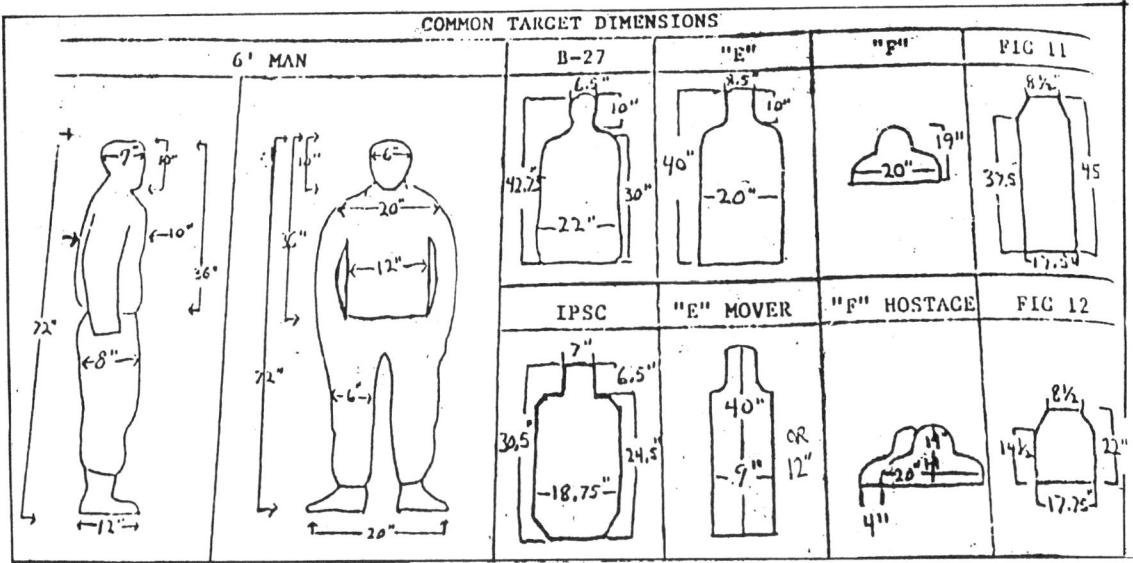

II. Target height charts.

Fig. 13.1. Common targets are redrawn and dimensioned by SFC Pete Carpentier for his own data/log book. The "E" is the M 1919 kneeling silhouette and the "F" is the M 1917 prone target. There is also a full height standing military target, the M 1913, which combines the "E" target and a trapezoidal section to go beneath the "E". Full height of "M" target assembled is 68". Fig. 11 and 12 are the European cartoon or "Hun" targets. (Courtesy Epitacio Carpentier)

CHAPTER 13: Managing The Log/Data Book

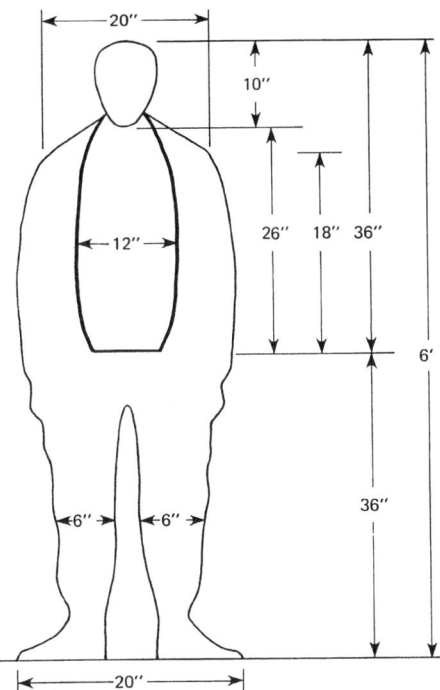

Fig. 13.2. Dimensions of 6' height human target from USMC Data/Log Book.

Aircraft Dimensions					Aircraft	Length	Wingspan	Tail Height
Aircraft	Length	Wingspan	Tail Height		Civilian			
Military					707	128 ft.	130 ft.	38 ft
SR-71	107 ft.	56 ft.	19 ft.		727/200	153	108	34
U2-A	50	80	13		737/300	109	95	37
C-130	98	133	39		737/400	119	95	37
E-3/E-6	145	146	42		737/500	101	95	37
KC-135	136	130	38		747/400	231	251	70
B1-B	147	137	34		747/500	250	251	70
B-2	69	172	17		747/600	279	211	64
B-52	157	185	48		(Crossection width on all 747's is 22 ft.)			
A-10	53	58	18		757-F	155	125	45
767	136	130	38		767/200	159	156	52
JSF	45	36			767/200ER	159	156	52
F-18	56	38	15		DC-8	150	142	42
F-15	63	43	17		DC-9	105	82	27
F-16	50	33	17		DC-10/10	182	156	58
F-4	63	38	17		DC-10/30	184	156	58
F-117	66	43	12		KC-10/A	184	166	58
Air Force 1	232	196	64		YC-15	125	110	44
					MD-11	200	170	58
					MD-80	148	108	30
					MD-90	153	108	30

Fig. 13.4. Military and Commercial Aircraft Dimensions. Tail height is measured from bottom of fuselage to top of tail. (Sam Chesnut)

Object	Ht. (in)	Width (in)	Lgth. (in)
MC/VISA decal			4 1/2
Door Frame (Glass/Alum or wood)	83	37	
Door	81	31	
Stop Sign (Texas)	24	24	
Stop Sign (Oklahoma)	30	30	
Fire Hydrant			
Telephone Pole (wood)			
Concrete Cinder Block (Wall/Trailer spt)	8	8	16
55 Gal Drum	36	24	
AK-47/AKM Full Stock			34.25
AK-47 Barrel Forward of Handguard			8.5
SVD (Dragunov)			48.2
SVD Barrel Forward of Handguard			14.75
M-1 Garand			43.6
M-1 Carbine			35.6
M16A1/A2			39.5
M14			44.3
M14 Barrel forward of Handguard			11.75
M40A1			44
M24			43
MP-5 Sub-machine gun (butt retracted)			19.4
Remington 870 w/18" bbl			40.5
RPG-7 loaded with grenade			52.6
RPG-7 unloaded			37.8

(Helicopter Ht:bottom fuselage to top of turbo; lgt: fuselge excluding tail upright)

Object	Ht. (in)	Width (in)	Lgth.
UH-1B or H (Blunt Nose)	83.5	98	36.5 (ft)
UH-1N (Pointy Nose)	110.5	108	41 (ft)
OH-6 (520N)	106	74.8	24.5 (ft)
M 206B (OH-58) JetRanger (1 pilot window and 1 passenger window each side)			
	110.2	71	39 (ft)
M 206L LongRanger (1 pilot window and 2 passenger windows ea side)			
	122	90.6	42.6 (ft)
AS 350B Ecureuil (ht. is bottom fuslg-top roto)	108.3		35.86 (ft,incl tail)
M24 Hind	168		56 (ft)

Aircraft height is given as bottom of fuselage to top of tail, width is wingspan, length does not include probe in nose.

Object	Ht.	Width	Lgth.
Super Etendard	12.3 (ft)	31.5 (ft)	47 (ft)
A-4 Skyhawk	15 (ft)	27.5 (ft)	40 (ft)
F-5A	11 (ft)	25.3 (ft)	47.2 (ft)
M151 Jeep	71 (w/roof)	64	11 (ft)
M998/M1025 HMMWV	72	84	15 (ft)

Fig. 13.3. Dimensions of Common Objects. (Mike R. Lau and Sam Chesnut)

CHAPTER 13: Managing The Log/Data Book

Photo 13.5. A-4 Skyhawk was used for aggressor aircraft at the old Dallas Naval Air Station. Numerous A-4s and F-5s were sold to foreign governments. Argentine Navy used A-4s against British Task Force during Falklands War in 1982. (Mike R. Lau)

TABLE OF MILS FOR OBJECTS

Inches =		9	12	16	18	20	22	24	28	32	36	60	66	69	72
Yards =		.250	.333	.444	.500	.556	.611	.667	.778	.889	1.000	1.667	1.833	1.917	2.000
	3/4	333	444	593	666	741	815	889	1037	1185	1333	2223	2444	2556	2667
	1	250	333	445	500	556	611	667	778	889	1000	1667	1833	1917	2000
	1 1/4	200	266	355	400	445	489	534	622	711	800	1334	1466	1537	1600
	1 1/2	167	222	296	333	371	407	445	519	593	667	1111	1222	1278	1333
	1 3/4	143	190	254	285	318	349	381	445	508	571	953	1047	1095	1143
	2	125	167	222	250	278	306	334	389	445	500	834	917	959	1000
	2 1/4	111	148	197	222	247	272	296	446	395	444	741	815	852	889
	2 1/2	100	133	178	200	222	244	267	311	356	400	667	733	767	800
	2 3/4	91	121	161	182	202	222	243	283	323	364	606	667	697	727
	3	83	111	148	167	185	204	222	259	296	333	556	611	639	667
	3 1/4	77	102	137	154	171	188	205	239	273	308	513	564	590	615
	3 1/2	71	95	127	143	159	175	191	222	254	286	476	524	548	571
	3 3/4	67	89	118	133	148	163	178	207	237	267	445	489	511	533
	4	63	83	111	125	139	153	167	195	222	250	417	458	479	500
	4 1/4	59	78	104	118	131	144	157	183	209	235	392	431	451	471
	4 1/2	56	74	99	111	124	136	148	173	197	222	370	407	426	445
	4 3/4	53	70	93	105	117	128	140	164	187	210	351	386	404	421
	5	50	67	89	100	111	122	133	156	178	200	333	367	383	400
	5 1/4	48	63	85	95	105	116	127	148	169	190	318	349	365	381
	5 1/2	45	61	81	91	101	111	121	141	162	182	303	333	349	364
	5 3/4	43	58	77	87	97	106	116	135	155	174	290	319	333	348
	6	42	56	74	83	93	102	111	130	148	167	278	306	320	333
	6 1/4	40	53	71	80	89	98	107	124	142	160	267	293	307	320
	6 1/2	38	51	68	77	86	94	103	120	137	154	256	282	295	308
	6 3/4	37	49	66	74	82	91	99	115	132	148	247	272	284	296
	7	36	48	63	71	79	87	95	111	127	143	238	262	274	286
	8	31	42	56	63	70	76	83	97	111	125	208	229	240	250
	9	28	37	49	56	62	68	74	86	99	111	185	203	213	222
	10	25	33	44	50	56	61	67	79	89	100	167	183	192	200

USING THE TABLE

1) Estimate the size of the object (to be range estimated) and locate across the top of the table.
2) Measure the height or width of the object in mils and locate down the side.
3) Move down from the top and right from the side to find the distance in yards.

MIL FORMULA

$$\frac{\text{Height or width of object (in yards)} \times 1000}{\text{Height or width of object (in mils)}} = \text{Distance (in yards)}$$

PREMIER RETICLES • 920 BRECKINRIDGE LANE • WINCHESTER, VA 22601
(540) 722-0601 • FAX (540) 722-3522

III. Range Estimating Chart.

*Fig. 13.5. MIL Range Estimation Table in YARDS for **Standard** size objects. (Premier Reticles)*

CHAPTER 13: Managing The Log/Data Book

RANGE ESTIMATION TABLE OF MILS FOR PERSONNEL – 6', 5'9" and 5'6"

MILS	6' = 2 YDS	5'9" = 1.9 YDS	5'6" = 1.8 YDS
1	2000	1900	1800
1¼	1600	1520	1440
1½	1333	1266	1200
1¾	1143	1085	1028
2	1000	950	900
2¼	888	844	800
2½	800	760	750
2¾	727	690	654
3	666	633	600
3¼	615	584	553
3½	571	542	514
3¾	533	506	480
4	500	475	450
4¼	470	447	423
4½	444	422	400
4¾	421	400	378
5	400	380	360
5¼	380	361	342
5½	362	345	327
5¾	347	330	313
6	334	316	300
6¼	320	304	288
6½	308	292	277
6¾	296	281	266
7	286	271	257
8	250	237	225
9	222	211	200
10	200	190	180

*Fig. 13.6. MIL Range Estimation Table For **Personnel** (USMC)*

Size of object in mils \ Height of target in inches	6	9	12	14	16	18	20	22	24	28	32	36	60	66	69	72	Range in meters
0.25	610	914	1219	1422	1626	1829	2032	2235	2438	2845	3251	3658	6096	6706	7010	7315	
0.5	305	457	610	711	813	914	1016	1118	1219	1422	1626	1829	3048	3353	3505	3658	
0.75	203	305	406	474	542	610	677	745	813	948	1084	1219	2032	2235	2337	2438	
1	152	229	305	356	406	457	508	559	610	711	813	914	1524	1676	1753	1829	
1.25	122	183	244	284	325	366	406	447	488	569	650	732	1219	1341	1402	1463	
1.5	102	152	203	237	271	305	339	373	406	474	542	610	1016	1118	1168	1219	
1.75	87	131	174	203	232	261	290	319	348	406	464	523	871	958	1001	1045	
2	76	114	152	178	203	229	254	279	305	356	406	457	762	838	876	914	
2.25	68	102	135	158	181	203	226	248	271	316	361	406	677	745	779	813	
2.5	61	91	122	142	163	183	203	224	244	284	325	366	610	671	701	732	
2.75	55	83	111	129	148	166	185	203	222	259	296	333	554	610	637	665	
3	51	76	102	119	135	152	169	186	203	237	271	305	508	559	584	610	
3.25	47	70	94	109	125	141	156	172	188	219	250	281	469	516	539	563	
3.5	44	65	87	102	116	131	145	160	174	203	232	261	435	479	501	523	
3.75	41	61	81	95	108	122	135	149	163	190	217	244	406	447	467	488	
4	38	57	76	89	102	114	127	140	152	178	203	229	381	419	438	457	
4.25	36	54	72	84	96	108	120	131	143	167	191	215	359	394	412	430	
4.5	34	51	68	79	90	102	113	124	135	158	181	203	339	373	389	406	
4.75	32	48	64	75	86	96	107	118	128	150	171	193	321	353	369	385	
5	30	46	61	71	81	91	102	112	122	142	163	183	305	335	351	366	
5.25	29	44	58	68	77	87	97	106	116	135	155	174	290	319	334	348	
5.5	28	42	55	65	74	83	92	102	111	129	148	166	277	305	319	333	
5.75	27	40	53	62	71	80	88	97	106	124	141	159	265	292	305	318	
6	25	38	51	59	68	76	85	93	102	119	135	152	254	279	292	305	
6.25	24	37	49	57	65	73	81	89	98	114	130	146	244	268	280	293	
6.5	23	35	47	55	63	70	78	86	94	109	125	141	234	258	270	281	
6.75	23	34	45	53	60	68	75	83	90	105	120	135	226	248	260	271	
7	22	33	44	51	58	65	73	80	87	102	116	131	218	239	250	261	
8	19	29	38	44	51	57	64	70	76	89	102	114	191	210	219	229	
9	17	25	34	40	45	51	56	62	68	79	90	102	169	186	195	203	
10	15	23	30	36	41	46	51	56	61	71	81	91	152	168	175	183	

Fig. 13.7. *MIL Range Estimation Table in METERS for Standard size Objects. (US Army NGSSS)*

MILS \ FEET	3	4	5	6	7	8	9	10	11	12	13	14	15	16	17	18
YARDS	1	1.3	1.7	2	2.3	2.7	3	3.3	3.7	4	4.3	4.7	5	5.3	5.7	6
2	500	650	850	1000	1150	1350	1500	1650	1850	2000	2150	2350	2500	2650	2850	3000
2.5	400	520	680	800	920	1080	1200	1320	1480	1600	1720	1880	2000	2120	2280	2400
3	333	425	566	666	766	900	999	1100	1230	1332	1433	1566	1665	1766	1900	1998
3.5	285	371	486	571	657	771	855	943	1057	1140	1229	1343	1425	1514	1629	1710
4	250	325	425	500	575	675	750	825	925	1000	1075	1175	1250	1325	1425	1500
4.5	222	289	378	444	511	600	666	733	822	888	950	1044	1110	1178	1267	1332
5	200	260	340	400	460	540	600	660	740	800	860	940	1000	1060	1140	1200
5.5	182	236	309	362	418	491	543	600	673	724	782	855	905	964	1036	1086
6	167	217	283	334	383	450	500	550	617	668	717	783	835	883	950	1000
6.5	154	200	262	308	354	415	462	508	569	616	662	723	770	815	877	924
7	143	186	243	286	329	386	429	471	529	572	614	671	715	757	814	858
7.5	133	173	227	266	307	360	399	440	493	532	573	627	665	707	760	798
8	125	163	213	250	288	338	375	413	463	500	538	588	625	663	713	750
8.5	118	153	200	234	271	318	351	388	435	468	506	553	585	624	671	702
9	111	144	189	222	256	300	333	367	411	444	478	522	555	589	633	666
9.5	105	137	179	210	242	284	315	347	389	420	453	495	525	559	600	630
10	100	130	170	200	230	270	300	330	370	400	430	470	500	530	570	600
10.5								314	352	381	410	448	476	505	543	571
11								300	336	367	390	427	455	482	518	545
11.5									322	348	374	409	435	461	496	522
12									308	333	358	392	417	442	475	500
12.5										320	344	376	400	424	456	480
13										308	331	362	385	408	438	462
13.5											319	348	370	393	422	444
14											307	336	357	379	407	429
14.5												324	345	366	393	414
15												313	333	353	380	400
15.5												303	323	342	368	387
16													313	325	356	375
16.5													303	321	345	364
17														312	335	353
17.5														302	326	343
18															317	333
18.5															308	324
19															300	316
19.5																308

1) Estimate height of target and locate across the top
2) Measure height of target in Mils and locate down the side
3) Move down from the top and right from the side to find the range

$$\frac{\text{Height of TGT (yards)} \times 1000}{\text{Height of TGT (Mils)}} = \text{Range}$$

Fig. 13.8. *MIL Range Estimation Table in YARDS for **Large** objects. (USMC)*

CHAPTER 13: Managing The Log/Data Book

$\frac{A}{4}$ = MPH (A = Angle)

$\frac{R \times V}{15}$ = ½ MINUTES (1 Click)

1 CLICK MOVES STRIKE OF BULLET APPROXIMATELY ½ " PER 100 YARDS/METERS

Observation Method — If the tactical situation prevents the use of the above methods, snipers can use the following guides:
 (a) A wind under 3 miles per hour can hardly be felt, but it causes smoke to drift.
 (b) A 3- to 5-mile per hour wind is felt lightly on the face.
 (c) A 5- to 8-mile per hour wind keeps tree leaves in constant motion.
 (d) An 8- to 12-mile per hour wind raises dust and loose paper.
 (e) A 12- to 15-mile per hour wind causes small trees to sway.

Circles represent wind velocity as indicated. Roman numerals indicate wind direction. Small numerals indicate **clicks** of windage for ½ minute sights.

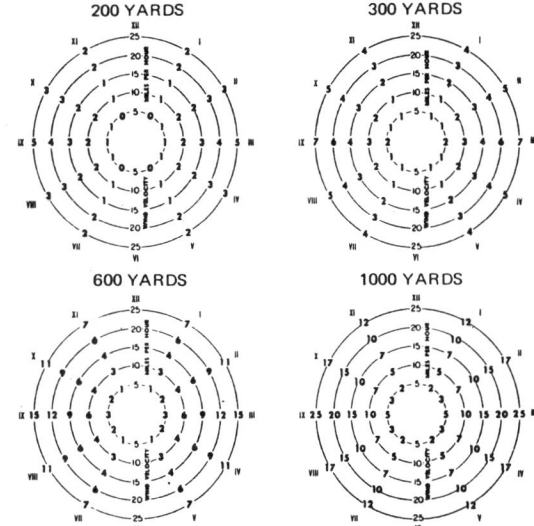

IV. Wind

Fig. 13.9. Rosette wind chart for YARDS and 7.62mm M118, M40A1. (USMC)

RANGE IN YARDS	WIND VALUE	2 MPH		4 MPH		6 MPH		8 MPH		10 MPH		12 MPH		14 MPH		16 MPH		18 MPH		20 MPH	
		Min	In	Min	In	Min	In	Min	In	Min	In	Min	In	Min	In	Min	In	Min	In	Min	In
100	½ FULL	0 ⅛	.073 .146	⅛ ½	.146 .292	¼ ½	.219 .438	¼ ½	.292 .581	¼ ¾	.365 .73	½ ¾	.484 .876	½ 1	.511 1.02	½ 1 ¼	.584 1.16	¾ 1 ¼	.657 1.31	¾ 1 ½	.78 1.40
200	½ FULL	0 0	.3 .6	0 ½	.6 1.2	¼ 2	.9 1.8	½ 1	1.2 2.4	1 1 ½	1.5 3	1 2	1.8 3.6	1 2	2.1 4.2	1 2 ½	2.4 4.8	1 2 ½	2.1 5.4	1 ½ 3	3 6
300	½ FULL	0 ½	.7 1.4	½ 1	1.4 2.8	½ 1 ½	2.1 4.2	1 2	2.8 5.6	1 2	3.6 7	1 ½ 2 ½	4.2 8.4	1 ½ 3	4.9 9.8	2 4	5.6 11.2	2 4	6.3 12.6	2 ½ 4.5	7 14
400	½ FULL	½ ½	1.25 2.5	½ 1	2.5 5	1 2	4 8	1 2 ½	5.25 10.5	1 ½ 3	6.5 12	2 4	7.75 15.5	2 4 ½	9 18	2 ½ 5	10.5 21	3 6	11 ¾ 23.5	3 6 ½	13 26
500	½ FULL	½ 1	2.0 4.5	1 1 ½	4 8.5	1 2 ½	6.5 13	1 ½ 3 ½	8.5 17	2 4	11 21.5	2.5 5	13 22	3 6	15 30	3 ½ 7	17 34.5	4 8	19 38.5	4 ½ 8 ½	21.5 43
600	½ FULL	½ 1	3.0 6.5	1 2	6.5 13	1 ½ 3	10 19.5	2 4 ½	12 26	2 ½ 5 ½	16 32.5	3 6 ½	19.5 39	3 ½ 7 ½	23 45	4 ½ 8 ½	26 52	5 9 ½	29 58.5	6 ½ 11	32.5 65
700	½ FULL	1 1	5 9.5	1 2 ½	9 18.5	2 4	14 28	2 ½ 5	18.5 37	3 6 ½	23 46.5	4 8	28 56	4 ½ 9 ½	32.5 65	5 10 ½	37 74.5	6 12	42 83.5	6 ½ 11	46.5 93
800	½ FULL	1 1 ½	6 12.5	1 ½ 3	13 25.5	2 4 ½	19 38	3 6 ½	25.5 51	4 8	32 63.5	5 9 ½	39 76	5 ½ 11	44.5 89	6 ½ 12 ½	51 101½	7 14	57 114½	8 16	63.5 127
900	½ FULL	1 2	8.5 17.0	2 4	17 34	3 5 ½	25 50.5	7 7 ½	34 67.5	4 ½ 9 ½	42 84.5	5 ½ 11	51 101½	6 ½ 13	59 118½	7 ½ 15	67 135	8 ½ 17	76 152	9 ½ 19	84.5 169
1000	½ FULL	1 2	11.0 21.5	2 4	22 43.5	3 6 ½	32.5 66	4 8 ½	43.5 87	5 ½ 10 ½	54 108½	6 ½ 13	65 130	7 ½ 15	76 152	8 ½ 17 ½	87 173½	10 19 ½	98 195½	11 21.5	108½ 217
1100	½ FULL	1 2 ½	18.5 27.0	2 ½ 5	27 54.5	4 7 ½	41 81.5	5 10	34.5 109	6 12	68 136	7 ½ 15	87.5 163	9 17	95.5 180	10 14 ½	104 217	11 22	122½ 245	12 24 ½	136 292
		2 mph		4 mph		6 mph		8 mph		10 mph		12 mph		14 mph		16 mph		18 mph		20 mph	

Fig. 13.10. Wind chart for YARDS and 7.62mm M118, M40A1. (USMC)

CHAPTER 13: Managing The Log/Data Book

ANGLE OF TARGET MOVEMENT

Angle of Target Movement. Shows the leads required for a target moving at a 90-degree angle to the sniper. In most cases, however, targets will be moving at some other angle toward or away from the sniper's position. A method for estimating the angle of movement is as follows:

Full Lead Target. When the target is moving across the observer's front and only one arm and one side are visible, the target is moving at or near an angle of 90 degrees and a full value lead is necessary.

Half Lead Target. When one arm and two-thirds of the front or back are visible, the target is moving at approximately a 45-degree angle and a one-half value lead is necessary.

No Lead Target. When both arms and the entire front or back are visible, the target is moving directly toward or away from the sniper and will require no lead.

Double Leads. The leads previously mentioned hold true for a right-handed shooter firing on a target moving from his right to left. If the target is moving from left to right, the lead must be doubled due to a natural hesitation in follow-through when swinging against the shooting shoulder. This hesitation is extremely difficult to overcome even by the most experienced shooters.

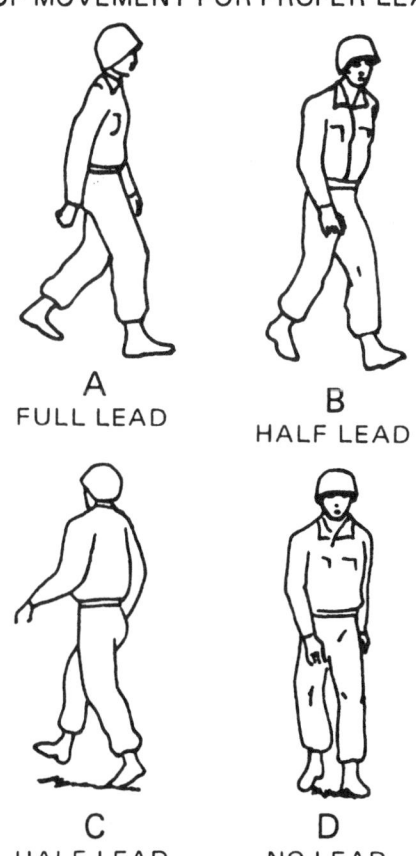

LEADS
METHOD OF DETERMINING ANGLE OF MOVEMENT FOR PROPER LEAD

A FULL LEAD
B HALF LEAD
C HALF LEAD
D NO LEAD

V. Moving Target Leads Adjustment Chart. (See Appendix II for Tables)

Fig. 13.11. Angle of Moving Target for Leads. Note the comment for "Double Leads." (USMC)

VI. Elevation and Slope Adjustment Charts

```
              POINT OF IMPACT RISE AT NEW ELEVATION (MINUTES)
     RANGE        2500 ft        5000 ft        10,000 ft

      100           .05            .08             .13

      200           .1             .2              .34

      300           .2             .4              .6

      400           .4             .5              .9

      500           .5             .9             1.4

      600           .6            1.0             1.8

      700          1.0            1.6             2.4

      800          1.3            1.9             3.3

      900          1.6            2.8             4.8

     1000          1.8            3.7             6.0
```

With a rifle zeroed at sea level, shooting at 700 yards at 5000 feet will result in rounds hitting the target 1.6 minutes high. The sniper will need to come down on his fine tune 1 1/2 minutes to correct for this.

VI. Elevation and Slope Adjustment Charts

Fig. 13.12. Altitude MOA elevation adjustment chart for 7.62mm M118 ammunition fired from 24" barrel M40A1 and M24. (USMC)

CHAPTER 13: Managing The Log/Data Book

Zero summary chart
Temperature (F/C)

Yards	Meters	20/-6	30/-1	40/4	50/10	60/15	65/18	70/21	75/24	80/26	85/29	90/32	95/35	100/37	105/40
100	90														
150	145														
200	180														
250	225														
300	270														
350	315														
400	360														
450	405														
500	450														
550	495														
600	540														
650	585														
700	630														
750	675														
800	720														
850	765														
900	820														
950	855														
1000	900														
1050	945														

Range

Fig. 13.13. Zero Summary Chart is for recording your elevation differences from a specific year round zero. Chart is given in yards and meters. Go back through your old data and record the adjustments to your zero for different temperatures. You can also obtain data from the Sierra Ballistics Program. (U.S. Army NGSSS)

Deviations from zero

Variable	Change from zero	Average correction required		
		@ 300m	@ 600m	@ 1000m
Temperature	+ 30 deg or more	- 2 moa	- 3 moa	- 4 moa
	+20 deg	- 1 moa	- 1.5 moa	- 2 moa
	- 20 deg	+ 1 moa	+ 1.5 moa	+ 2 moa
	- 30 deg or more	+2 moa	+ 3 moa	+ 4 moa

Fig. 13.14. Zero deviations from zero with large changes in temperatures for M118 type ammunition. Adjustments shown may vary with your own local weather conditions, altitude, weapon and ammunition, etc.. You need to check this out for yourself and your own ammunition. (U.S. Army NGSSS)

VII. Firing Record Page.

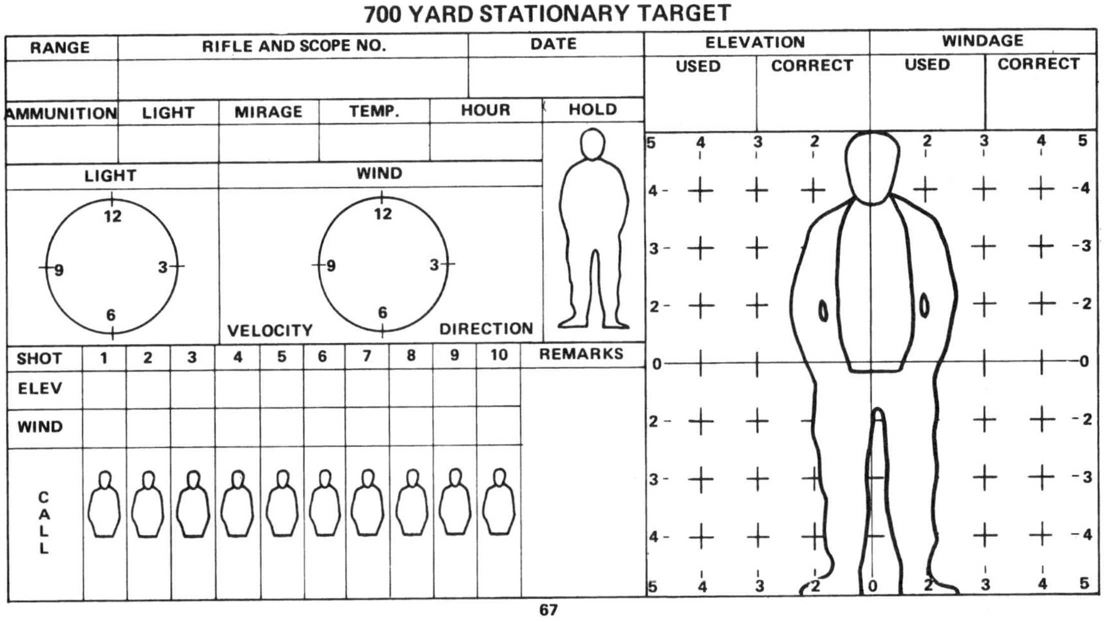

Fig. 13.15. *Firing record for stationary target from Marine Corps Scout/Sniper data book. There is a different page for each 100 yard increment giving the exact elevation and windage adjustment figures around the full height target. This particular page is for 700 yards range showing 10 MOA (approximately 73 inches) height. (USMC)*

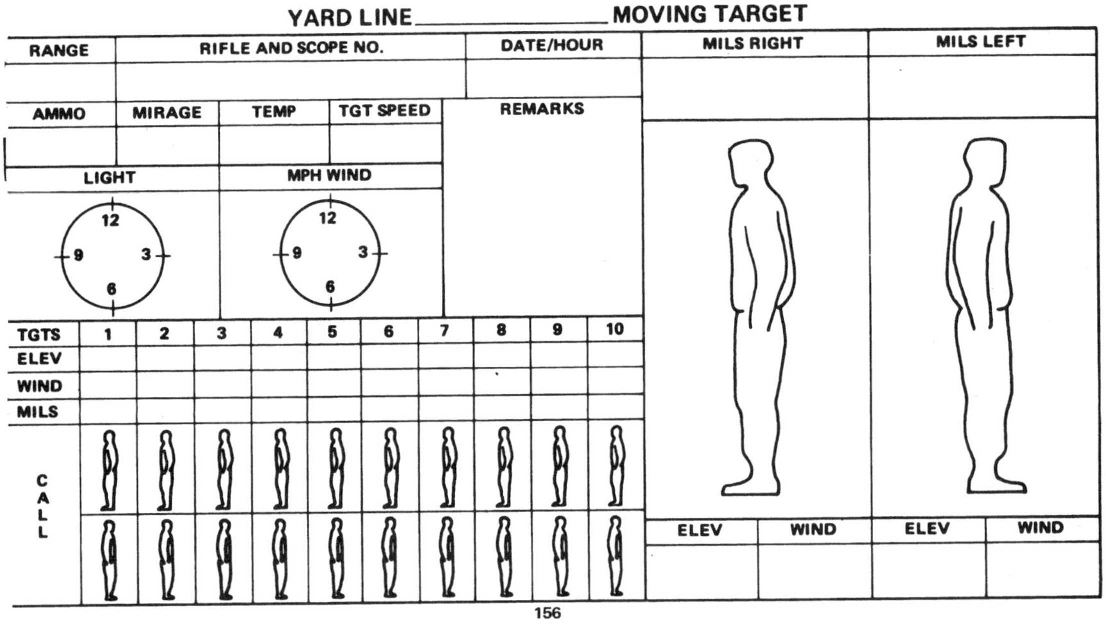

Fig. 13.16. *Firing record for moving target at any range. (USMC)*

CHAPTER 13: Managing The Log/Data Book

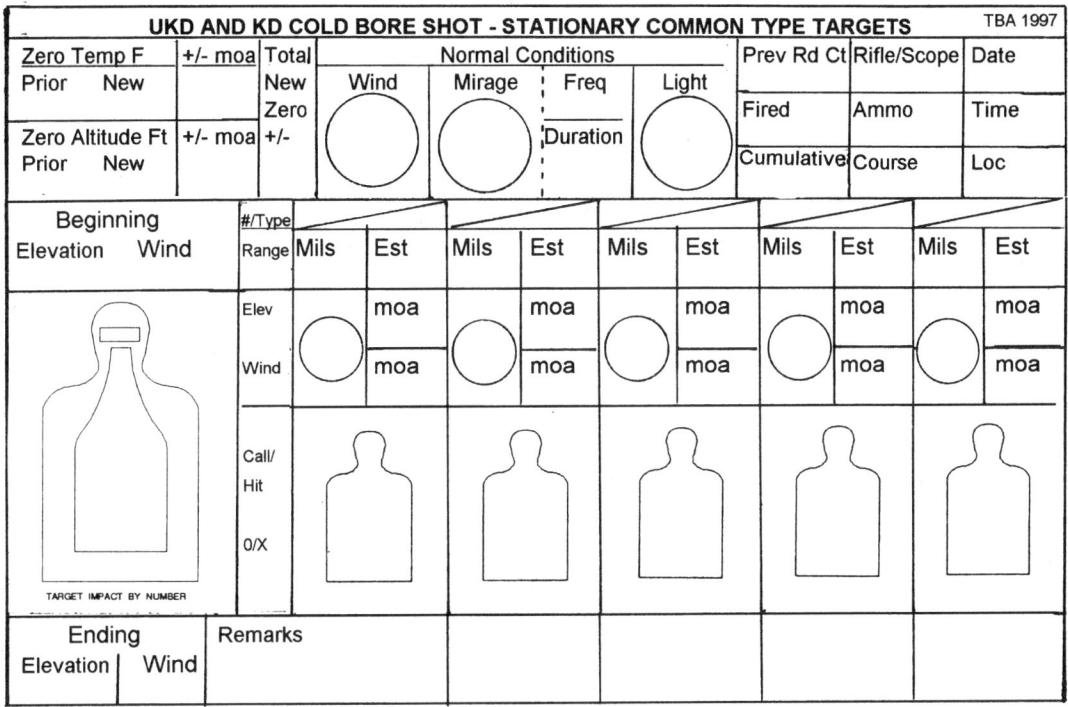

Fig. 13.17. Firing record for any type UKD and KD stationary targets at any distance. Can be used for Fig. E, F, 11, or any full size or half size silhouette target. Use for multiple targets at different ranges or will walk you through for single cold bore shot Data collection for normal condition also included. Note in spaces: target number, type, MIL height, range, mirage or wind direction/speed. (Mike R. Lau)

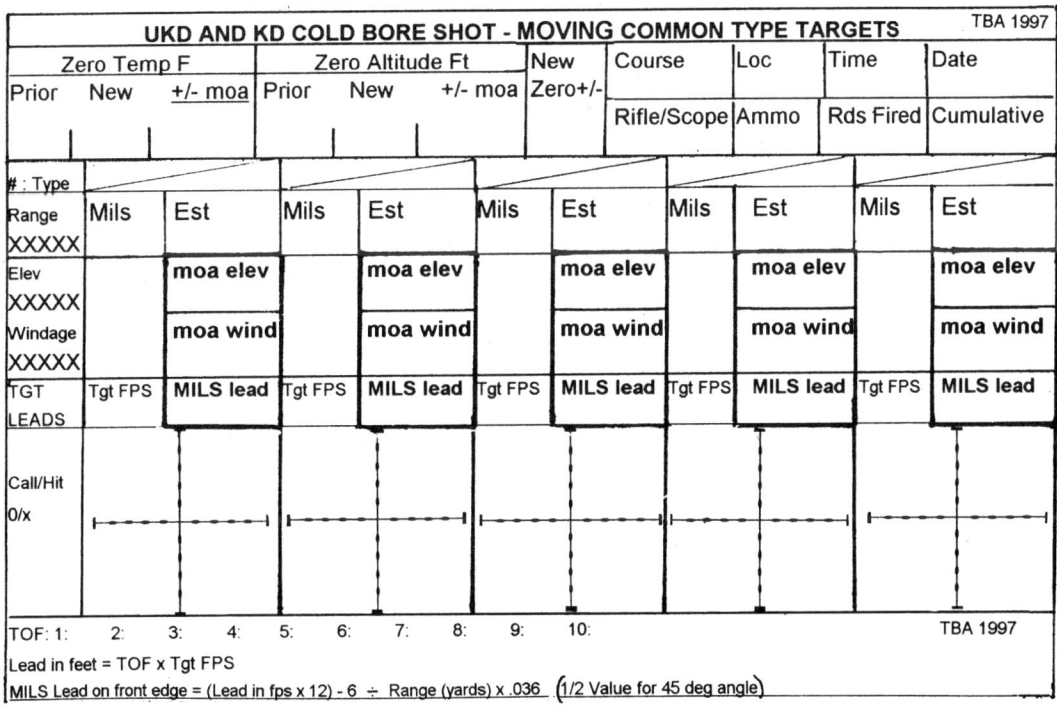

Fig. 13.18. Firing record for any UKD and KD moving target at any distance. Indicate lead in MILs by drawing a single line representing target's front edge and height observed. Fill in TOF at bottom for yards or meters range for your particular rifle/ammunition. (Mike R. Lau)

Fig. 13.19. Firing record for stationary targets at any yard/meter distance. (US Army National Guard Scout-Sniper School)

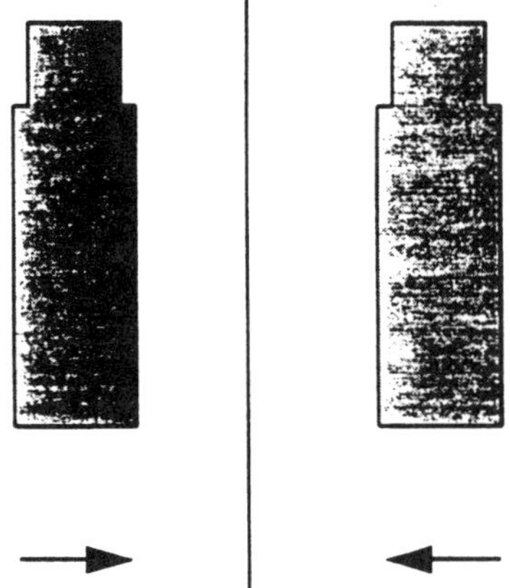

Fig. 13.20. Firing record for moving targets at any yard/meter distance. (US Army National Guard Scout-Sniper School)

CHAPTER 13: Managing The Log/Data Book

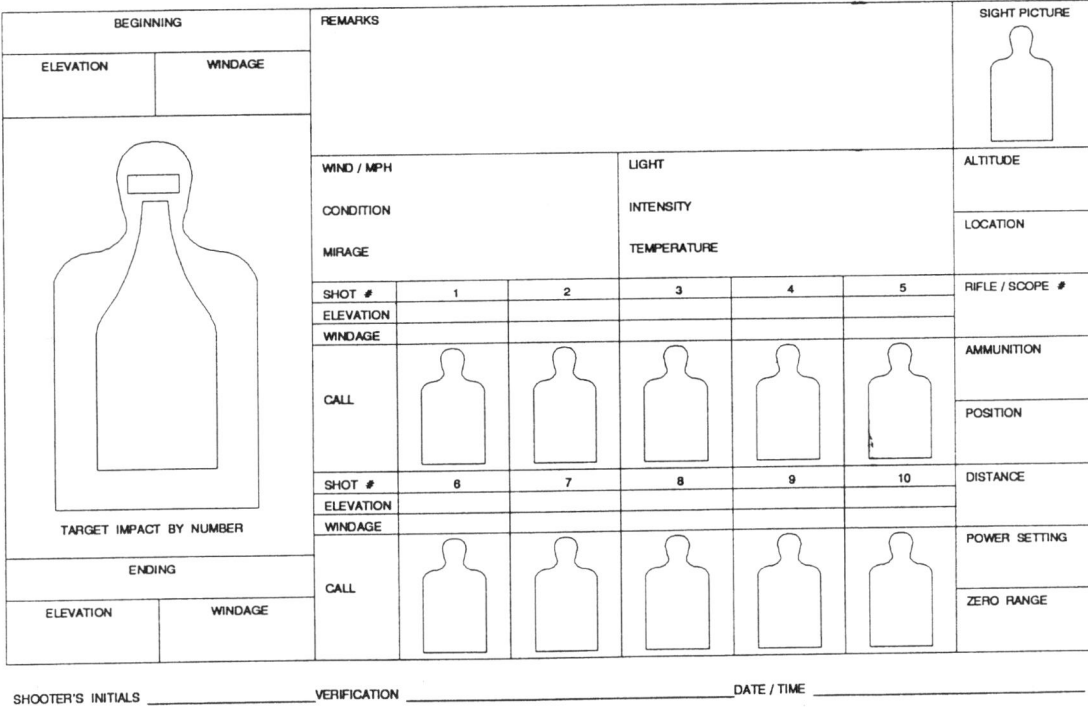

Fig. 13.21. Firing record for qualification. Use for single distance. Record data and location of each round fired on each target. Good example of 10 target "E" type. (unknown source)

MOVING TARGET DATA

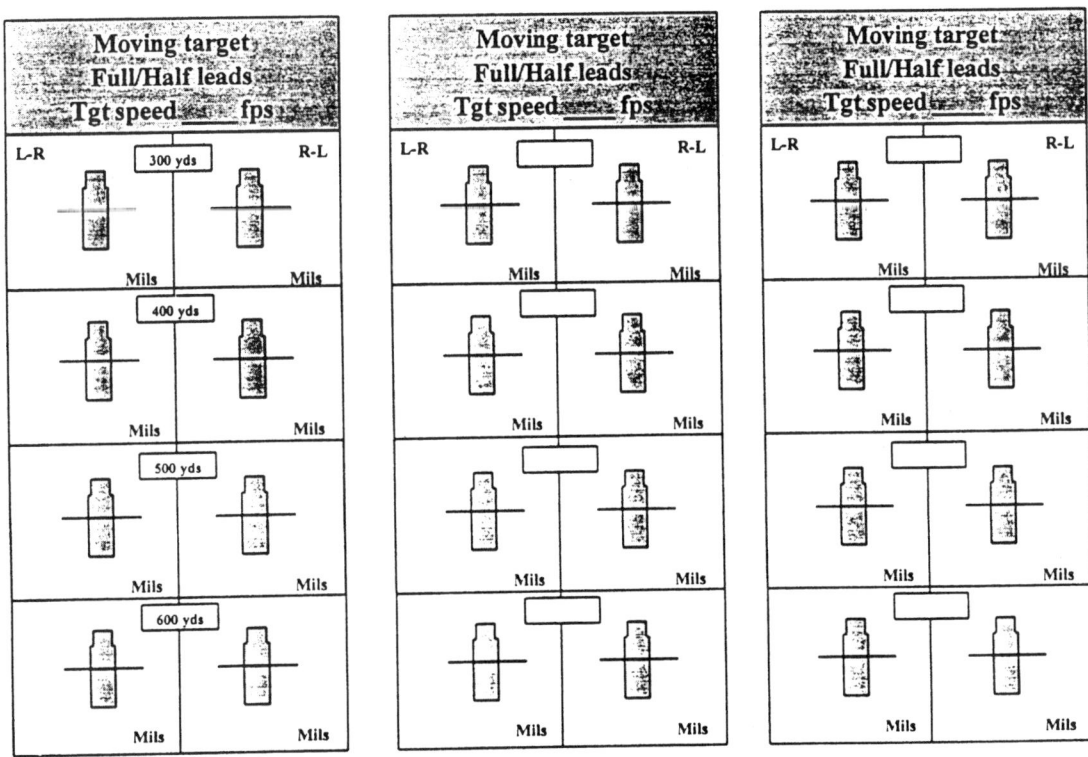

Fig. 13.22. Firing record for multiple moving target at different distances. Use for any height target. (US Army National Guard Scout-Sniper School)

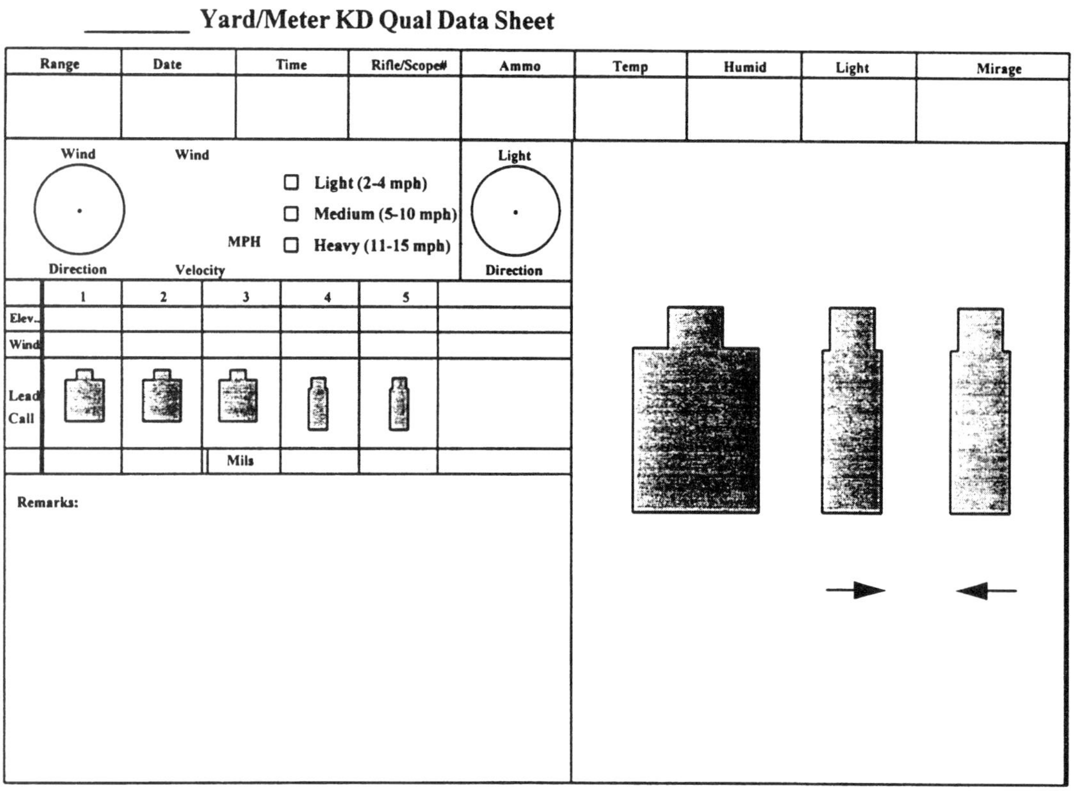

232 The MILITARY and POLICE SNIPER

Fig. 13.23. Firing record for both stationary and moving targets at any yard/meter distance. (US Army National Guard Scout-Sniper School)

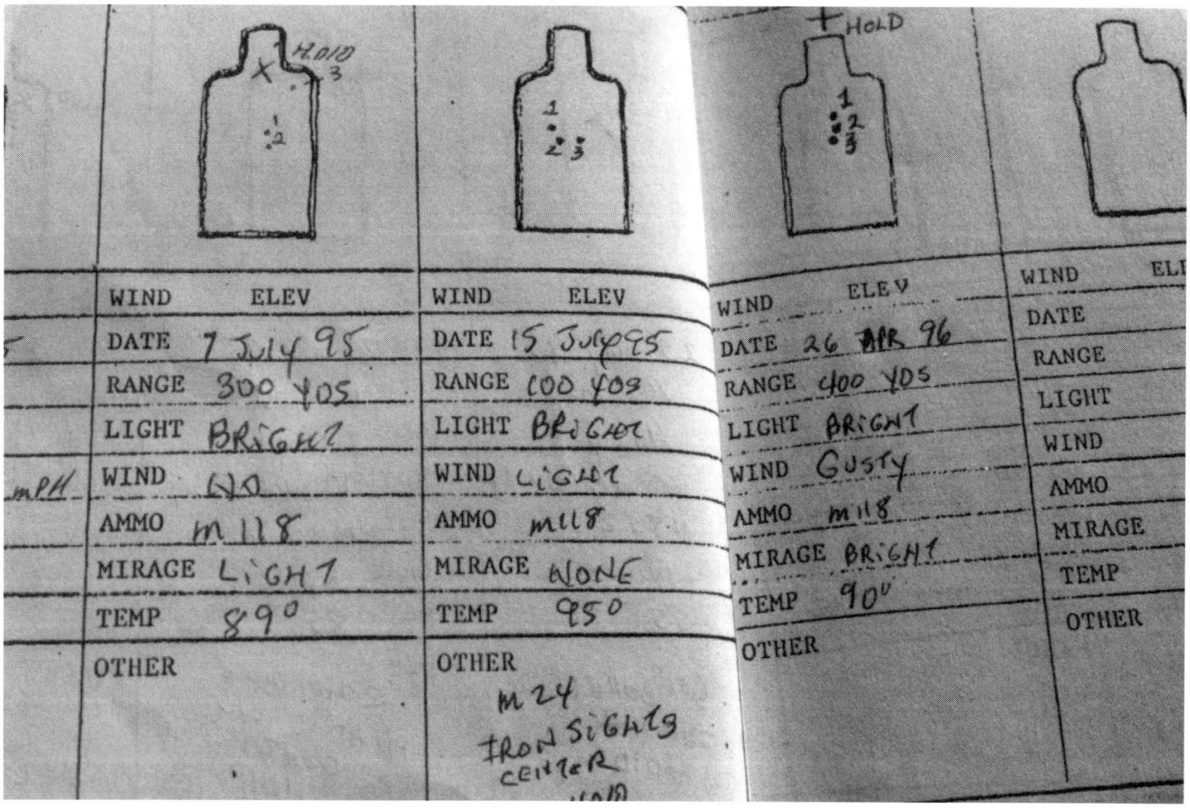

Firing record from SFC Pete Carpentier's log book page. Simple, easy to use and find, does not waste paper or clutter book with pages that have little data on them. (Courtesy Epitacio Carpentier)

CHAPTER 13: Managing The Log/Data Book

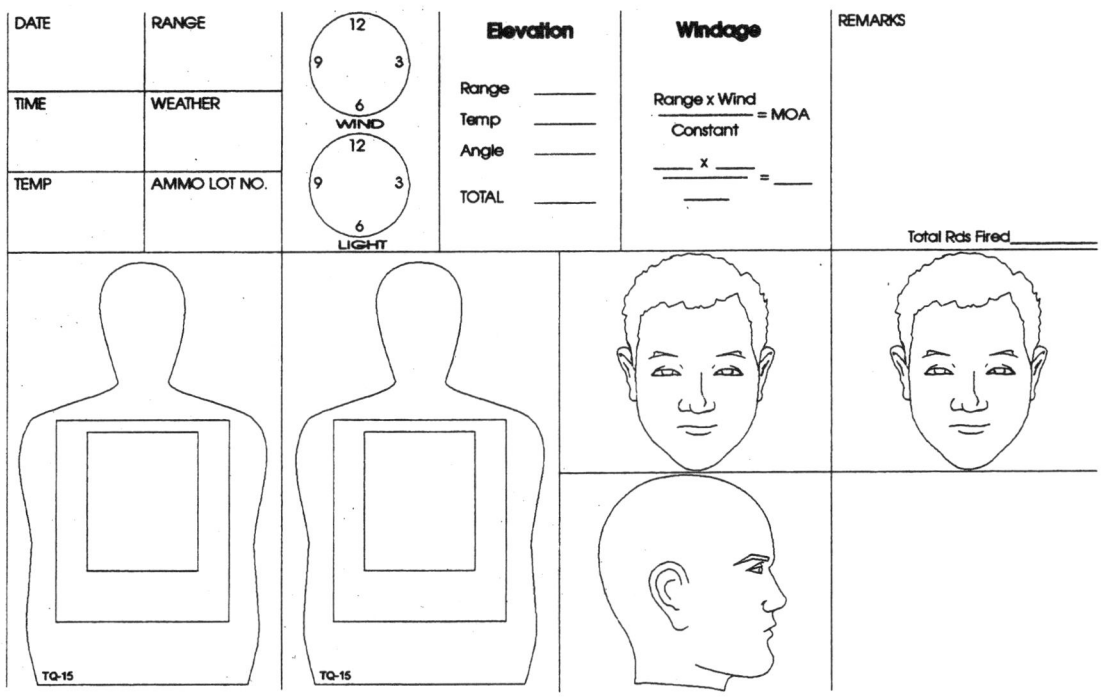

Fig. 13.24. Firing record for Arlington Police Department. For urban and field use, this page depicts the type of targets used by law enforcement. Pertinent information only is required for use in your local area keeps page simple and easy to use. Classic simplicity designed by Keith Scullins. (Arlington, Texas, Police Department)

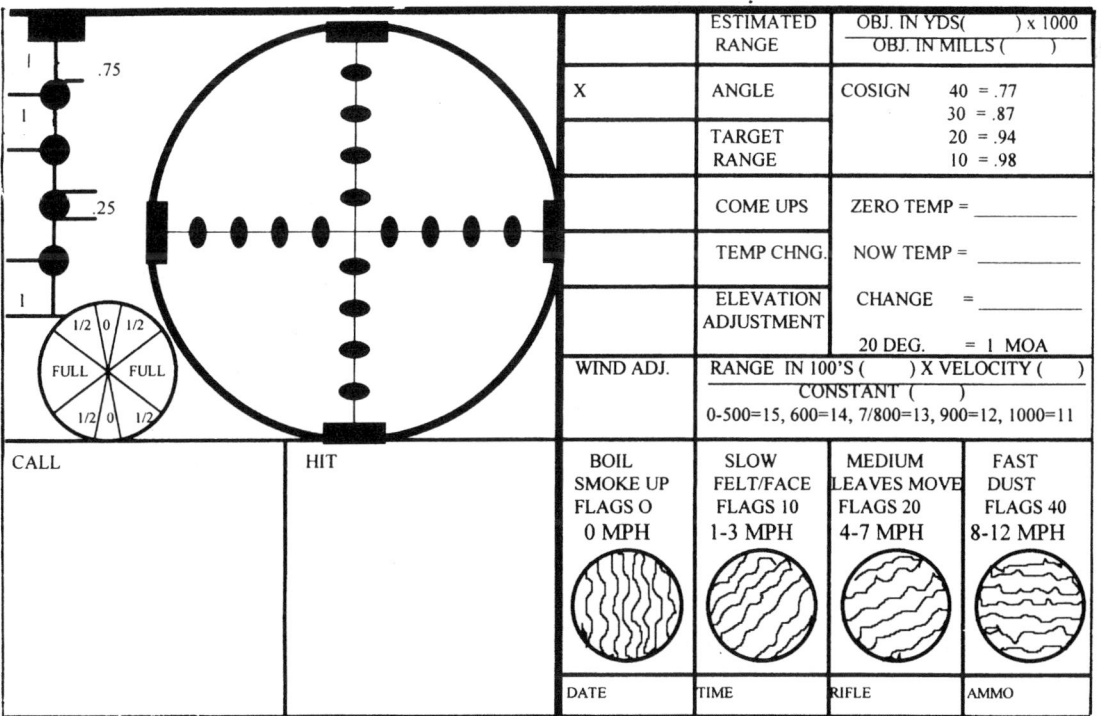

Fig. 13.25. Firing record designed by SGT Bob Newton, Dallas PD. Another good example of law enforcement style data record page collecting only required information with minimum required formulas to make shot. (Robert Newton)

234 *The MILITARY and POLICE SNIPER*

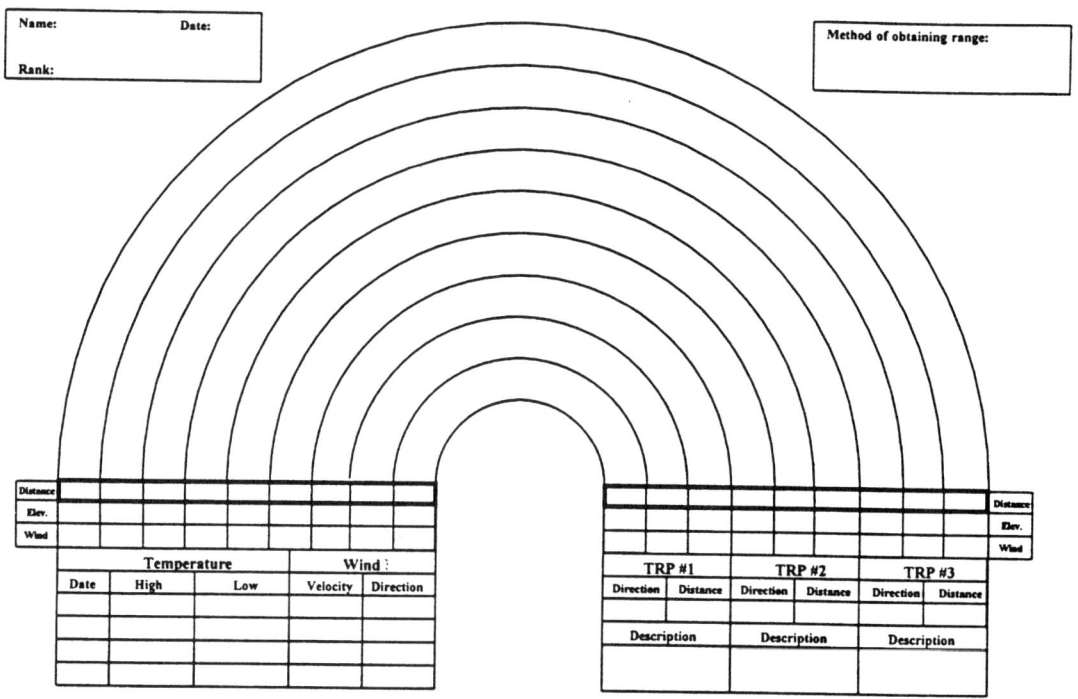

VIII. Range Card.

Fig. 13.26. Range card. TRP is "target reference point" and targets are numbered when drawn in location. (US Army NGSSS)

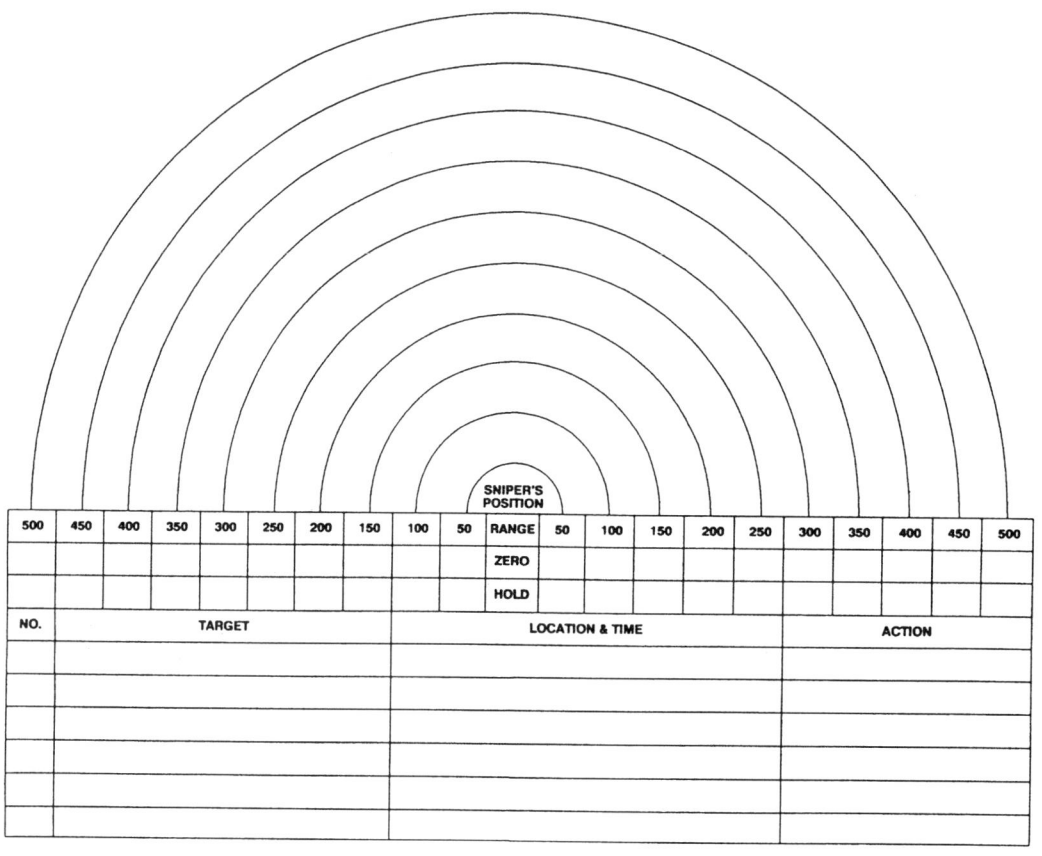

Fig. 13.27. Range card. (US Army)

CHAPTER 13: Managing The Log/Data Book

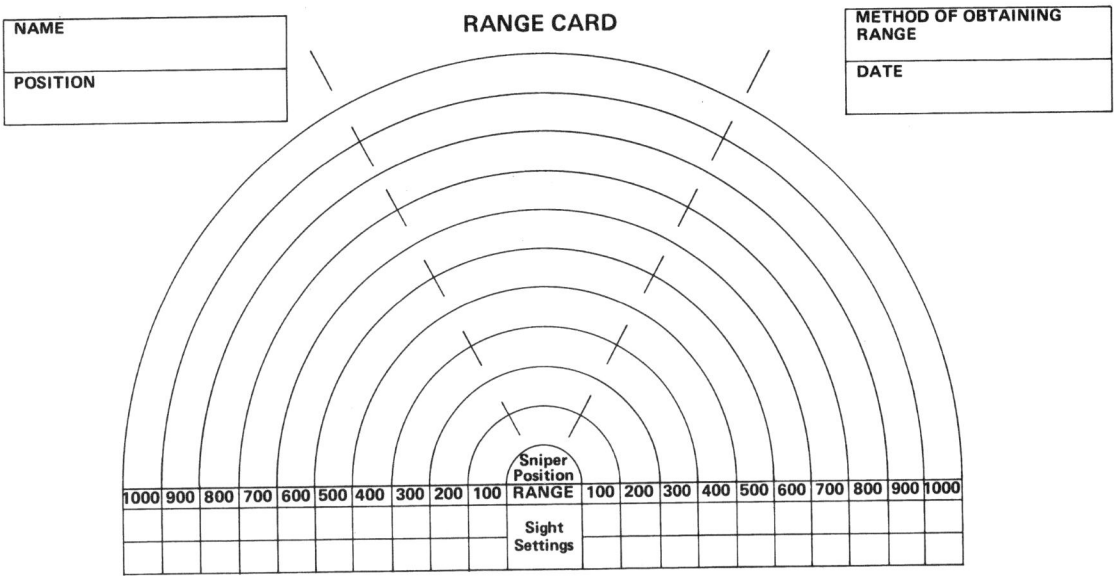

Fig. 13.28. Range Card. (USMC)

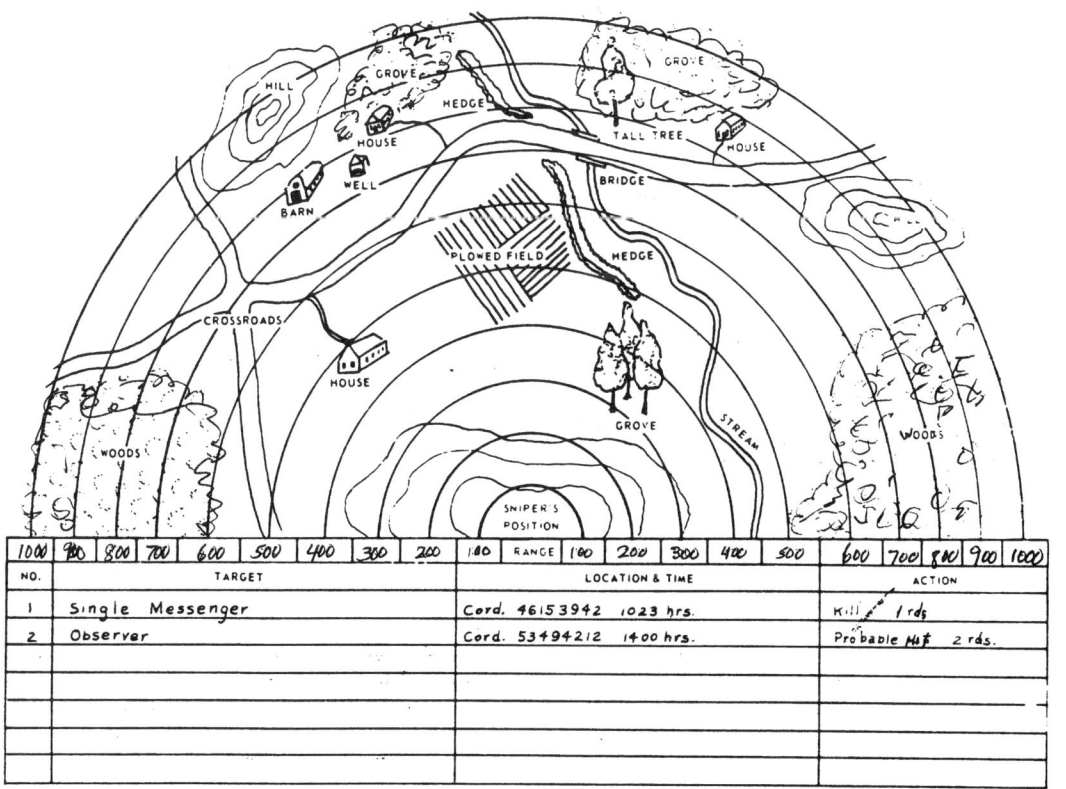

Fig. 13.29. Range card filled out. (US Army)

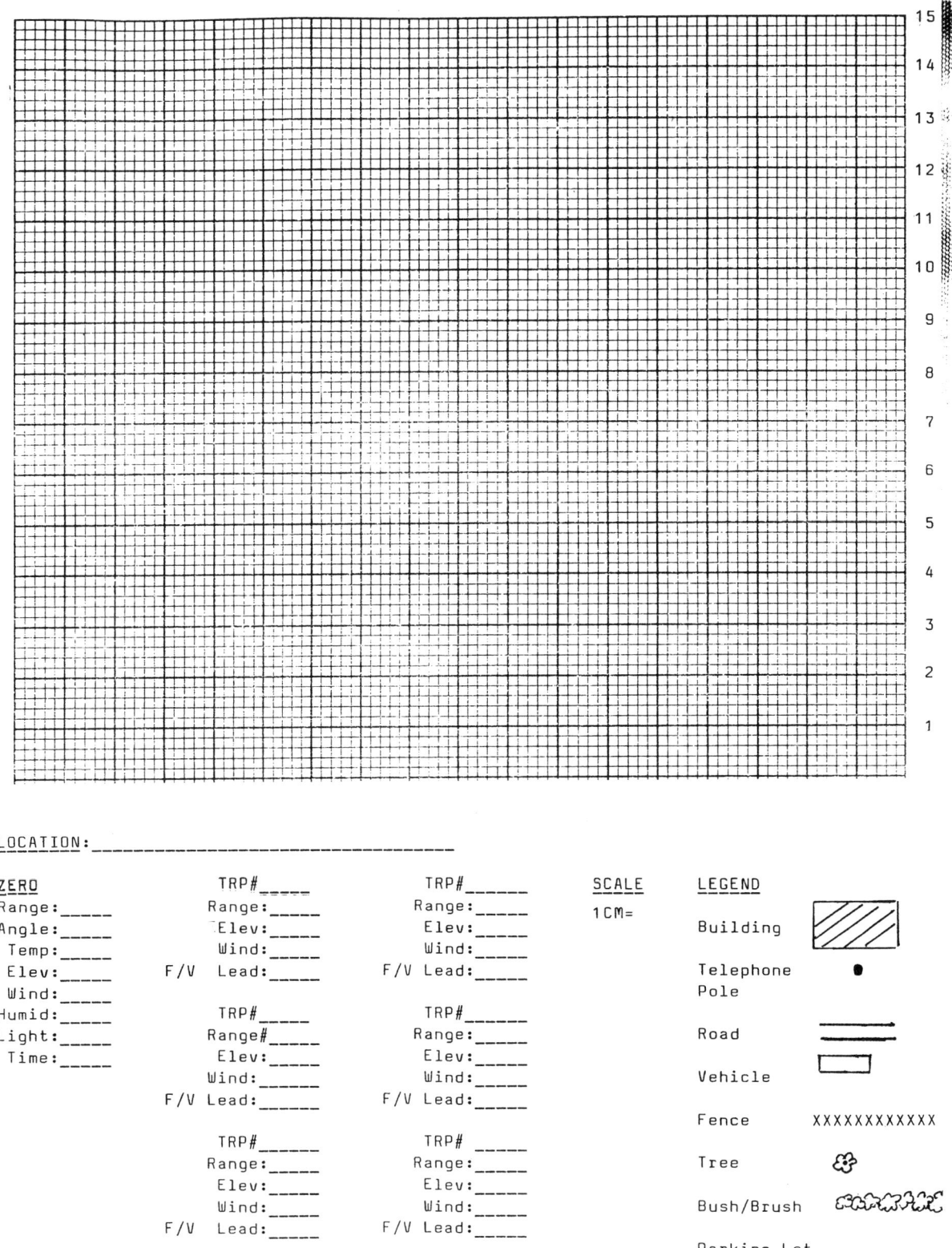

Fig. 13.30. *Centimeter Scale Range Card used by Marine Corps and Law Enforcement.
(Sam Chesnut)*

COMMON FORMULAS
RANGE, WINDAGE, LEADS, ANGLE FIRING
(IF YOU DON'T USE TABLES OR CHARTS)

Range Estimation

Range in *METERS* = Target Size in Inches x 25.4 / Mil Size of Target
or
Range in *METERS* = Target Size in Meters x 1000 / Mil Size of Target
Range in *YARDS* = Target Size in Inches x 27.77 / Mil Size of Target
or
Range in *YARDS* = Target Size in Yards x 1000 / Mil Size of Target
Size of Target in *METERS* = Range in Meters x .001 x MIL size
Size of Target in *YARDS* = Range in Yards x .001 x MIL size
Size of Target in MILs = Size of Target in Inches x 25.4 / Range in Meters
Size of Target in MILs = Size of Target in Inches x 27.77 / Range in Yards

Windage (For M118 SB, LR, & 175 gr .308 Match, 2600 fps)

$$\frac{\text{Range (YARDS) in hundreds} \times \text{Effective Wind Speed}}{\text{Constant}} = \text{Wind Adj (MOA)}$$

Constant: 100-500 = 15;   600 = 14;   700-800 = 13;   900 = 12;   1000 = 11

Moving Targets

Lead in Feet from Center of Mass = TOF (sec) x Tgt Speed (fps)
Lead in MILS from Leading Edge of Target =
$$\frac{(\text{Lead in Ft from Ctr of Mass} \times 12) - 6}{\text{Range (yards)} \times .036 \text{ OR Range (meters)} \times .039}$$

Slow Patrol = 1 fps; Slow Stroll = 2 fps; Walk = 4.4 fps; Fast Walk = 6 fps
Trot = 8.8 fps or 6 mph; Run = 14.6 fps or 10 mph

Angle Firing

Slope Range x Cosine of Slope Angle = Horizontal (Zero) Range

Angle:	30	40	45	50	60	70
Cosine:	.87	.77	.70	.64	.50	.34

IX. Formulas and Miscellaneous Data

Fig. 13.31. Sniper Math Formulas (USMC, US Army)

CONVERSION TABLE I: METERS and YARDS

To Convert	Multiply By
Inches to millimeters	25.4
Inches to centimeters	2.54
Inches to decimeters	.254
Inches to meters	.0254
Feet to centimeters	30.48
Feet to decimeters	3.048
Feet to meters	.3048
Yards to meters	.9144

To Convert	Divide By
Millimeters to inches	25.4
Centimeters to inches	2.54
Decimeters to inches	.254
Meters to inches	.0254
Centimeters to feet	30.48
Decimeters to feet	3.048
Meters to feet	.3048
Meters to yards	.9144

Yards to Meters
75 yards = 68.48 meters
100 yards = 91.44 meters
200 yards = 182.88 meters
250 yards = 228.60 meters
300 yards = 274.32 meters
400 yards = 365.76 meters
500 yards = 457.20 meters
600 yards = 548.64 meters
1,000 yards = 914.40 meters

Meters to Yards
75 meters = 82.02 yards
100 meters = 109.36 yards
200 meters = 218.72 yards
250 meters = 273.40 yards
300 meters = 328.08 yards
400 meters = 437.44 yards
500 meters = 546.80 yards
600 meters = 656.16 yards
1,000 meters = 1,093.60 yards

CONVERSION TABLE II: MOA and MILs

1 MOA ≅ 1"   @ 100 yards   1 MOA ≅ 1.13" @ 100 meters
1 MOA ≅ 10" @ 1000 yards   1 MOA ≅ 11.3" @ 1000 meters
1 MIL ≅ 3.6" @ 100 yards   1 MIL ≅ 3.9"   @ 100 meters
1 MIL ≅ 36" @ 1000 yards   1 MIL ≅ 39.37" @ 1000 meters

1 MOA ≅ .28 MIL
1 MIL ≅ 3.6 MOA

CONVERSION TABLE III: GRAMS and POUNDS

To Convert	Multiply by
OZ. to Grams	28.00
LB. to Kilograms	.45
Grams to OZ.	.35
Kilograms to LB.	2.2

CONVERSION TABLE IV: TEMPERATURE

To Convert F° to C°
subtract 32 and multiply by .55

To Convert C° to F°
multiply by 1.8 and add 32

Fig. 13.32. Unit of Measures Conversion Tables (USMC, US Army)

CHAPTER 13: Managing The Log/Data Book

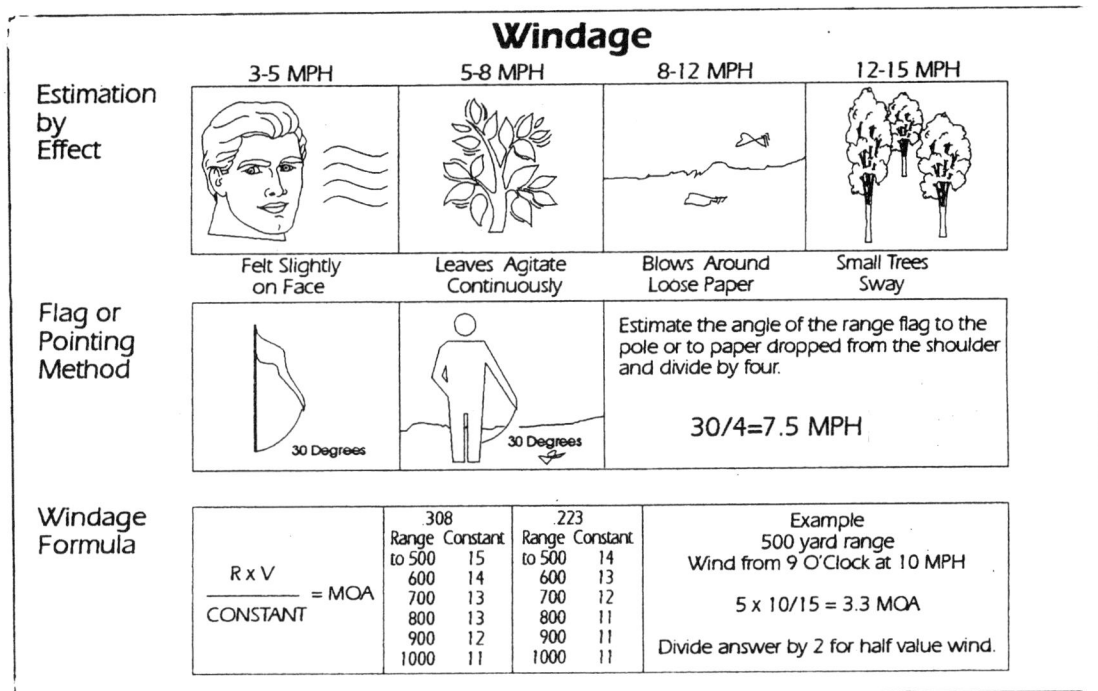

Fig. 13.33. Data page from Arlington Police Department Log Book created by Keith Scullins. Note range constants for .223/5.56mm ammunition for calculating windage. (Arlington Police Department)

Rifle Round Count Data Sheet
Rifle serial number_____

Date	Type	Lot #	Daily total	Total	Date	Type	Lot #	Daily total	Total

Fig. 13.34. Round Count Log is sheet in Log Book. (US Army NGSSS)

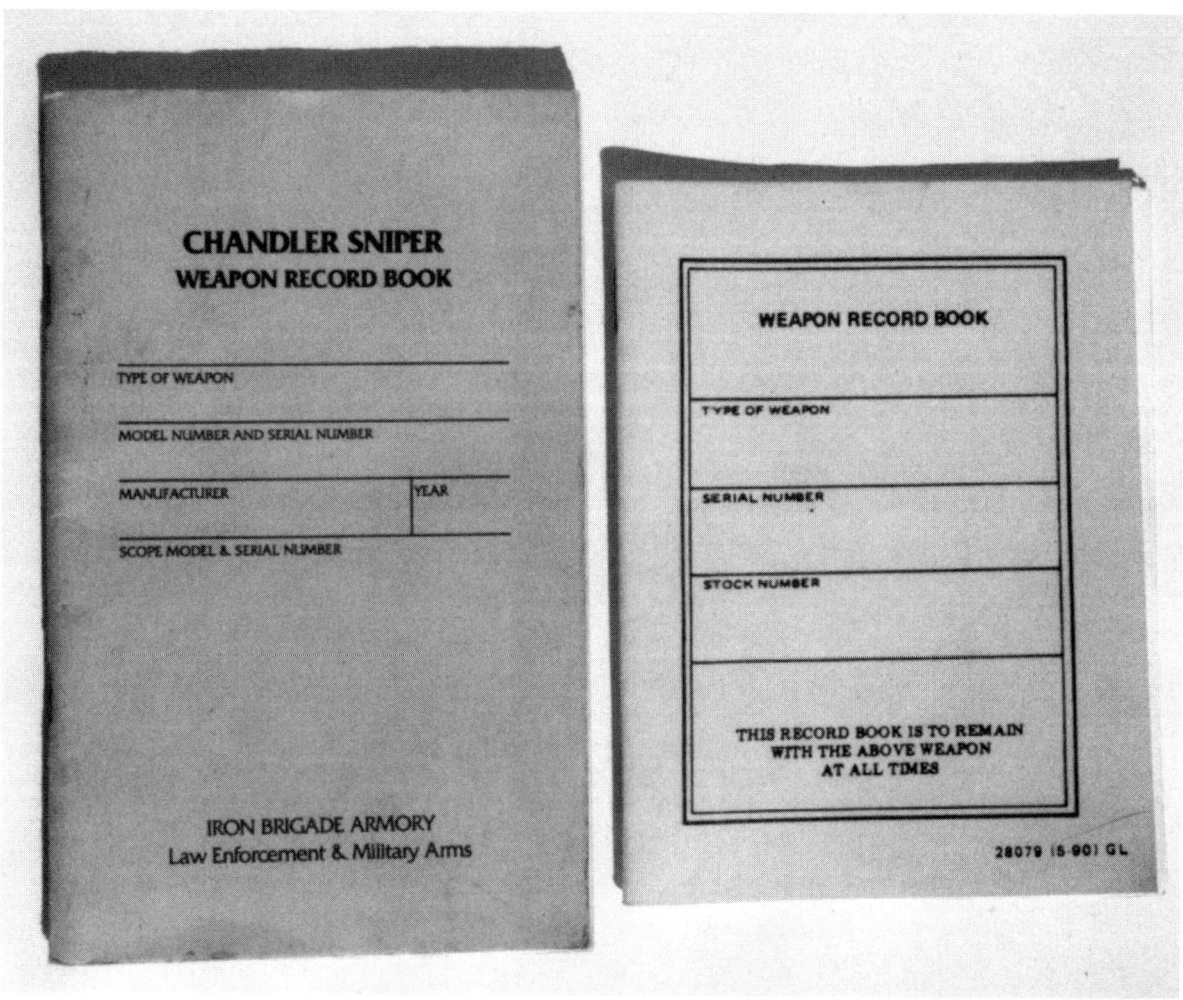

Weapon Round Count and Maintenance Record Books. Left is Norm Chandler's Sniper Weapon Record Book which has both round count and weapon maintenance record in it. Earlier Navy type on right has round count with comments. Separate books are usually required for weapons that are unit owned so that it follows the rifle when turned in for maintenance or passed to another shooter.

CHAPTER 13: Managing The Log/Data Book

Texas National Guard, SFC Pete Carpentier (right) and SPC Alan Donaldson. (Mike R. Lau)

CHAPTER 14: TACTICAL SCENARIOS

In the previous chapters, we covered weapons and necessary data required by the military and LE sniper to engage targets at different distances under different conditions. The data required by the sniper to engage a target seems overwhelming and tedious at times and in some situations it is. Only through practice will you eventually develop a system for obtaining the information quickly and eliminate what you don't need. A lot of LE agencies and military units develop priority lists of info needed before the shot can be made. Besides shooting data, you should also include safety to bystanders, hostages, other team members, and some basic tactics. Gather info according to priority until you are forced to take the shot then you know you have at least covered the highest priorities. In this chapter we are going to detail several scenarios to show you how sniper teams use their weapons and data in engaging targets.

Numerous books, magazine articles, and videos have explained the use of the MIL-dot reticle and how to estimate range. For someone who has never used the MIL-dot range finding telescope, it's use appears complicated and difficult to imagine. Up to now, the only way to learn how to use it is by actually using it. It is hoped that the examples of sniper target engagements in this chapter will give you a better idea of how to use the MIL-dot and how the sniper considers some of the other factors in his shooting. For my money, the MIL-dot reticle is the most useful of all range finding reticles to use because it will handle most situations easily and accurately.

All photos were taken by author except as noted.

If you have sniper tips, trivia, stories, etc., and you would like to share these with others, please send them to author at Texas Brigade Armory, 906 Middle Run, Duncanville, TX 75137. Any comments on the material in this book, corrections, or new information would be appreciated and we will update these in a future edition.

244 *The MILITARY and POLICE SNIPER*

SCENARIO 1: The Long Range UKD Cold Bore Shot

Every practice session and most sniper matches begin with the cold bore shot. For military sniper matches it is usually the long range unknown distance on a type E or F silhouette. For police practice it is usually the known distance at 100 yards with the face target or a small 1/2" - 1" paster. The target must be hit with one or two rounds. At distances less than 600 yards the first round should be a hit. For police, first round hit is mandatory for instant incapacitating shot.

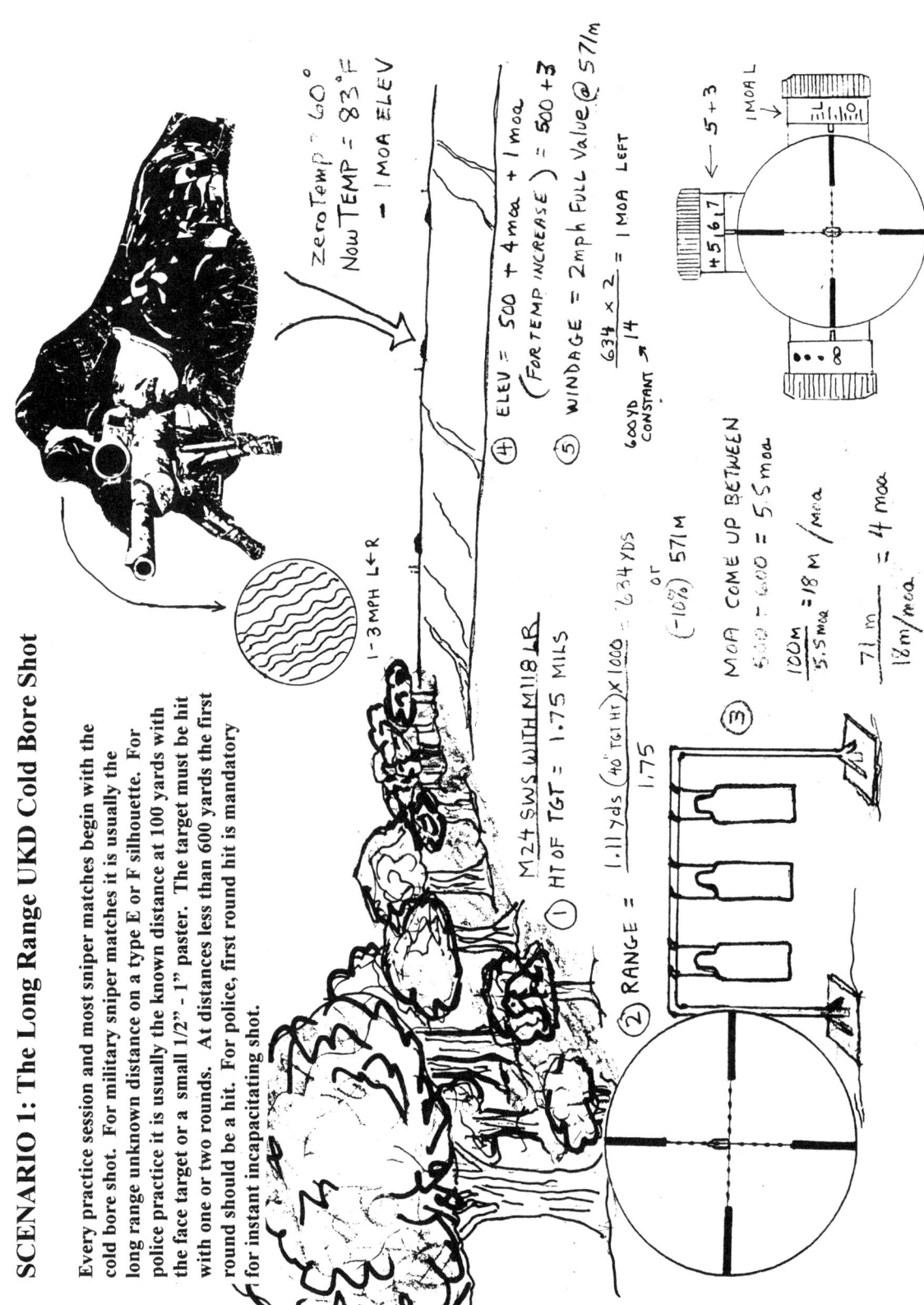

CHAPTER 14: Tactical Scenarios   245

"Iron Maidens" are used for long range UKD because you can see the splash of the bullet with a very high powered spotting telescope. Paint the metal a light color and put a 4" "V" circle or square in the center chest and a 1"x4" rectangle on the face area. This will break ties and cause shooters to be more precise. Hang the "maiden" from a stand that will hold-up in wind and bullet impacts, but still can be moved by two persons in a vehicle to other locations.

Sniper Tip: The Supported Prone Position. Marines and Army snipers learn to (1) Keep as low as possible. (2) The body should be kept in line with the rifle and not angled off to present less of a target to the enemy and gives more body mass to absorb recoil better. (3) The insides of the feet are flat on the deck. (4) Adjust the stock height with the non-shooting hand under the rear sling swivel.

246 *The MILITARY and POLICE SNIPER*

SCENARIO 2: High Altitude, Low Temperature, and Bracketing

A small UN force, working a border checkpoint in a small southern European country, has come under harassing fire by a guerrilla sniper. The sniper only makes irregular visits and never fires more than two rounds. Three days ago he wounded one of the Peacekeepers. A Marine Scout/Sniper team from Camp Pendleton has just been assigned to a carrier task force in the Mediterranean. Early in the morning they are choppered in to the remote site, high on a mountain range, to support the UN team. After moving into a position overlooking the checkpoint they observe for activity where shots had been fired from before. They brought along a Leica laser range finding binoculars, GPS, and a recent satellite photo map.

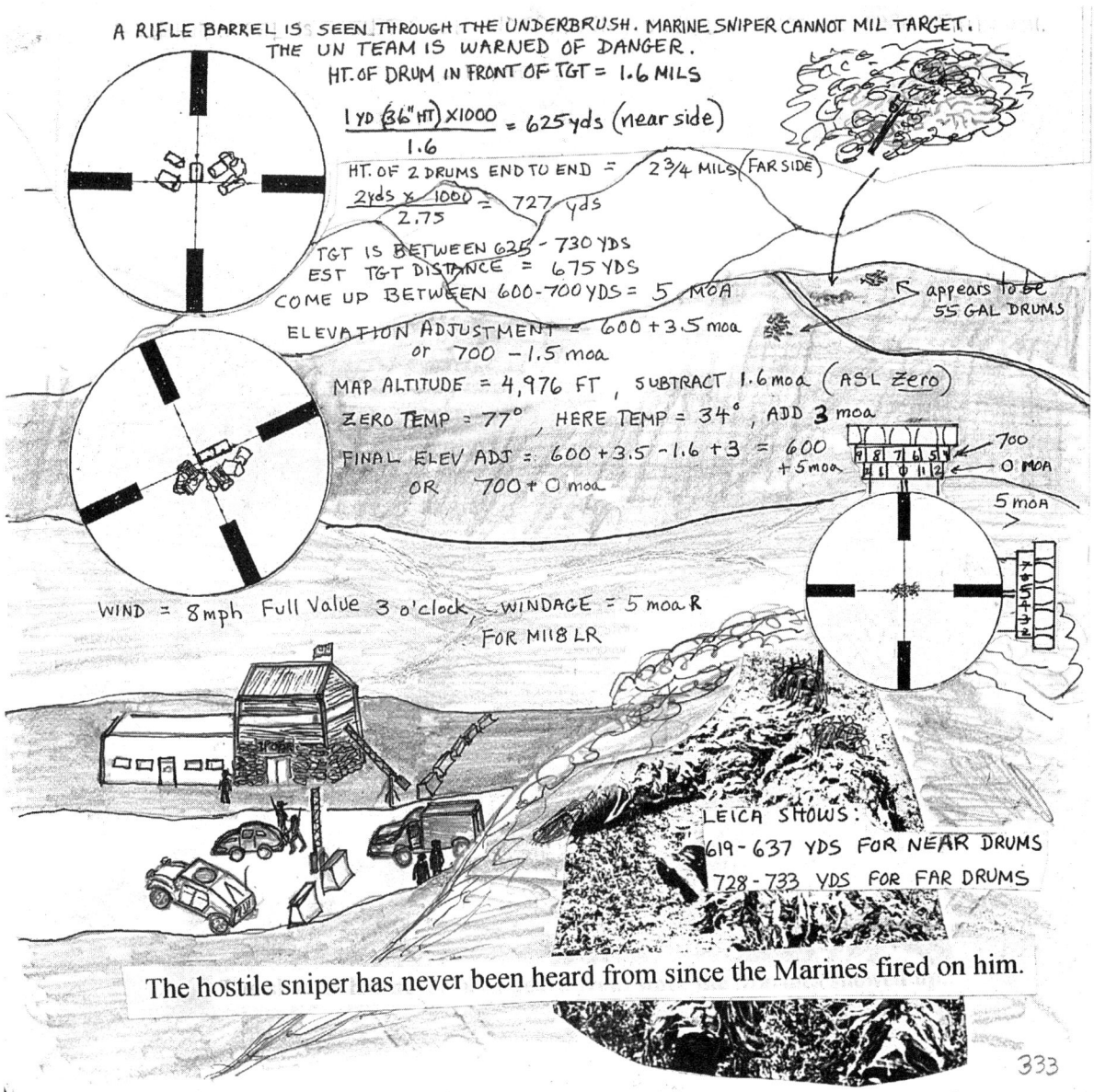

The hostile sniper has never been heard from since the Marines fired on him.

CHAPTER 14: Tactical Scenarios

Sniper Trivia: The First Marine Division Scout/Sniper School at Camp Pendleton, California, teach a prone position known as the ***Hawkins Position.*** This prone position was taught to British snipers during World War II. It involved placing the toe of the rifle butt on the ground with the forearm of the stock being placed over the hand, arm, or object. This firing position provided a very stable one for the rifle and also allowed the shooter to assume a very low profile. One drawback to this position is that you cannot get enough elevation on level terrain to sight on target. It works better if you have a low bank to raise the forend or a shallow depression for you to lay in. If on fairly level terrain the view to target must be fairly open with short grass, sand, gravel, etc.

Sniper Tip: "The Right Environmental Attitude." Military snipers are more affected by weather than the competitive shooter. The sniper, cannot decide what days he is going to go to the "range" and he cannot head for the dugout if the weather gets bad. He does not have a shooting season. His "game" requires playing against the elements and the rules include any kind of weather. He must always be in total awareness of his environment. Besides changes in weather, he must be aware of animals, insects, leaves, and any other activity around him that can distract him. He will have to sit motionless for long periods of time with no radio, TV, books (except data book), cigarettes (snipers don't smoke), or other creature comforts. He will curse and get "pissed" a lot. His mind will wander and he will lose his concentration. But, he will remind himself of his mission, his own personal survival, and maintain a positive mental attitude. When the time comes to shoot he will block out all distractions and concentrate on the shot.

SCENARIO 3: UKD, GLASS SHOOTING, ANGLE FIRING, and DIGNITARY PROTECTION

A famous Italian film star is attending his daughter's wedding in a small Mid-West USA town. The local police chief has been told by the film star that he got a death threat letter stating that he would be killed if his daughter got married. He believes the threat came from the daughter's ex-lover who is a member of a mob. The police chief assigns his entire force, four officers, to cover the wedding. Two officers cover Main St. at opposite ends from roof tops because they own rifles. The police chief and the other two provide security on the ground along with the film star's body guards.

As the people are leaving the church, inexperienced Officer Duffey notices what looks like a rifle barrel slowly inching out a partially open sliding glass window from the top floor of the Cheap Motel. Then it disappears back inside. He can't see the shooter because the curtain is completely closed. His 2-way doesn't work so he can't warn anyone. To him, distance to target appears to be anywhere from 200 to 350 yards. He has no idea what shooting through glass will do. He does have a mil-dot 3.5x10 scope and he did attend a 1 hour class on police sniping a week ago and he remembered a couple of things.

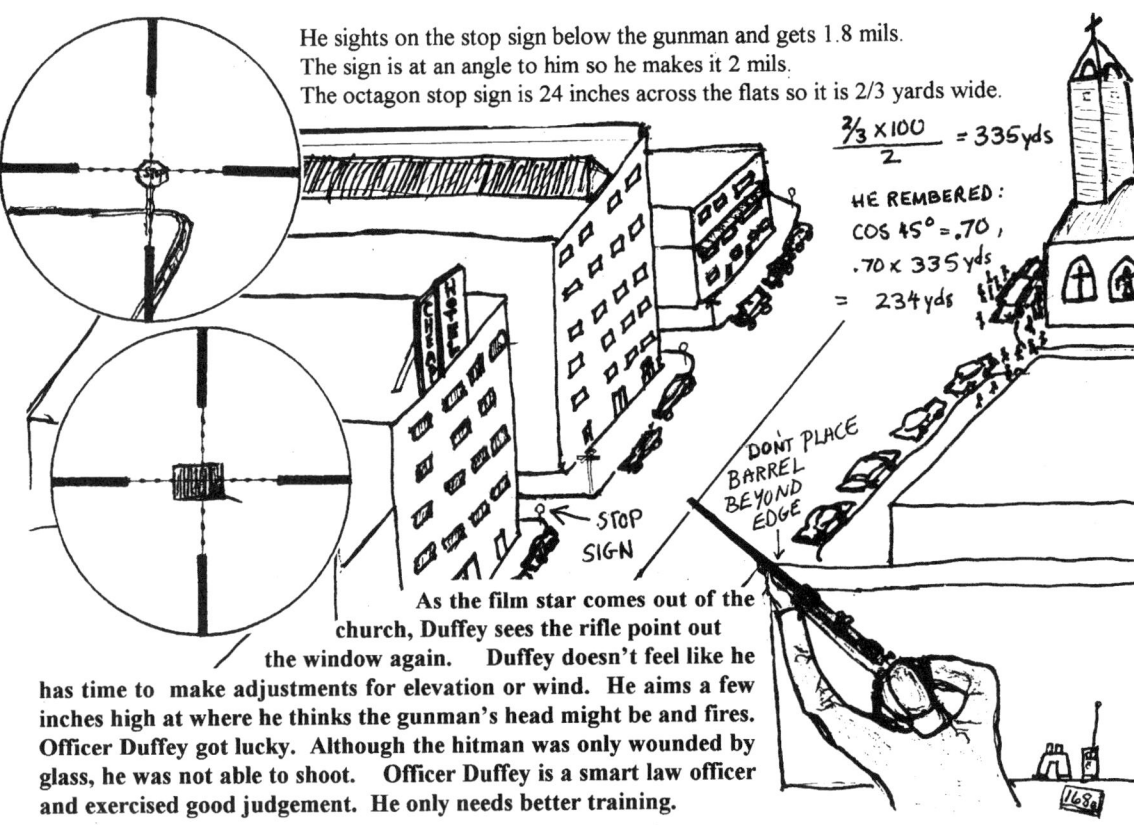

He sights on the stop sign below the gunman and gets 1.8 mils.
The sign is at an angle to him so he makes it 2 mils.
The octagon stop sign is 24 inches across the flats so it is 2/3 yards wide.

$$\frac{2/3 \times 100}{2} = 335 \text{ yds}$$

HE REMBERED:
COS 45° = .70,
.70 × 335 yds
= 234 yds

DONT PLACE BARREL BEYOND EDGE

STOP SIGN

As the film star comes out of the church, Duffey sees the rifle point out the window again. Duffey doesn't feel like he has time to make adjustments for elevation or wind. He aims a few inches high at where he thinks the gunman's head might be and fires. Officer Duffey got lucky. Although the hitman was only wounded by glass, he was not able to shoot. Officer Duffey is a smart law officer and exercised good judgement. He only needs better training.

CHAPTER 14: Tactical Scenarios   249

Sniper Tip: Marine and Army Sniper Bolt Operation. Ejection of fired cases not caught by the shooter can be easily seen by the sudden movement and flash of the brass case. This is a major reason why semi-auto rifles are not used by military snipers. After firing, maintain your natural point of aim. Keep the butt of the stock in the pocket of the shoulder. With the body and elbows in place, slowly lift the bolt handle and slowly bring the bolt handle back. Slowly tilt the rifle to the ejection side to let the case fall directly below. Do this under your veil if you have one. Another method is to unlock and begin to move the bolt to the rear with your thumb. As the bolt moves rearward, use your index and middle finger to pinch the case when it pivots out of the receiver. This is called "bolt manipulation" and is the preferred method.

Sniper Tip: The Bad Habit of Not Seeking Cover. Military training manuevers that use blank ammunition conditions the soldier to **_not_** seek cover. No one gets hit, no bullets whiz by, and it takes some effort to hit the ground or get down and come back up if you are wearing a lot of equipment. It is not instinctual, so make it a habit to get down or get behind cover when you hear gunfire (except on the range) even if the aggressor is firing blanks in a tactical exercise. Experienced troop leaders will yell at you if you don't, and in real situations you can get yourself killed.

250 *The MILITARY and POLICE SNIPER*

SCENARIO 4: ESTIMATING TARGET RANGE WITHOUT USING COSINE and USE OF COVER.

Two SWAT officers respond to gunman in tower shooting at people below.

Officer 1 cannot see gunman at top of tower. Officer 2 can see target from where he is and needs target distance. Officer 1 sees the concrete bench near him and the other officer and also one at the base of the tower building.

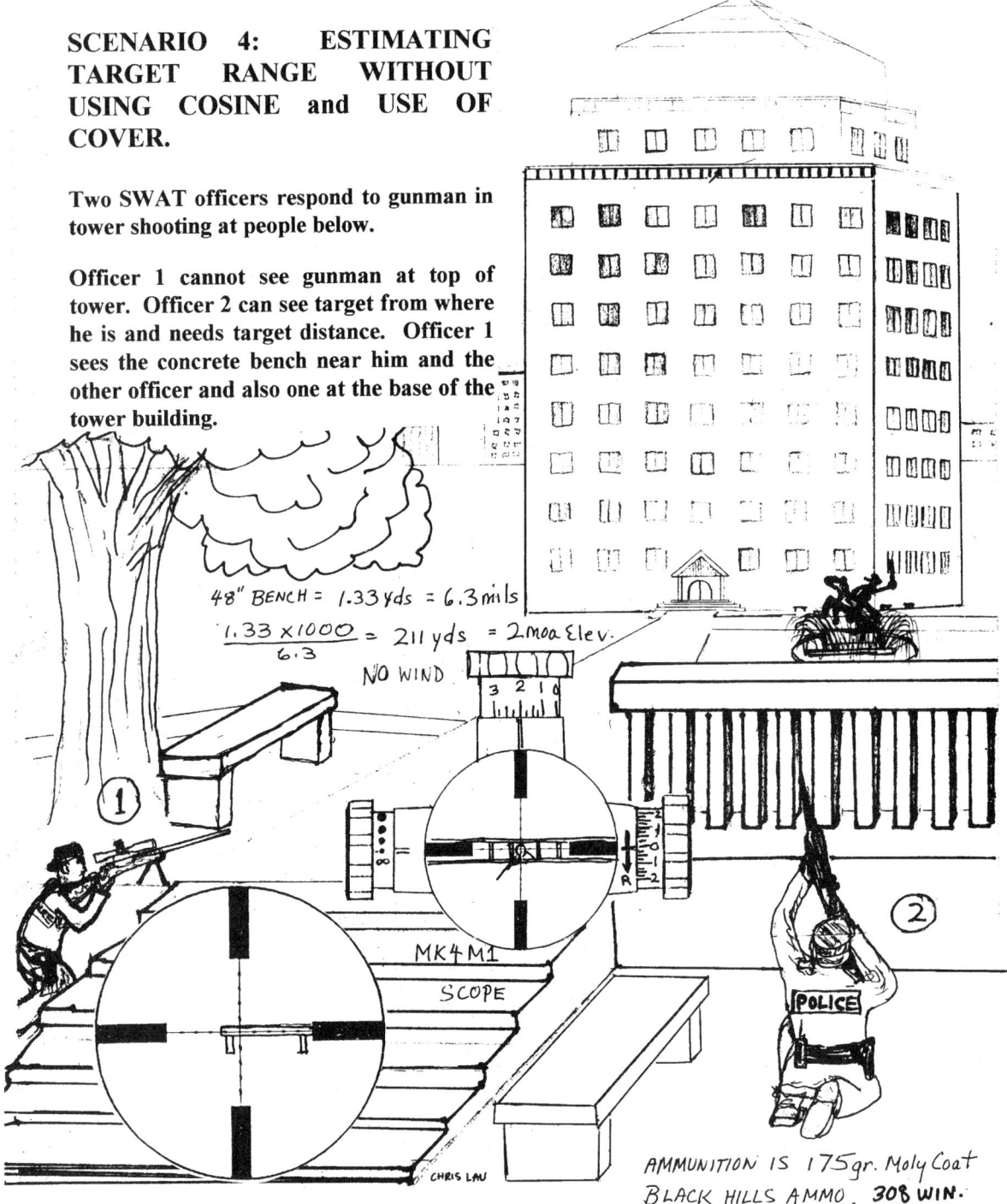

Sniper Field Craft: Field expedient method for camouflaging your rifle. (Sean Little) Wrap the blousing rubbers (elastic band for holding the bottom cuffs of your trousers to the top of the boots) around the barrel and stock and stick grass and twigs under the band.

CHAPTER 14: Tactical Scenarios

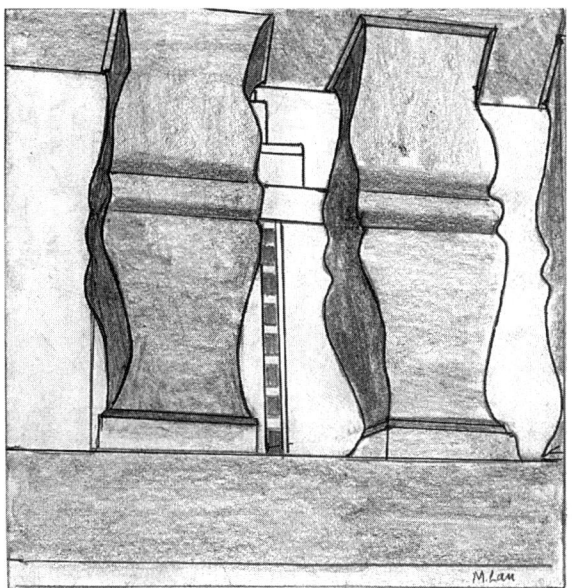 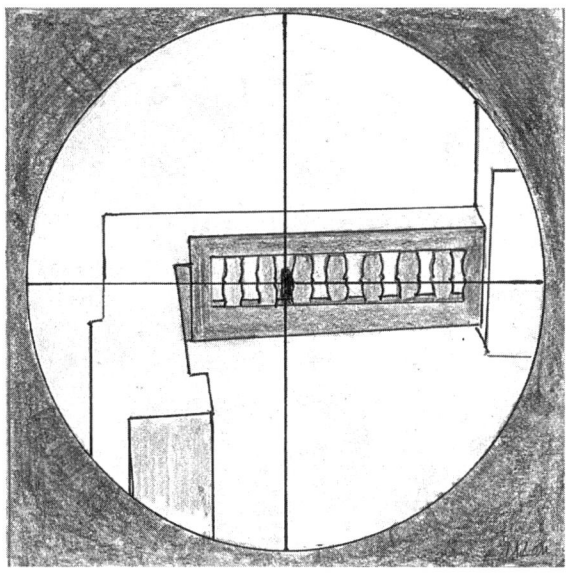

The drawing on the left is a view of the UT Austin tower as seen from behind a concrete railing. An Austin police officer took cover behind this railing believing he was safe from Charles Whitman's deadly gunfire. The drawing on the right is the view Whitman had from the observation tower as seen through a rifle telescope. The officer was killed with one shot from Whitman's rifle. These drawings depict the vulnerability of someone trying to take cover from a gunman with a scoped rifle. "Cover" means protection from small arms and artillery fire. "Concealment" means you cannot be seen, but not protected by direct or indirect fire. A large tree trunk may give you protection from direct fire, but the leaves and small branches may only conceal you from overhead gunfire. Law officers are taught not to get careless when moving about behind walls and other cover with a barricaded suspect in a high position. It is easy to forget that you can be seen by the gunman if you move just a little ways from the wall because you are not looking up while you are moving about.

Sniper note: Shooting with the sling. The Marine Corps teaches basic recruits how to use the sling with the M16A2 rifle. The Marine scout/sniper is taught to use either the "one point suspension" (sling attached only to the front swivel), or the "two point suspension" technique which is the hasty sling. The leather sling is preferred because the web and nylon slings are too light and less stable in supporting the shooter. When the sniper decides to use the sling, he should make these considerations: (1) Sling manipulation must be done without excessive movements and may have to be done slowly when in a hide. (2) If the sniper changes his shooting position, he must recheck the sling position, tension, and adjust. (3) The sniper must minimize the heart beat which can be transferred to the scope through the sling. (4) Sniper needs to practice loading and reloading techniques with the sling being used. (5) The sling may have to be readjusted between shots. (6) Make sure the rifle's zero is the same for using with and without use of sling or know difference. Carlos Hathcock used position shooting without support except for the sling because he learned how to do that as a long range target shooter and of course, was deadly effective.

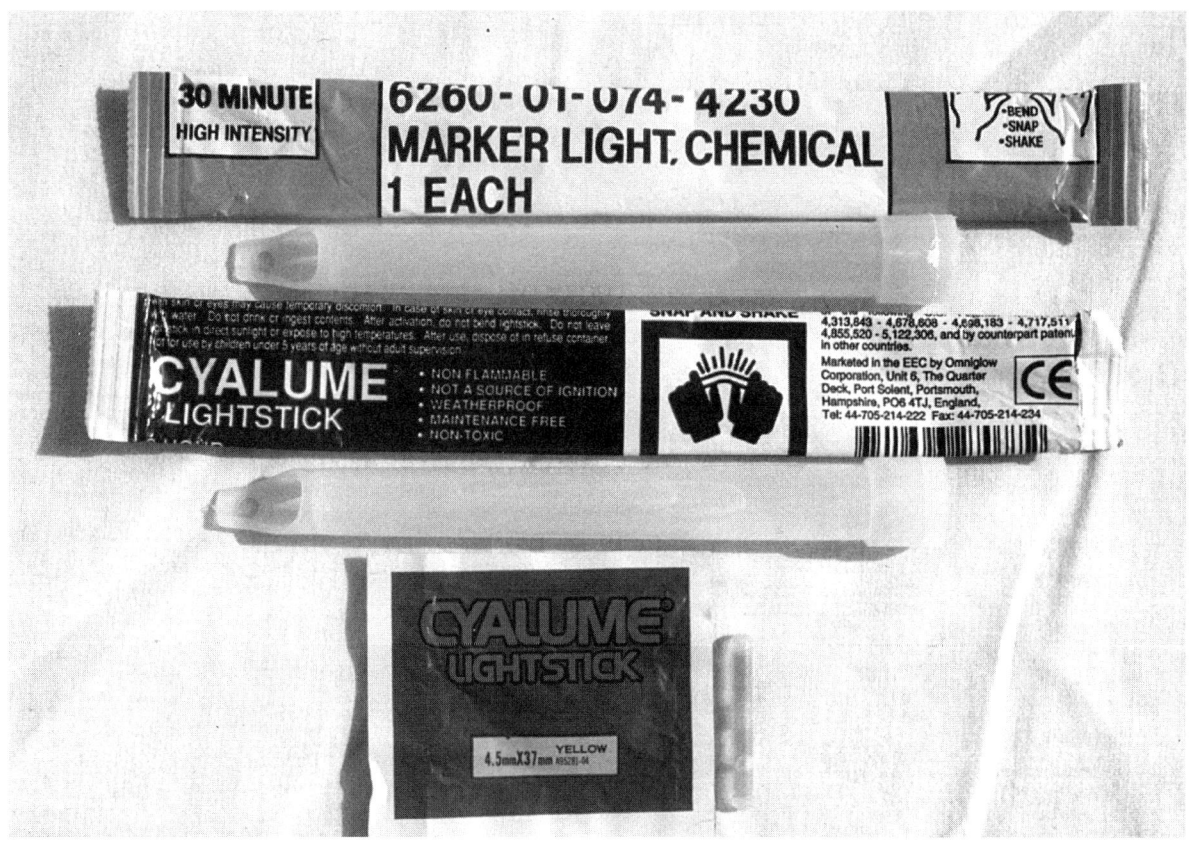

Sniper Tip: Night Reticle Illumination. (Submitted by Sean Little.) The large Cyalume Lightsticks are now available to the Army and the Marine Corps troops. The plastic container of the Lightstick is filled with two separate non-toxic liquid chemicals. One of the chemicals is inside another smaller plastic container surrounded by the other chemical inside the main outer container. When you bend the container, the inner case breaks open and allows the two chemicals to mix, creating a light source that will last for 1/2 hour to 8 hours. Shown at the top is the 30 minute High Intensity Marker Light. Below it is the 8 hour standard light. Marine snipers carry the small Lightsticks, aka "peanuts", with them to light the Unertl's reticle at night. Any kind of sticky tape is wrapped around the peanut to allow only a very small amount of the container to be exposed. After bending the peanut, it is placed into the eyepiece housing so that the small amount of light shines toward the reticle to illuminate it. The light will last several hours and is not visible to anyone else but the user. The "peanut" can also be used to illuminate the elevation and windage dials at night. For military use make sure the amount of light being used to reflect on the reticle is very small. If not done carefully, the light can be observed by someone with a night vision device or binoculars looking directly into the scope.

CHAPTER 14: Tactical Scenarios

SCENARIO 5: AWKWARD SHOOTING POSITIONS and THE PRECISION TOWER SHOT.

This is the angle shot requiring very high accuracy. Sniper schools and SWAT Team matches usually have this exercise. In some cases, where the equipment or safety doesn't allow it, the angle may be a shallow 20-30 degrees (targets 1 and 2) and at far or short ranges. Adjusting for the angle on these may not be necessary to still make a precision surgical shot. To get a steep angle like 40-50 degrees, the target is usually within 150 yards (targets 3 and 4). Like in Practical Pistol Matches, to increase the difficulty of the shot, the shooter may have to crawl through a small space, either height or width wise, and poke his rifle through a firing port. Sometimes the shooting position will be awkward, and is neither sitting, kneeling, full height standing, or prone. Sometimes the only apparent support for the rifle is 6" from the muzzle. Because the target may be a half inch paster, the awkwardness is designed to disadvantage the shooter with the 26" barreled 1/2 moa rifle. A firing space may favor short barreled "Scout" rifles or AR-15s, but the decrease in accuracy will put these snipers right back at equal or less advantage with the other guys. To make it even more difficult, the stage of fire is usually timed and the shooter must calculate distance, angle, and wind adjustments in a hurry. Time may be between 2-7 minutes including climbing the stairs, reloading, and catching your breath. You and your partner should divide the data tasks.

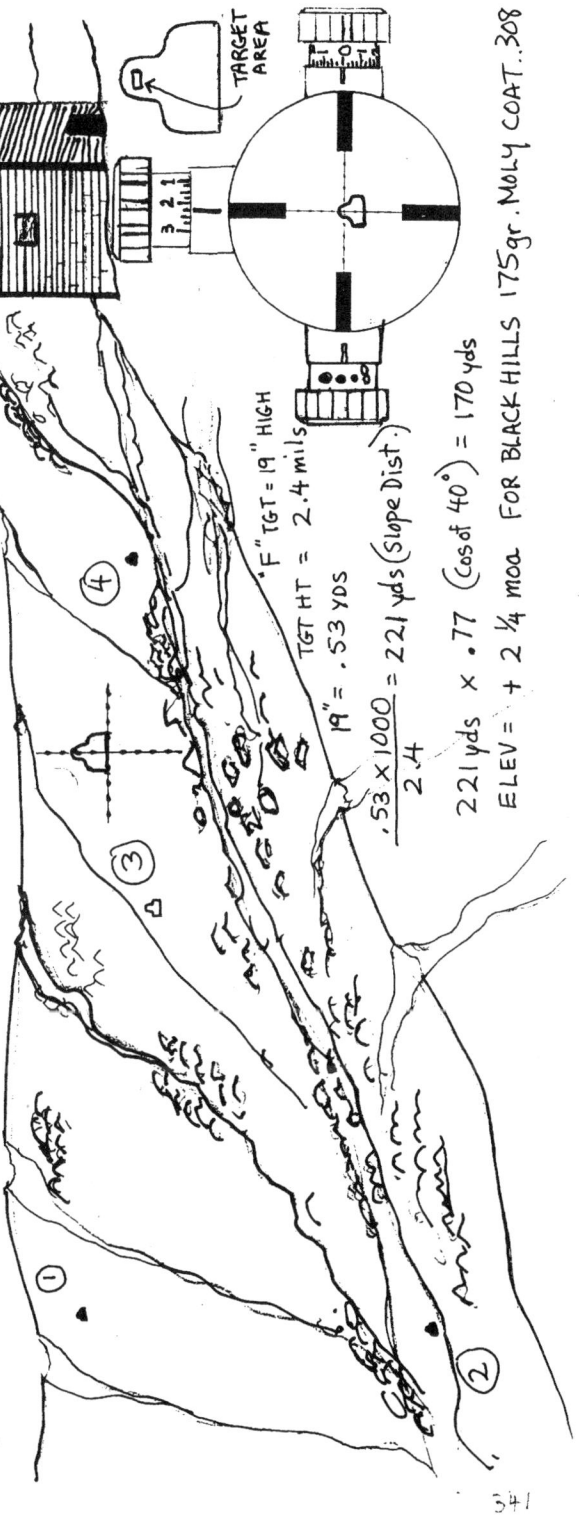

"F" TGT = 2.4 mils
TGT HT = 19" HIGH
19" = .53 YDS

$$\frac{.53 \times 1000}{2.4} = 221 \text{ yds (Slope Dist.)}$$

221 yds × .77 (Cos of 40°) = 170 yds
ELEV = + 2 ¼ moa FOR BLACK HILLS 175 gr. MOLY COAT .308

Sniper Fieldcraft: Another Use for Toilet Paper. (Sean Little) Inside the military MRE (Meal, Ready to Eat) is a small wad of toilet paper. And I mean small amount. Troops save it until they acquire enough to do nature's bidding. Marines snipers have found another use for it. Sometimes when you pull your rifle out of the drag bag, the scope lenses will fog up even if you have scope caps on. This usually happens when the temperature changes and there is a lot of moisture in the air. Use the toilet paper to dab the moisture off the lens. Care should be taken to not rub the lens roughly or hard with the paper. Water and dust will scratch the lens when rubbed and can remove the magnesium fluoride or HELR coating on the glass lenses.

SWAT TEAM MATCH BONUS POINTS

SWAT Team matches usually require two man teams. Shooting events usually require the use of the semi-auto service pistol, police 12 Ga. shotgun, assault rifle, sub-machine gun, and the sniper rifle. Each team member should be proficient in all and have a mastery of one or two of the weapons. There will be scoring and trophies for each individual weapons competition as well as for the team efforts. All of the events are usually timed and teams add time with each event to reduce their total match score To help teams that have had problems during the match, a final "bonus point" event will usually take place. This event usually requires precision so is done by the most proficient sniper with the sniper rifle.

Bonus Point Shoot Example No. 1: Time for this event may vary. You are allowed 5 shots at 100 yards from the prone using support. Targets are 1" circles worth 5 points per shot or 1/2" circles worth 20 points per shot. The way Sam Chesnut engages this event is to fire the first shot at one of the 1" circles to check body position and zero. Make adjustments, then engage the four 1/2 inch circles with one round each.

Bonus Point Shoot Example 2: Time var-

ies, but is usually short. You are allowed 5 shots at 100 yards and can use support. Targets are five 1/2 inch circles worth 10 points each or a price tag target hanging from a string and moving in the wind. This target is about 3/4"x1" in dimension and the string is 6 or 7 inches long. Hitting the tag is worth 25 points. Sam fires this event by firing at four of the 1/2" pasters to check zero and body position. Between shots at the 1/2" pasters, check to see what kind of pattern the hanging price tag makes in the wind. Whether the wind is blowing hard steadily, or gusting lightly, the tag will have a position that is stable for maybe a couple of seconds. Shoot the tag like a bobbing target. With your scope's crosshair, pick the spot that the tag is most stable at and wait for it to come back to that position. Firing at the lowest or hanging/rest position requires no

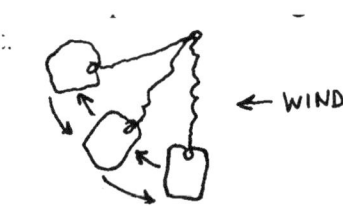

wind adjustment.

In a well publicized police sniping incident, a man had threatened to kill himself with a handgun while sitting on a chair in his front lawn. A police sniper watched the man raise the pistol to his head. When the man's arm got tired, the arm holding the pistol was lowered back to the lap. This occurred several times so the sniper aimed for where the pistol would be at when the man lowered it back to his lap. The sniper supposedly used frangible ammunition so that ricochet danger posed no problem to the man. The sniper shot the pistol out of the man's hand and ended the incident with no one injured.

CHAPTER 14: Tactical Scenarios

Any hits outside outer scoring ring is +90 seconds.

Scoring ring must be cut to score in next lower penalty.

LE Practice Target also used for SWAT sniper matches.

256 *The MILITARY and POLICE SNIPER*

SCENARIO 6: High Supported Position

This position is quite common in the field and is practiced in sniper courses, schools, and in actual field operations. Many times this happens because of tall grass. This position makes shooting difficult especially if the shooter and observer are exhausted from low crawl or running. If you think about it there are many ways to provide a high steady support. Use your imagination.

(1) Improvise a long bipod or tripod from branches before the stalk or at the FFP.
(2) Carry an extra long extendible bipod like the Harris 25 inch.
(3) Have your spotter create a tripod using himself and his M16A2.

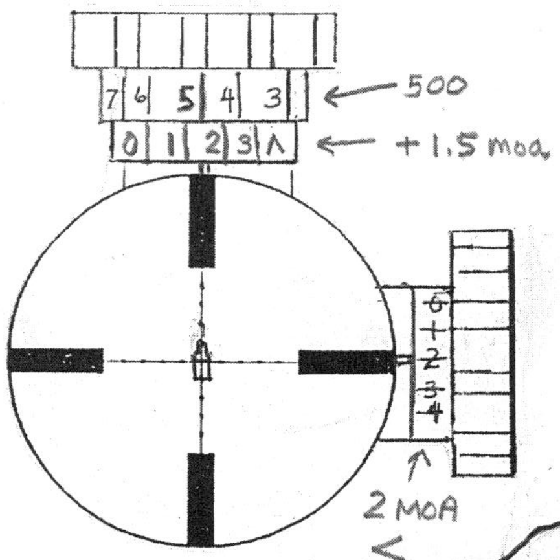

To aim and fire, hold your breath together on the count of "3".

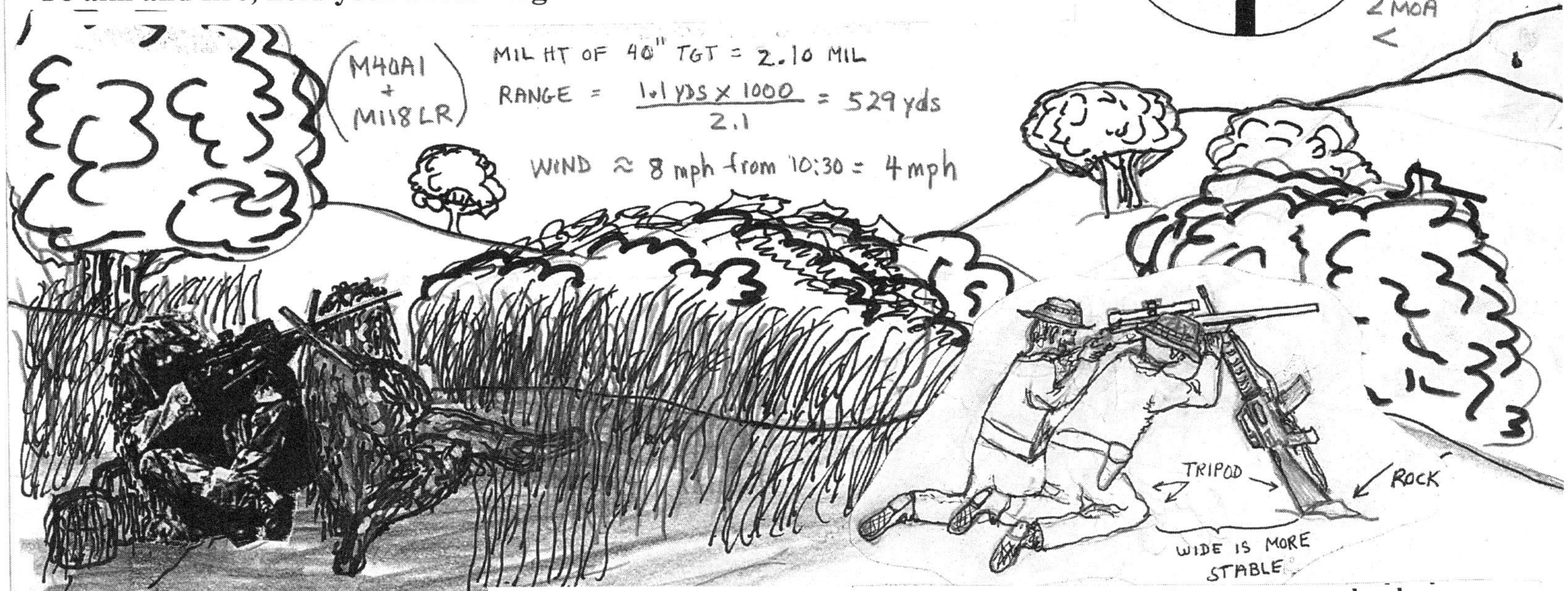

Place your hands over the front sight bracket. Leaning your head on your arms or hands removes some of your body movement and breathing impulses from the position. On the count of "3" both hold your breath. Aim and fire.

(M40A1 + M118LR)

MIL HT OF 40" TGT = 2.10 MIL

RANGE = $\frac{1.1 \text{ YDS} \times 1000}{2.1}$ = 529 yds

WIND ≈ 8 mph from 10:30 = 4 mph

500
+1.5 MOA
2 MOA

CHAPTER 14: Tactical Scenarios

Sniper Tip: Engaging very close targets. When on the move in hostile territory do not store your rifle in the drag bag. Have it loaded and put your scope power on low or turn your focus knob to under 50 yards. Snipers with scopes fixed at 10X, that have focus set to long range infinity, will have some difficulty engaging targets that suddenly appear within 100 feet. Don't attempt to engage the target by refocusing the scope. The enemy soldier or terrorist will engage you faster with his AK-47 than you can dial the focus. Don't try to locate him with the fine cross wires and find an aiming point. If he is wearing a dark uniform, focusing your eye to locate the fine cross wires will take too long. Trying to focus the target in the unfocused scope is almost impossible. He will have emptied his magazine at you while you are trying. Try centering him up in the scope with the wide outer posts. Focus on either the top or the bottom post. I like to focus at the bottom of the top post because I find it easier to keep the target centered between all four posts. If you focus on the top of the bottom post, like an M16A2 or M14 front sight, the center of the cross hair may be above his head and you might not see this when you fire. Keep the target's unfocused image centered between all four posts and try not to jerk the trigger. You can only fire one shot. If he turns on you with his rifle and you are not even sighted on him, you'd be better off to dive for cover before engaging him. Practice this technique and see what works best for you. It can happen during a stalking exercise or during movement to your final firing position (FFP).

Marine Scout/Sniper Brian Gauthier with TBA M40A1.

Sniper Shooting Tip: Harris Bipods. (Mike Lau and Sam Chesnut) Army snipers and many LE snipers use the Harris Engineering bipod because it is inexpensive, light weight, and allows for quick artificial support. The Harris bipod can be had in either the fixed or the tilt swivel version. Be aware, however, that the bipod can cause problems with your shooting if not used correctly. Nearly every place you go prone in the field the ground will be uneven because of grass, rocks, depressions, or slope. When using the non-tilting fixed bipod there is a tendency to level the rifle by twisting the rifle to raise one side because you don't have the time to adjust the leg heights. There is also the tendency to put torque on the bipod base laterally when you follow a target moving across your view. Both of these conditions can cause you to not be in a natural shooting position. To correct the torque on the bipod, sight on the target with a natural point of aim. Pick up one leg of the bipod with the rifle and let it swing or snap back into an unrestrained position. In addition, if you don't level the rifle, you can put an additional accuracy error which can be great at long range. There are bipods that actually tilt from one side to the other such as the Harris swivel version, the Versa Pod, and the Parker Hale. These bipods allow you to tilt the rifle on a narrow hinge type pin so you must be careful not to use too much sideways pressure in trying to hold the rifle level.

CHAPTER 14: Tactical Scenarios

SCENARIO 7: LONG RANGE MOVING TARGET

A drug king pin in a small foreign country is responsible for the deaths of several police chiefs and political and military leaders. As a result, the country's armed forces are slow in locating and capturing or killing the drug leader. El Presidente has asked the US for help in disrupting the drug organization. Several US Special Ops units have been checking out possible drug transferring sites located with satellite photos. When learning about the US operation, the drug leader decides to temporarily leave the country. A US Army Scout-Sniper team observing one of the sites has just been informed that the drug czar is going to leave by plane at their site.

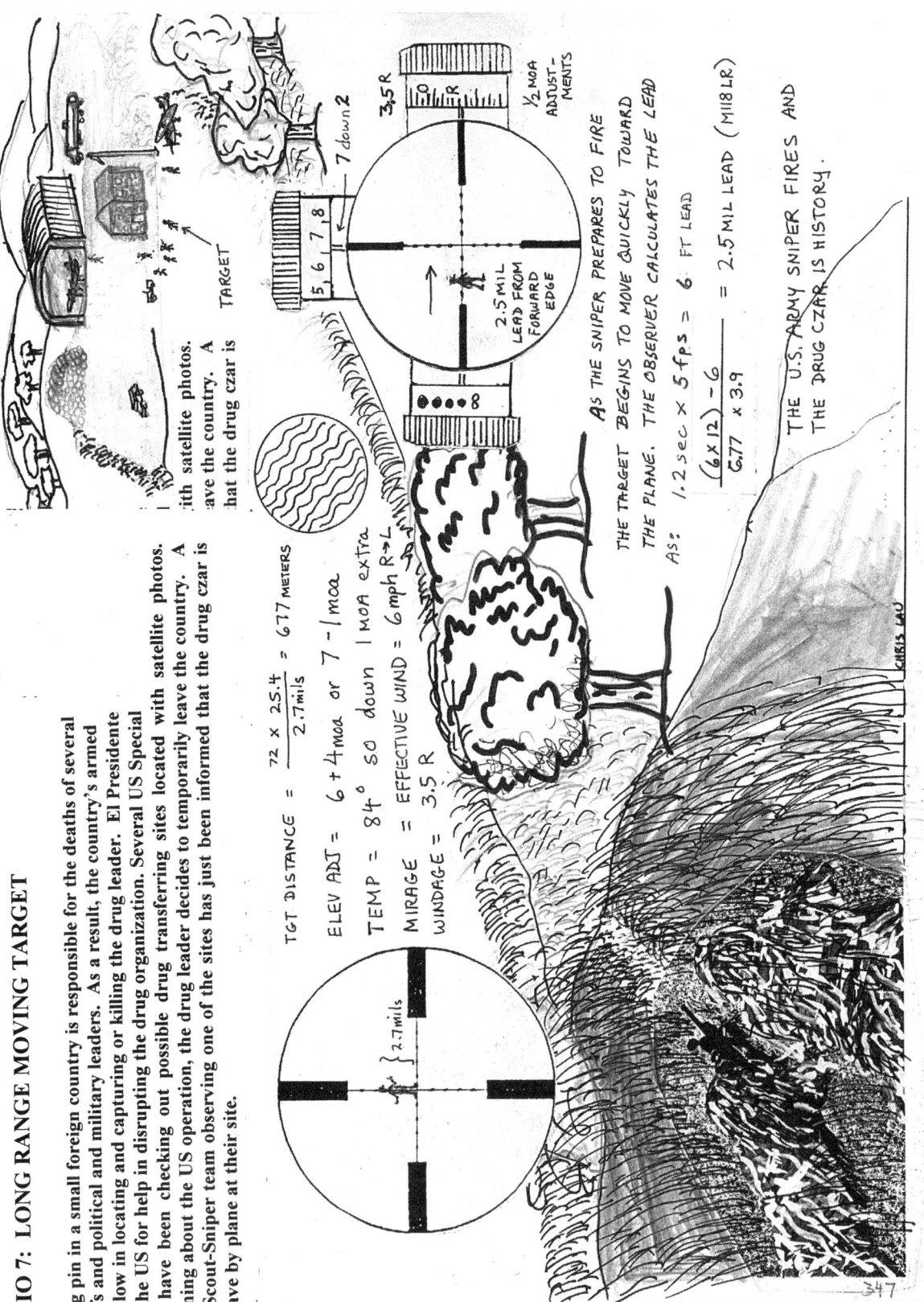

TGT DISTANCE = $\dfrac{72 \times 25.4}{2.7 \text{mils}}$ = 677 METERS

ELEV ADJ = 6 + 4 moa or 7 - 1 moa

TEMP = 84° SO down 1 MOA extra

MIRAGE = EFFECTIVE WIND = 6mph R→L

WINDAGE = 3.5 R

AS THE SNIPER PREPARES TO FIRE THE TARGET BEGINS TO MOVE QUICKLY TOWARD THE PLANE. THE OBSERVER CALCULATES THE LEAD AS: 1.2 sec × 5 fps = 6 FT LEAD

$\dfrac{(6 \times 12) - 6}{6.77 \times 3.9}$ = 2.5 MIL LEAD (M118 LR)

THE U.S. ARMY SNIPER FIRES AND THE DRUG CZAR IS HISTORY.

CHAPTER 14: Tactical Scenarios

Above is some of the basic equipment a Marine Scout/Sniper would carry for most missions. At left is Sean Little posing with his TBA M40A3. Sean builds complete ghillie suits for commercial sale in several different styles. He also offers a complete or modified USMC Ghillie suit kit with all correct items as issued in the Marine Corps to snipers. Shown is the modified kit which leaves out some items that are costly and also not totally necessary to build the suit with.

Eagle Drag bags are very useful pieces of equipment. They appear to be designed by persons who have a lot of field experience and it is no wonder why these bags are the choice of the military and LE sniper. The two items at the top are the large Eagle drag bag and outer cover as preferred by the Army, Marine Corps, and SOCOM. The bag can be used as a backpack as it has provisions for straps and has numerous pockets. It will also stow a one piece cleaning rod. The next item below these is Eagle's assault rifle case. At the bottom and also in lower photo is author's favorite, the Eagle tri-fold drag bag. It has 550 parachute cord sewn on the outside and has a pocket and locations for attaching other pieces of equipment with alice clips. Best of all the bag opens up as a shooting mat. Author's drag bag has been spray painted desert tan. TBA M40A1 rifle with adjustable cheekpiece belongs to a TBA customer.

CHAPTER 14: Tactical Scenarios

SCENARIO 8: WORKING THE PITS, AKA BUTTS, FOR BUTTRESSES.

Hostage target. Which target is the gunman? Wrong, it is number 2. Make sure you know the ID of the target before firing. Be careful when firing at number 2, because there is a good guy standing behind him.

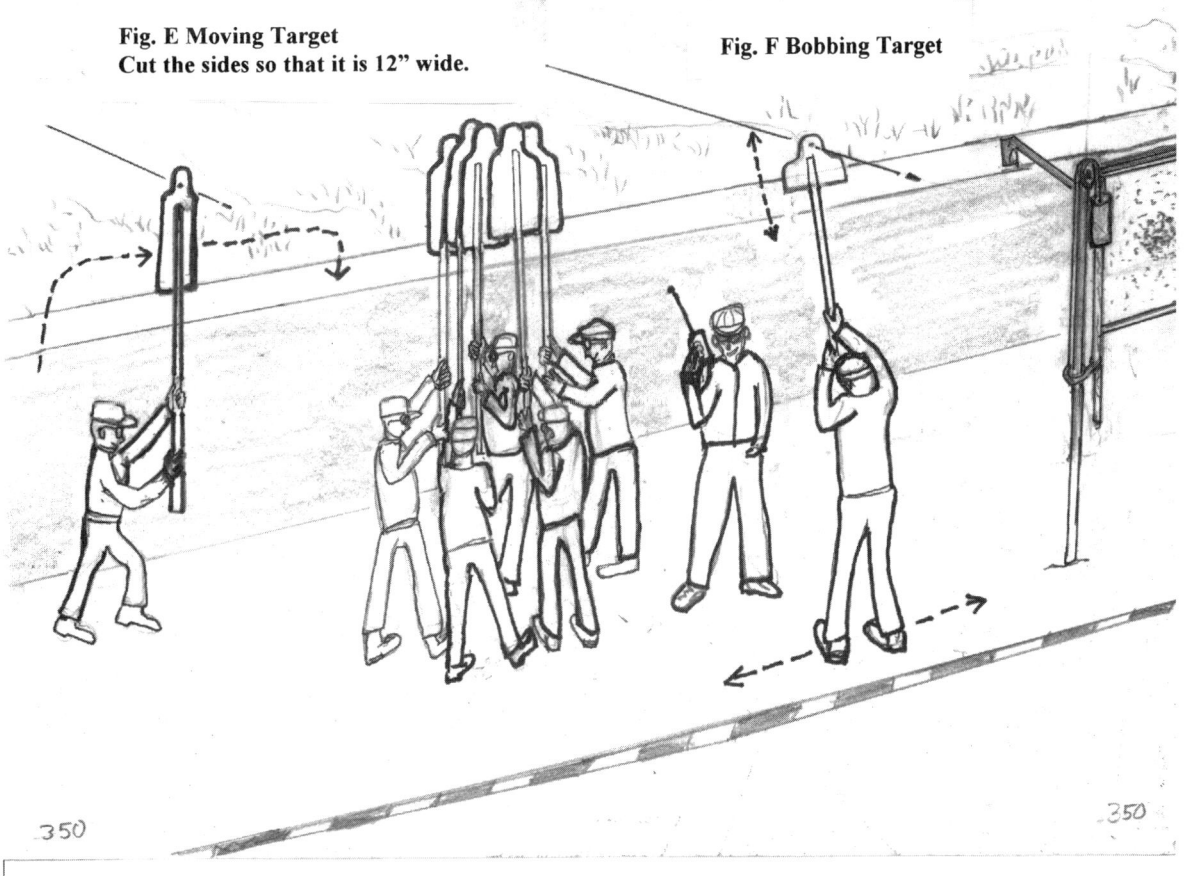

Fig. E Moving Target
Cut the sides so that it is 12" wide.

Fig. F Bobbing Target

Some safety tips and comments about using the pits.

(1) First of all, you will be safe in a certified, well constructed pit. The military has been using this type of firing for years and most military installations have them. If you go to Camp Perry then you know they are safe. The pit on military ranges are deep and wide to allow for adequate distance, reducing ricochet/ fragmentation from back burm and top of pit.
(2) The standard bullseye frame with the pivot or weights can be used for tactical type shooting with silhouettes stapled to the cardboard backing. They cannot be raised or lowered as fast as a person holding a stick with the target on it. You also cannot do moving targets unless you hold the target on a stick.
(3) Use staples instead of nails for attaching your target board to the stick. Less chance of bullet breakup when it hits the staple. Or you can use glue.
(4) When you use the target on a stick, wear safety glasses. Helmets are not necessary, but wear a hat in case of wood splinters and bullet fragmentation. Limit the number of times and amount of time you look up at the target you are holding. You will be able to feel the bullet hit the target board.
(5) Check to make sure no one is using special type ammunition, such as tracer, explosive, frangible, etc.
(6) Pistols fired from long range, 100-150 yards, require a very high trajectory and may fall inside the marked safety zone. Pistol bullets also fall just short of the target frequently and will kick up a lot of dirt on the people in the pits. For long range pistol shooting it is better to use iron targets on the back burm with no one in the pits.
(7) Do not fire rifles at iron targets on the back burm with people in the pits.
(8) Use high grade soft woods, like pine without a lot of knots, for exposing targets. They will not throw as much splinters as harder wood.
(9) Clean up the range better than you received it. Volunteer to repair or help with the expenses for maintaining the range.

CHAPTER 14: Tactical Scenarios

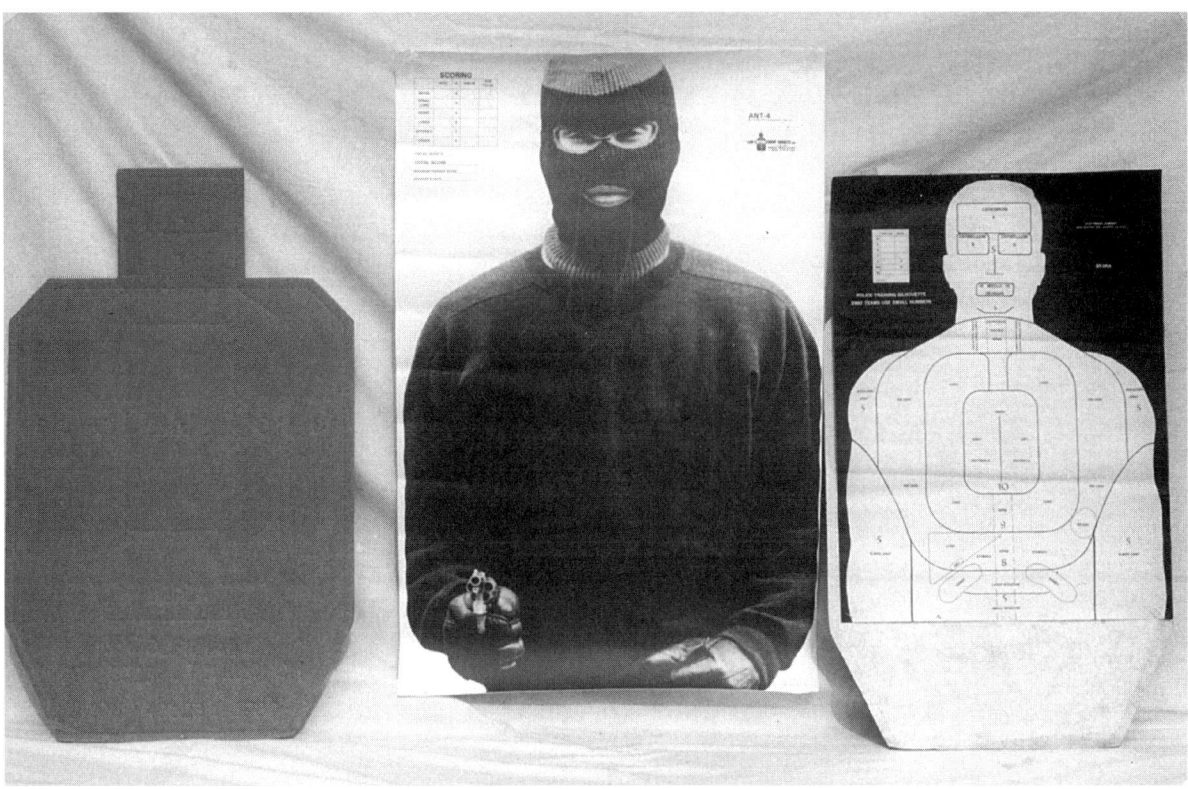

At top, from left to right, IPSC target has scoring areas depicting incapacitating areas. Terrorist target also has very faint scoring lines on it that outline the different incapacitating zones. LE type target with many different scoring zones showing other anatomy. Photo below shows the standard military targets. From l to r: Fig. 11 Cartoon or "Hun"; Fig. E with sides cut to be used for moving target; Fig. F standard prone target; and Fig. F "hostage" target. White portion is the 4" aiming area.

266 The MILITARY and POLICE SNIPER

SCENARIO 9: KD MOVING TARGET IN BOTH DIRECTIONS WITH WIND

Moving targets are difficult to shoot with wind and target moving at varying speeds. What can also compound the problem is a target that changes directions. If you want to add more difficulty, put the target at an unknown distance.

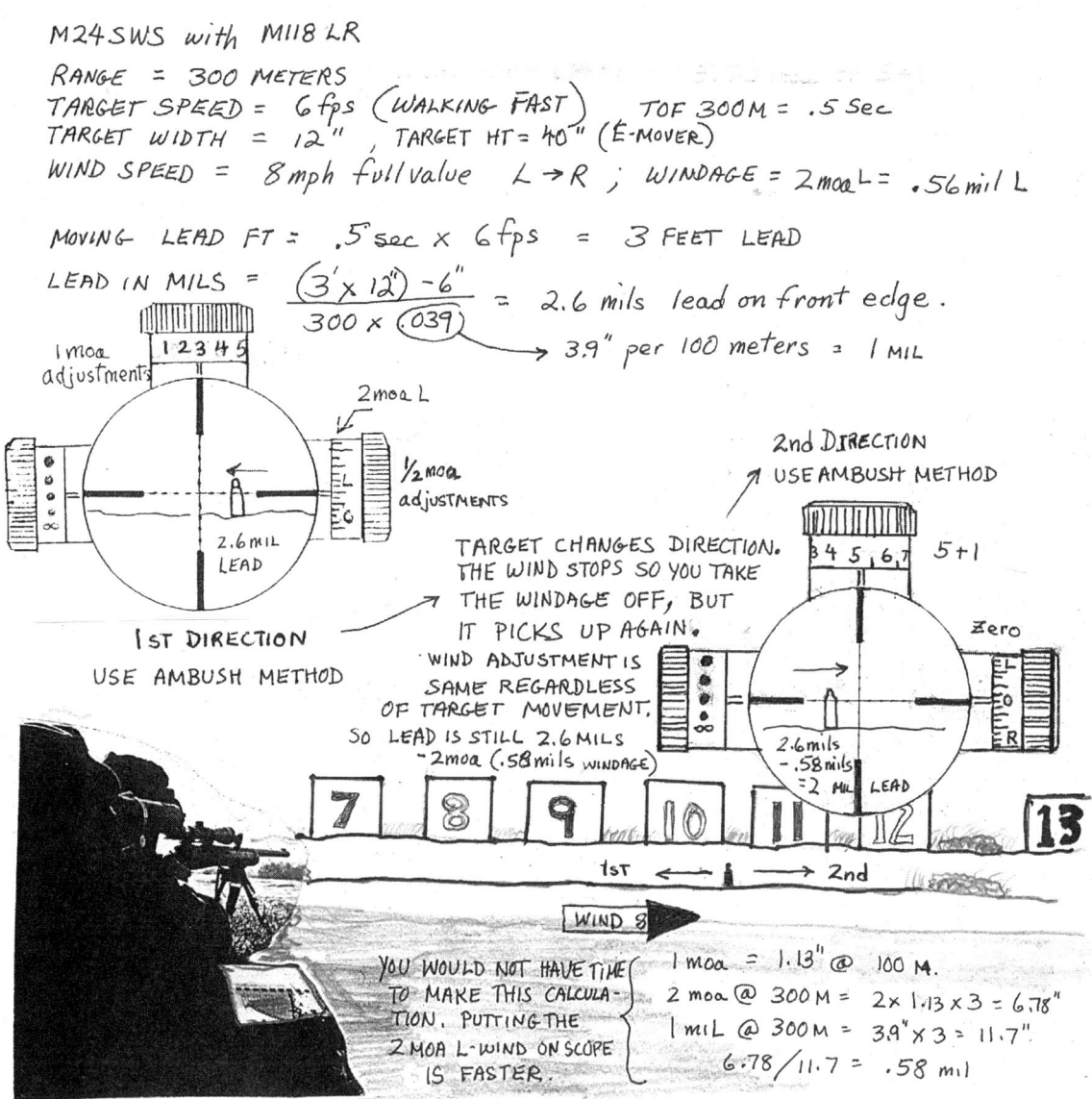

CHAPTER 14: Tactical Scenarios

Sniper Tip: Opposite Rotation of Windage Dial. Target engagement may look easy but it can get very complicated if you run into unforeseen difficulties with equipment. At the Proskopathlon 1997 Sniper Competition, in Wyoming, one of Jerry Lane's friends was not able to hit a target because he lost track of the additional clicks he put on his M3's BDC for a specific range. That happened to me one time when shooting moving targets with some of our local PD snipers. I normally use the Leupold M1 scope on my M40A1 because the elevation and windage knobs are easy to read and clearly marked. This time, however, I used an M3 scope on a customer's rifle. The BDC was calibrated for the 168 gr. Federal Match load. At first I was doing OK. At 100 yards I either tracked just inside the leading edge or ambushed the narrow target at that same point of aim. Sometimes the person moving my target inside the pits walked slowly and I made hits holding center of mass. As we moved back another 100 yards, the wind picked up. At first I made hits. Then the guy pulling my target realized he was being to easy on me and started walking faster. I missed the target. I started correcting for windage and more lead and missed again. Tony Black, yelled out, telling several others on the line to put on 1 minute left windage because the wind picked up. I looked at the windage knob of my M3 and then I realized I may have turned the knob the wrong way the first time. As I looked at the windage knob closer (I wear contact lenses and have a hard time reading up close without my granny glasses) I paid greater attention to the "L" and the "R" above the zero tick mark on the dial. Since the "L" was above the "0" mark, I had turned the knob in the direction of the letter "L". I had actually put on right windage. By the time I figured all this out and readjusted my windage, the moving target exercise was over. The windage knob on the M3 scope works just the opposite of the Leupold M1.

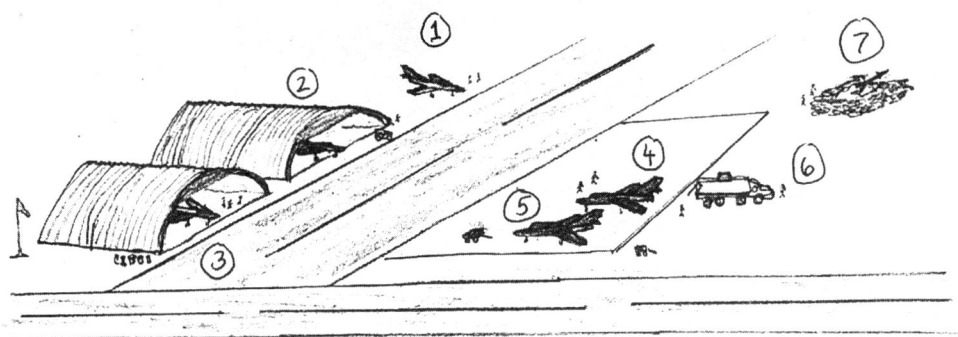

Scenario 10: Long Range Target Engagement with Cal..50 Barrett

The Aggressor Country has flown most of it's aircraft
into a neighboring neutral country to safeguard them
from US air attack pending breakdown of negotiations

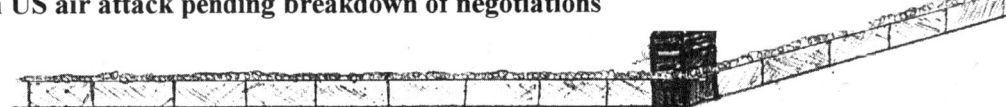

and imminent war. The Aggressor has five French made Super Etendard attack aircraft capable of firing the AM.39 Excocet anti-ship missile. A coordinated air strike and ground invasion by offshore US forces against the Aggressor Country has already been given the OK. A US Marine recon team, reinforced with three additional Scout/Sniper teams with two M40A1s and one Cal..50 Barrett, had been sent into the neutral country two weeks earlier to locate either the Etendard aircraft and/or the missiles. The team has been given the directive to destroy them at 7 minutes prior to the invasion kick off time. The mission is clandestine and air attack by US Navy planes into the neutral country is ruled out. The Marine team has located the aircraft. The sniper teams deploy to destroy primary and secondary targets. The remaining team members provide flank and rear security.

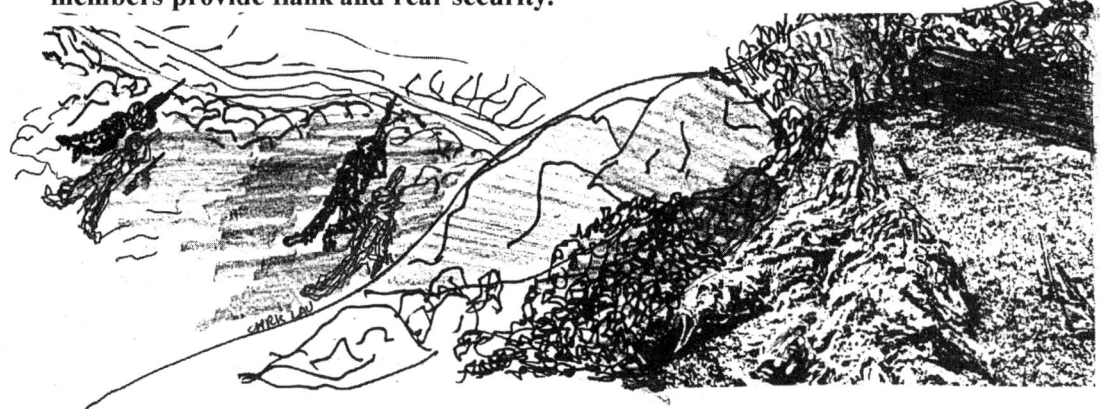

CHAPTER 14: Tactical Scenarios

TARGET #1, FARTHEST AIRCRAFT, TAIL HT = 12.3 FT = 4.10 yds
HEIGHT OF TAIL = 3.00 MIL

$$RANGE = \frac{4.10 \times 1000}{3} = 1367 \text{ yds}$$

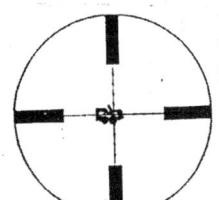

FROM TRAJECTORY TABLE FOR .50 RAUFOSS

	DROP	COMEUP	DIFFERENCE
1300 yds	.570"	44.0 MOA	} 6.5 MOA ELEVATION
1400 yds	.705"	50.5 MOA	between 1300-1400

TARGET IS BEYOND 1300 yards BY 67 yds WHICH IS 2/3rds of 100 yds
SO 2/3 × 6.5 MOA ≈ 4.5 MOA , ELEV ADJ = 1400 - 2.0 MOA
(NOTE: LOWER-FINE ADJUST-CAM HAS 3 MOA UP AND DOWN)

TARGET NO. 2, FARTHEST AIRCRAFT IN HANGAR, CANNOT SEE TAIL
LENGTH OF M151 JEEP = 2.8 MILS
LENGTH OF M151 JEEP = 11 FT = 3.67 YDS

$$RANGE = \frac{3.67 \times 1000}{2.8} = 1311 \text{ YDS}$$

AIRCRAFT IS ABOUT 10 YARDS CLOSER THAN M151 SO USE 1300 YDS ELEV.

TARGET #3 CLOSEST HANGAR AIRCRAFT
HT. OF TAIL = 3.4 MIL = 1206 YDS   ELEV = 12

TARGET #4 FARTHEST AIRCRAFT ON FUELING PAD
HT. OF TAIL = 3.6 MIL = 1139 YDS
5.4 MOA BETWEEN 1100 + 1200 YDS SO 18 YDS PER MOA
2 MOA ≈ 36 YDS,   ELEV = 11 + 2 MOA

TARGET #5 = 3.7 MIL = 1108 YDS ,   ELEV = 11

SECONDARY TARGETS
#6 FUEL TRUCK = SAME AS TGT #4 = 11+2 ELEV
#7 SAM MISSLE = LOOKS LIKE 100 YDS BEYOND TGT #4
   ELEV = 12+2

WIND CONSTANT:
4-5 MPH

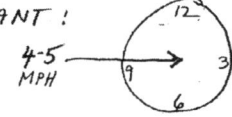

FROM SPCL TABLE. 5 MPH FULL:

YDS	DRIFT	MOA
1367	66.5"	5.0 L
1311	61"	4.5
1200	49"	4.0
1139	44"	4.0
1108	41"	3.5

TEMP: Z_{ero} = 78° HERE = 86° NO ADT.
ALTITUDE OK NO CHG, ANGLE, LESS THAN 10° = NO ADJ

270 *The MILITARY and POLICE SNIPER*

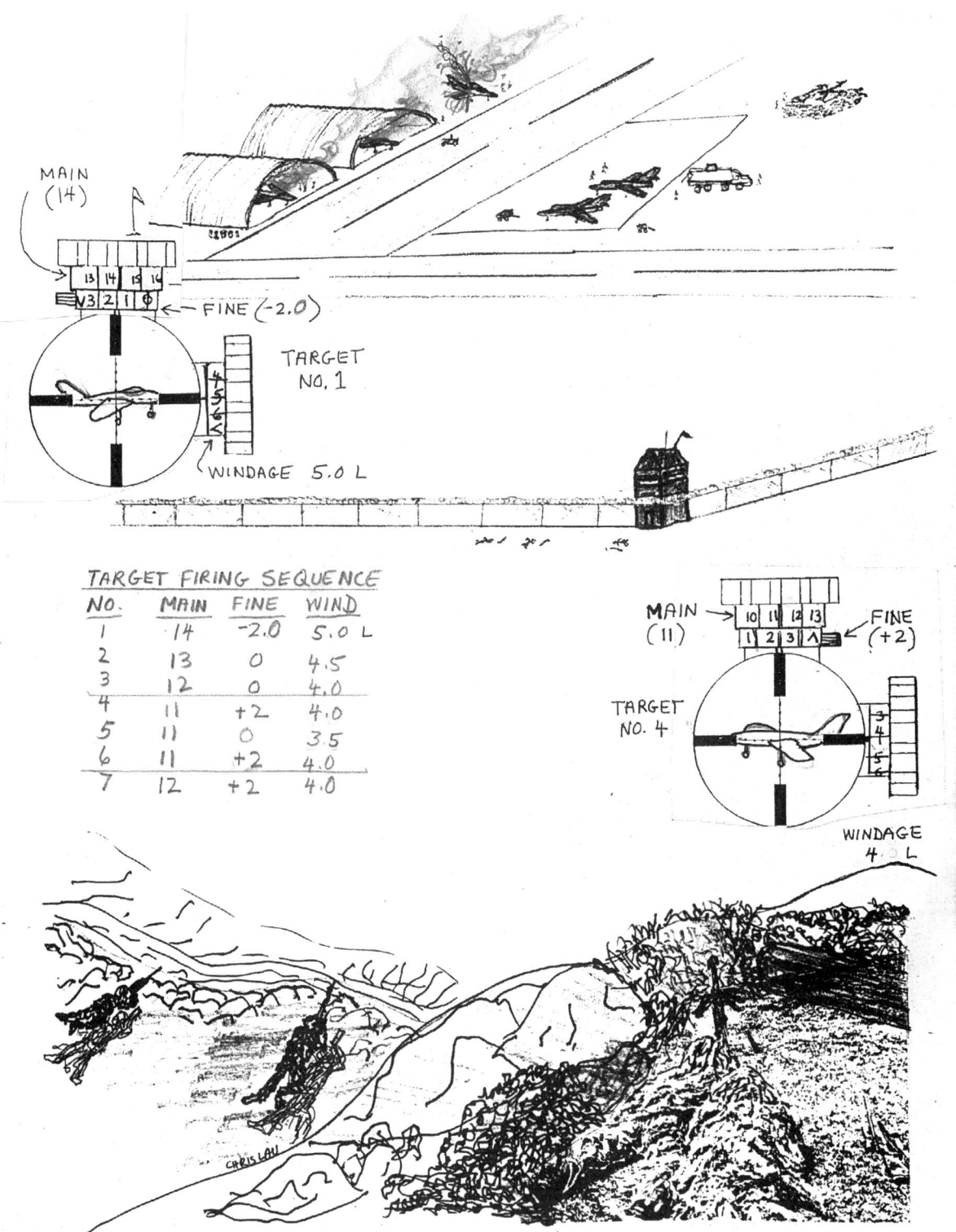

TARGET FIRING SEQUENCE			
NO.	MAIN	FINE	WIND
1	14	-2.0	5.0 L
2	13	0	4.5
3	12	0	4.0
4	11	+2	4.0
5	11	0	3.5
6	11	+2	4.0
7	12	+2	4.0

Special Operations to Destroy Critical Material

Numerous situations occur where Caliber .50 sniping rifles could be employed on the battlefield. During Desert Storm, the Marines used Barretts against Iraqi armored vehicles and heavy equipment.

The previous scenario in which Marines take out aircraft with their Barrett, never actually happened. A similar event, however, did take place during the Battle for the Falkland Islands in 1982. On 4 May, the British destroyer, **HMS Sheffield**, was hit by an AM.39 anti-ship Exocet missile fired from a French built Super Etendard jet fighter. The ship was gutted and 20 British seamen were killed. On 25 May, an Argentine Navy Super Etendard attacked the Merchant Marine ship **Atlantic Conveyor**. Fire engulfed the ship and she was abandoned. The **Atlantic Conveyor** lost all of it's cargo which consisted of Harrier jump jets and most of the Task Force's helicopters. Other British ships were also hit with Exocets, but did not sustain as much damage. To counter this threat, numerous actions were taken by the Royal Navy. The Exocet air threat was put to rest when SAS (Special Air Services, the British counterpart to our Green Berets) Commandos went into the Argentine mainland, located, and destroyed the Argentine Super Etendards on the ground. These French built aircraft were the only ones capable of firing the Exocets. SAS assaulted the airbase at night and destroyed the aircraft with explosives. (Note: initial information referring to this incident was only evidenced by a short news report that a British helicopter went down in Chile. What was not reported was that the SAS had conducted the raid successfully and escaped to Chile where the helicopter went down. Not being friendly to Argentina, Chile helped the SAS get to the coast where they were picked up by submarine.)

Marine Scout/Snipers of STA Platoon, 3/1, in Somalia. Black bullet proof vests are normally issued to the Recon Platoon because they have anti-terrorist and hostage rescue responsibility. Marine snipers here are wearing the vests because these were more comfortable to wear in the heat and also less confining when shooting than the standard armor vest. Brian Gauthier is on the left. (Brian Gauthier)

Outpost was built by Pakistani UN troop overlooking Mogadishu. US Marines occupied the bunkers as rear guard while UN troops debarked Somalia at the airport and in the harbor. As the Marines abandoned the outposts, rebels moved into them, but they kept their distance. Several fire fights broke out, but their were no casualties. (Brian Gauthier)

CHAPTER 14: Tactical Scenarios 273

A view of the city from the outpost. On the left of the outpost is no man's land. Behind the outpost is the airport. (Brian Gauthier)

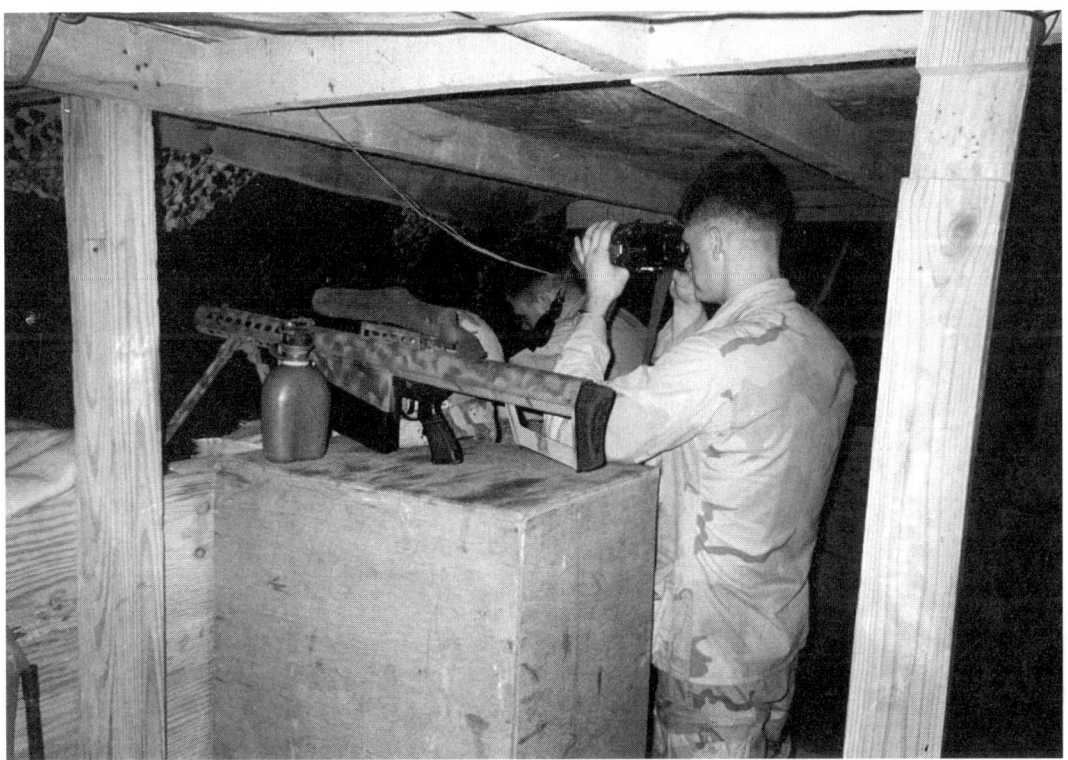

Night vision. Note sock over Unertl scope mounted on Cal..50 Barrett. Special cams were installed in the Unertl scopes for the .50 Caliber trajectory. (Brian Gauthier)

SCENARIO 11: THE SECOND MOST IMPORTANT SHOT OF THE DAY.

Almost as important as the cold bore shot, is the last shot fired at the end of a training day for law enforcement personnel. Before you leave the range to head back to the office or go home, confirm your zero by actual firing. The Dallas, Arlington, Texas, and Oklahoma City SWAT snipers do. Use the face target or small paster. It can mean saving someone's life.

LE Snipers usually check zero monthly and fire for record. Targets are reviewed by team leader and commander to insure there are no weapons problems and sniper maintains skill. Targets are kept for record that can be used in court to prove competency in shooting skills of LE sniper. Target at left was shot by Sam Chesnut at 100 yards. Three shot group measured .240 inches and is typical of Sam.

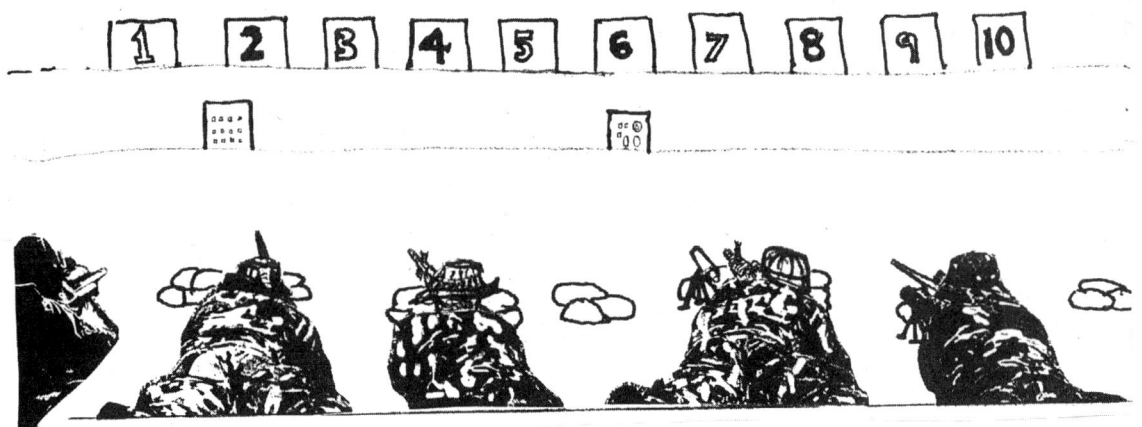

CHAPTER 14: Tactical Scenarios

When going through these scenarios, you should have noticed several things:

(1) I am prejudiced in that I favor the MIL-Dot system. Not just any MIL-Dot system, but the Marine Corps' and Army's mil dot.

(2) Army and Marine MIL-Dots reticles have fine wires and small dots. This makes estimating mil heights and widths easier. It also allows for pinpoint aiming on small targets.

(3) Army and Marine MIL-Dot reticles have wide outer posts on the crosswires. This allows for some dim light shooting and faster target pick up at close range, while still allowing for narrow center cross wires.

(4) The scope needs to have target type knobs that are easy to read and adjust without you having to get off the rifle. Turret caps are OK, but plan on losing them some day in the field.

(5) The more accurate your rifle, the better all around use you will get out of it. You'll have more fun practicing with it also. Remember Townsend Whelen, "only accurate guns are interesting."

(6) The .308 Win is still the best all around sniper cartridge.

(7) The simpler the weapons system the better off you will be because of speed and ease of handling.

(8) Stay in shape. Remember Proskopathlon Sniper Match. First day is 8 miles.

Final words: Keep things simple, yet maintain accuracy.

Use the simplest, fastest, yet most accurate method to get your data, adjust your sights, and engage the target. If you miss the target, make sure your sight adjustments on the scope dials are correct. If your spotter noted the impact, quickly readjust your sights or aim with hold-offs and refire. Or, if you have time, estimate the target distance and wind again, readjust sights and fire again. If you don't have time to make any calculations and you have to make the shot because you will lose the opportunity, like game hunting, you need to quickly decide whether to wait for a better opportunity or fire with hold-offs. Sometimes your best guess hold-offs and firing may be better than not firing at all.

SGT Sam Chesnut displays trophy for 1st place Sniper in the 1995 Southwestern Law Enforcement Emergency Response Team Competition. Over 25 teams participated and included Regular Army, National Guard, FBI, and numerous Law Enforcement. Sam is a dedicated competitor and practices a lot. He enters sniper and SWAT competitions all across the country when his schedule allows and he usually takes 1st Place. Sam competes with his duty sniper weapon, a Texas Brigade Armory M40A1 in .308 caliber with a Leupold 3.5x10 Tactical scope. (Mike R. Lau)

CHAPTER 15:
THE SNIPER MATCH:
Sustainment Training Through Practical Precision Shooting

During one week in October, 1997, over 970 men and women from all across the country and from foreign lands, converged at a place just north of Little rock, Arkansas. They brought with them assault rifles, sniper rifles, pistols, and light machine-guns. The men and women were members of Army National Guard units representing almost every state in the union and from foreign military units. The place was Camp J.T. Robinson and the event was the 1997 Wilson Matches. For one week they displayed their skills at arms, competing for trophies and increasing their knowledge. Thirty-two teams, sixty-two Army National Guard snipers, competed with their M24 SWSs for the honor of being Top Gun.

To maintain their skills, military and LE snipers participate in sniper matches. For some civilian type matches there are no national rules like in NRA and DCM sanctioned rifle matches. Rules may apply strictly to the specific event or course of fire to provide restrictions making the course favorable to all shooters and for safety reasons. Organizations and agencies that hold these matches can design content and make up rules for the competition in any way they want. This is what makes sniper competition interesting. Like any sniper mission or callout, the competitive sniper gets exposed to new situations and problems at these matches and must be prepared for anything.

Civilian sniper competition is usually open to anyone including military and law enforcement officers. Military and law enforcement sniper matches are usually closed to civilians. This is unfortunate since many civilian competitors are ex-military or ex-law officers. Even if they weren't, they wouldn't hesitate to back you up if a real threat situation occurred. Like IPSC and Practical Pistol, sniper competition is becoming more popular and professional. The winners of the 1997 Wyoming Proskopathlon Sniper Competition in June took home several thousand bucks and some nice rifles to boot.

A drawback to sniper matches is that for the civilian competitor, they are expensive. They are usually held on private property and the facilities cost money to run. The entry fees can run several hundred dollars and you would have to take a week off from your job. In addition there is the driving and most likely you will have to find your own place to stay at night. Military and law enforcement matches are usually on military and law enforcement installations so they are paid for by you, the tax payer. Military installations have barracks accommodations, meals, and transportation costs that could be charged to your city or Uncle Sam.

Sniper competition usually requires two man teams. Many of the recent matches the author has heard about, or attended, had anywhere from 10 to 32 teams, with the civilian matches having the less participants because of the cost. Police SWAT competition usually requires two man teams also, and tactical events usually involve the sniper rifle, assault rifle or submachine gun, combat shotgun, and pistol shooting.

Be aware of what the course might be like before you get there so you can bring the right kind of equipment. Otherwise bring everything. As you all get more skilled and familiar with the sniper courses around the country, the tactical problems will have to be made more difficult so they won't be "cake walks."

Some courses may be weighted more for short range law enforcement sniping rather than the long range military type because of the land area limitations and safety restrictions. You may find that a shorter 20 inch barreled rifle would have been more handy especially if you had to crawl through a small space and suddenly have to engage a target from that small space. On the other hand, a match may separate the winners and non-winners by long range shooting skills, like wind reading, engaging small targets, and shooting small shot groups at long range.

Many of the competing Army National Guard snipers competing in the 1997 Wilson Matches were senior non-commissioned officers that have experience in NRA and DCM competitive shooting. Many are Distinguished Rifle. Many compete in numerous law enforcement and civilian matches and are familiar faces. Increase your skills and have fun with competition matches and practice. However, be aware of the sniper shooting next to you, he may be a hard-core, dedicated, professional long range shooter.

A Photo-Journalist's Diary of the 1997 Wilson Sniper Matches.

CHAPTER 15: The Sniper Match: Sustainment Training Through Practical Precision Shooting

Day 1. Army National Guard MTU Headquarters, Camp Robinson, Little Rock, Arkansas. (Mike R. Lau)

Here's where the Scout-Snipers put their equipment (except weapons), clean up, and sleep for the next 5 days. (Mike R. Lau)

The first event is the unknown distance cold bore shot. The sniper is given 2 rounds of M118 SB. A one minute prep time is allowed after which the sniper will signal he is ready to fire to allow a scorer to observe the hit or miss on the iron maiden. He is allowed 3 minutes to fire the 2 rounds. A first round hit is 100 points and a second round hit is worth 50 points. Today's temperature was around 58 deg.

SSG John D. Hamel gives thumbs up to his spotter, SPC Allen J. Corey III, as he hit iron maiden target, estimated to be 600 meters, with first round. SSG Hamel won the Cold Bore UKD Match with a first round hit on Day 1 and again on Day 5. Team represents Co. C3/127 Infantry, Mountain, from New Hampshire. (Mike R. Lau)

CHAPTER 15: The Sniper Match: Sustainment Training Through Practical Precision Shooting 281

Team is given 1 minute to change over and SPC Corey is now the shooter with SSG Hamel spotting. Each team consists of an experienced and a new shooter. In this type of team match, the experienced shooter engages target first so that unknown distance range and wind correction data can benefit new shooter. This allows the new shooters to score as high a score, if not higher, than the experienced shooters. The team's rifle is painted OD green and chartreuse. (Mike R. Lau)

Second match, Day 1, is 10 UKD targets, type "E", "F", and iron maidens from 250 meters to 900 meters. Targets were laid out in a fan across the range so that they were not in line with each other. This also made wind estimation difficult. Targets were estimated to be at or near 250, 350, 400, 450, 550, 630, 700, 735, 800, and 850 meters. (Mike R. Lau)

Each sniper is given 20 rounds and 20 minutes. Each target must be engaged with 2 rounds and each hit is worth 5 points. SFC Tim Weber, HQ STARC, Minnesota National Guard, fired a perfect score with a number of "V"s on the iron maidens. His spotter is Sgt. James T. Kringlie. SFC Weber prefers the short 6-9" Harris to get low to the ground. (Mike R. Lau)

Day 2. Stalking competition. The ability of the Scout-Sniper to move on the battlefield undetected is a skill requiring patience, discipline, and the mastery of "field craft". Team Puerto Rico, Guillermo Aviles Ghiliotty, told author that camouflaging the M24 rifle is difficult because it is black. It is easier to make a light colored rifle dark than to make a dark rifle blend in with tan, green, and brown vegetation.

CHAPTER 15: The Sniper Match: Sustainment Training Through Practical Precision Shooting

The art of camouflage for the sniper is the ghillie suit. Assembling a suit like this takes 40 to 60 hours of your own time. Each is unique and may have many modifications suited to the individual wearer. Scout-Snipers apply camouflage paint to face and hands and then get into their suits.

About two to three gallons of water will be consumed during the 2 1/2 hour stalking exercise. Although not an issue item, the CamelBak™ is very useful because it allows the sniper to suck water from a tube without having to reach for a canteen. This is the large version and will hold 70 ounces of water. Cost is around $40, it is available from U.S. Cavalry, Ranger Joe's, AWG/ Dub Ball, etc. (Mike R. Lau)

Natural vegetation is blended in with shredded burlap to give the mix of colors as well as to make the suit look natural to the surroundings. The front of the suit is covered with OD tarp or nylon canvas to allow the sniper to slide easily over grass, small rocks, thorns, and other types of terrain without tearing up the suit or injuring the body. A suit can weigh up to 20 lb. or more when all decked out. (Mike R. Lau)

Suited up and ready to go. Gentlemen, you are under observation beginning now! (Mike R. Lau)

CHAPTER 15: The Sniper Match: Sustainment Training Through Practical Precision Shooting

Observation is done by SFC Gooch and an assistant from the back of two Humvees. The Sniper team must traverse 600-800 meters over open terrain with tall grass and within a few hours. Left and right boundary markers keeps the sniper team within the open area. When a team is within 300 meters of the observers and are not detected, they receive 40 points. Sniper teams that are located and identified are withdrawn from the exercise and accumulate no further points. (Mike R. Lau)

The team must position within 200 yards of the Humvees and fire a blank round for an additional 10 points. The M24 has to be in a supported position and both team members must be in the final firing position (FFP). To verify that the sniper team has a legitimate firing position without being detected, each member must identify one of two small placards with either a number or an alphabet on it. These can only be read with their scopes or binoculars. (Mike R. Lau)

SFC Gooch has several "walkers" in the area of the stalk that he uses to locate teams. If an observer on the Humvee detects movement, or believes he sees the sniper team, a "walker" is directed to the location. The "walker" is directed to put his hand on what the observer believes may be one of the team members. If the "walker" verifies, the team has been "spotted" and they will get up and leave the area. If the observer is wrong the exercise continues. 1st Lt. Chuck Bereen, MTU, acts as a "walker". He is a North Little Rock SWAT sniper, and graduate of NGSSS. (Mike R. Lau)

Final firing position of the North Carolina team, SSG Murphy Riggan, spotter, and SSG Kevin Griffin, shooter. When the snipers successfully identify the placards without being detected, they receive 10 more points. A second blank round must be fired from this position. If not detected during the second shot, the team receives another 10 points. The NC team was spotted after the second shot so accumulated 70 points. (Mike R. Lau)

CHAPTER 15: The Sniper Match: Sustainment Training Through Practical Precision Shooting

Final firing position of the Oklahoma Team. Second shot did not give away position. "Walker" comes over and indicates body direction of sniper to the observer in Hummer and cannot be seen. Team gets 10 more points. "Walker" then reaches and touches sniper on the head and observer still cannot make out body form. Sniper is asked to ID the number or alphabet on the placard again and correctly does so. The "walker" verifies the firing position and proper elevation on scope to be correct and the team earns 100 points. (Mike R. Lau)

Coming out of their FFP, the Oklahoma National Guard Team, SGT Sean Brownsen, shooter, and SGT Brent Newton, spotter, 1/158 FA (MLRS). This team placed second with a time of a little over 2 hours and 25 minutes. (Mike R. Lau)

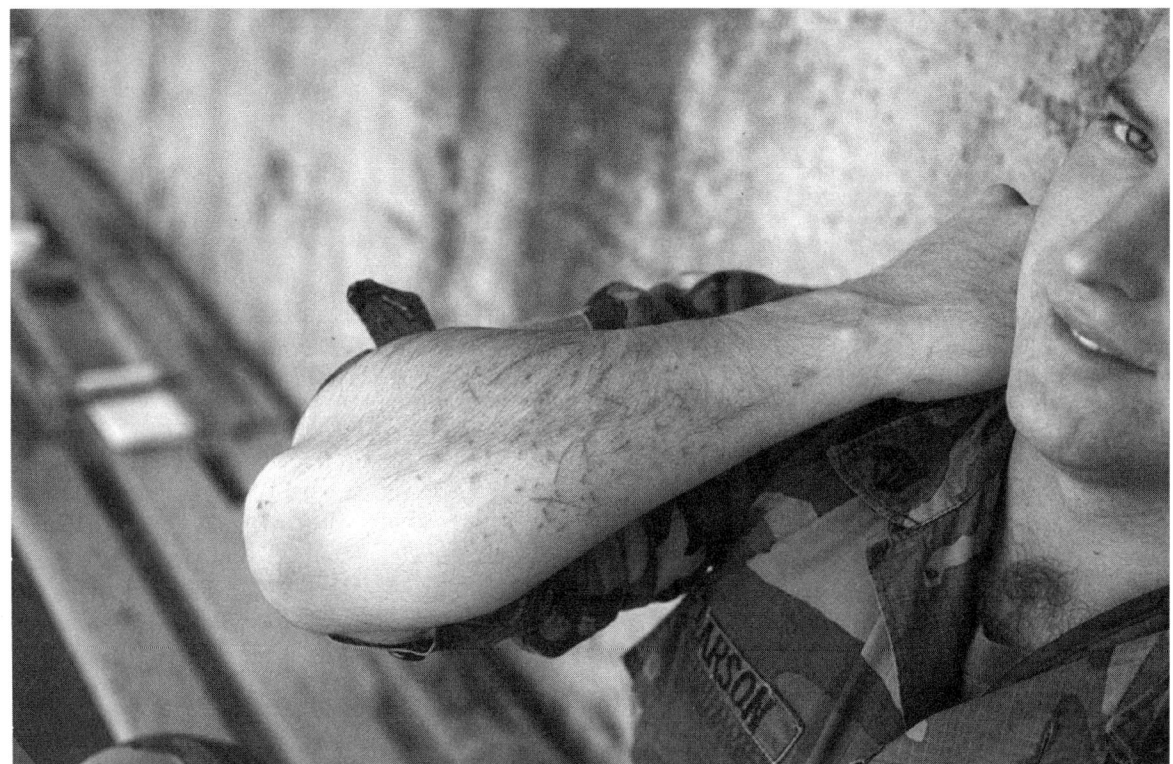

SGT Chris Larson shows me where the ants bit him during the stalk. He was also stung by wasps on the arm and calf, but did not get up or make any unnecessary movements to give the team's position away. His partner, SFC Tim Sowa, could not get into his firing position because he got a cramp in his leg. SAGT Larson became the shooter and the team scored 100 points in 2 hours and 20 minutes. The team from Montana took first place. Nice going! (Mike R. Lau)

Later on Day 2, the snipers continue with field craft exercises. Besides the stalking competition, there are a number of field craft exercises that can test the Scout-Sniper's skills. A few of these are range estimation, observation, call for artillery fire or air support. 15 teams are lined up, each to be given a map and field data to negotiate a land navigation course. (Mike R. Lau)

CHAPTER 15: The Sniper Match: Sustainment Training Through Practical Precision Shooting

Scout-Snipers prepare their route before moving out. Land navigation is considered one of the most fundamental skills of the US soldier and can only be learned by experience. Without this skill most field missions can't even begin. Expert land navigators like Army Special Forces, Rangers, and Scout-Snipers, can find a small objective area about 10 meters wide, over several miles away, using several different compass headings, over very rough terrain, and at night. (Mike R. Lau)

Scout-Snipers are examining pieces of equipment in an exercise called KIM's game. Named after a similar game used to teach the character "Kim" in Rudyard Kipling's novel, about a boy who was trained to become a British spy. Also acronym for "keep in memory". After observing numerous objects for a short period of time, the sniper is made to do other tasks which would distract his memory. Some time later he has to describe as many of the objects he observed and recall all pertinent information regarding it. (Mike R. Lau)

To observe, identify, and report accurately allied and enemy equipment, unit positions, uniforms, unit markings, rank and other intelligence data is a valuable field skill for the sniper. The last event of "field craft" is to identify armored vehicles of foreign countries. (Mike R. Lau)

The winning team of the land navigation, KIM's game, and armor ID events goes to Green Beret CPT Ken Chavez and SGT Darrell Brooks Jr., 5/19 Special Forces Group (A). CPT Chavez will rank third in individual scoring at the end of the matches. (Mike R. Lau)

CHAPTER 15: The Sniper Match: Sustainment Training Through Practical Precision Shooting

This guy must have been in the Navy, or worse yet, maybe the Marine Corps. He likes to speak from the "bridge", observe from the "crow's nest", and uses words like "deck" and "aye, aye" often. Day 3 starts with SFC Kent Gooch giving instructions on the next event, the Special Reaction Match. This event is similar to some HRT or Special Operations type firing. It can be done with or without urban environment obstacles such as firing through windows, negotiating obstacles, or wearing the gas mask. This time the event will be done without obstacles, but with rapid fire and running between stages. (Mike R. Lau)

Stage 1 is at 600 yards. From the prone position, you will have one Fig. 11 target exposed 5 times for 6 seconds each time with 8 seconds between exposures. One round per exposure. After you fire 5 fire rounds, change shooters. You will have 1 minute to prep and we will begin again. (Mike R. Lau)

Each relay has five teams. Five teams are in the pits. When the first relay finishes the entire course they will replace the team in the pits who will then await their relay at the starting position. (Mike R. Lau)

After all of the first shooters fires his 5 rounds the command to switch places is given and the 1 minute prep for the new shooter begins. At the end of the 1 minute the targets appear. (Mike R. Lau)

CHAPTER 15: The Sniper Match: Sustainment Training Through Practical Precision Shooting

When the second snipers complete their firing the rifles are loaded with safety on. The command to move to the next position is given. Snipers and safety personnel sprint 100 yards to the 500 yard line. Time allowed to run the distance with equipment is 30 seconds between 100 yard markers. (Mike R. Lau)

On right side of range, sniper teams get down in their positions and adjust for new range. Same course of fire on Fig. 11 targets. 5 exposures, 5 rounds, and then switch shooters. (Mike R. Lau)

Major Wigger checks safety and insures correct procedures on left side of range. (Mike R. Lau)

Teams are on their feet and sprint to the 400 yard line. This time they stay on line with their weapons loaded and safety on. (Mike R. Lau)

CHAPTER 15: The Sniper Match: Sustainment Training Through Practical Precision Shooting

At 400 yards, snipers engage a moving target 5 times with 1 round on each exposure. Target is cut down Fig. E to simulate 12" side body width. Moving targets are mounted on sticks and moved by Scout-Snipers operating the pits. After both team members complete their firing they are off and running again. (Mike R. Lau)

Five bobbing F silhouette targets are engaged with one round each at 300 yards. Each exposure is 3 seconds with 8 seconds between. You gotta be quick on this one. Get up and run another 100 yards to the 200 yard line. Engage the same target in the same manner as the 300 yard bobbing target. (Mike R. Lau)

Get up and run one more time to the 100 yard line. (Mike R. Lau)

The hardest to engage is the hostage target. An OD Fig. F is stapled in front of a partially exposed white Fig. F with an aiming zone of only a few inches wide. By this time you are somewhat exhausted and now comes a target that you cannot afford to miss. A hit on the hostage, OD Fig. F, will cost you negative 10 points. Exposure time is only 3 seconds per exposure for five exposures and five rounds. (Mike R. Lau)

CHAPTER 15: The Sniper Match: Sustainment Training Through Practical Precision Shooting   297

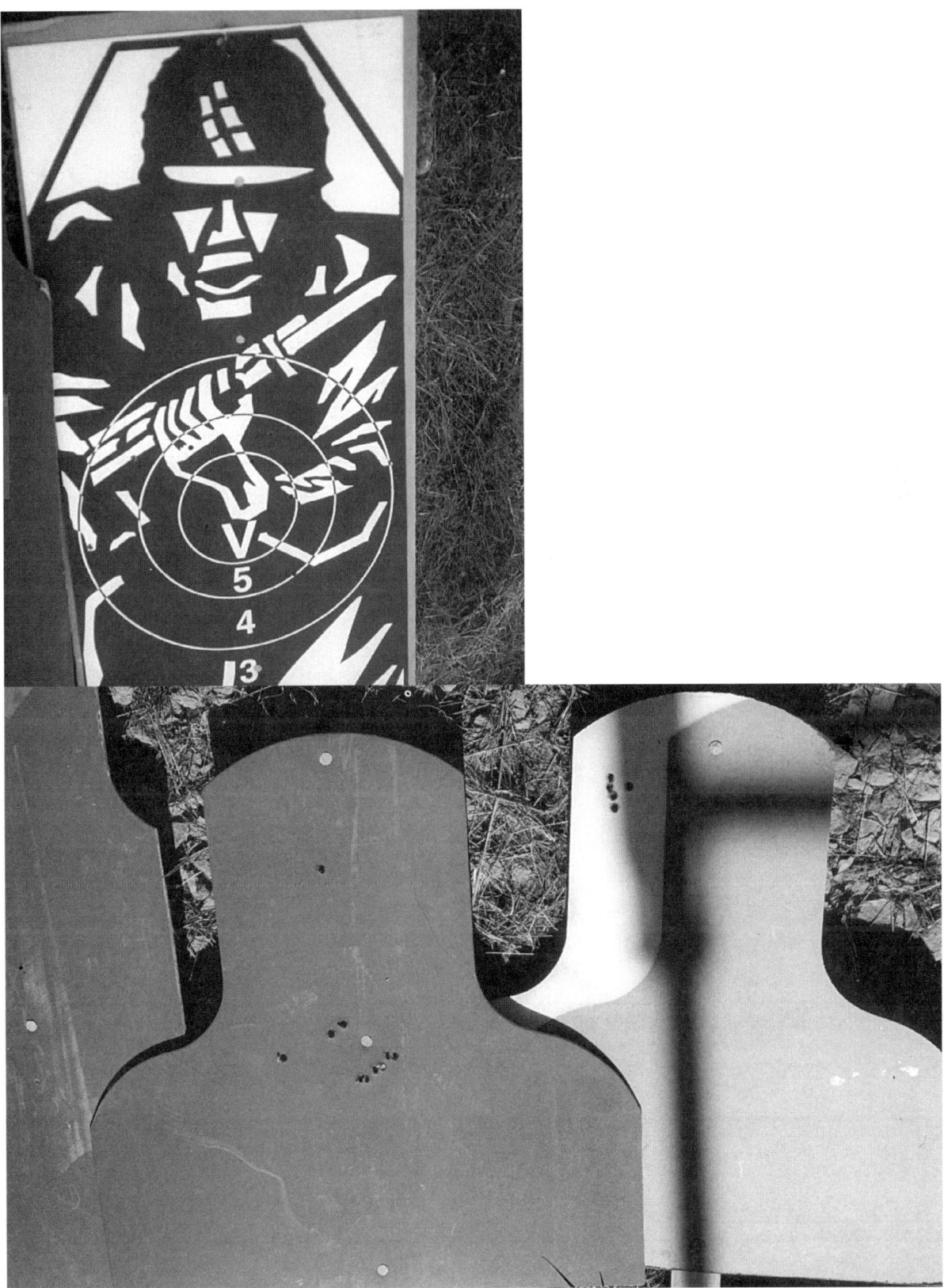

Not bad. Most of the snipers look like they practice a lot. At top is the Fig. 11 "cartoon" or "Hun" target with 10 bullet holes and 5 "V"s shot at 600 and 500 yards. At bottom left is the narrowed Fig. E moving target with five holes in it fired at 400 yards. Middle target is Fig. F bobber shot at 300 yards and 200 yards with all 10 rounds on it. Nice 5 shot group on the hostage target with all rounds in the "bad guy's" incapacitation area. Testimony to the precision and deadliness of the Army Scout-Sniper team and the M24 SWS. (Mike R. Lau)

Not until early evening did the relays finally finish. Major Wigger and SFC Gooch did a super job of coordination and planning and the event went smoothly. Placing 1st in the Special Reaction Match was SGT Sean Brownsen of Oklahoma. Second place went to MSG Mike Strasburger of Nebraska and third went to 1SG Greg Neiderhiser. (Mike R. Lau)

The day ain't over till it's over. Day 1's UKD 10 target event had a three way tie for second place and a two way tie for third and a three way tie for fourth. A shoot off match ensued from the 600 yard line as the sun was getting low. When it was over, SGT Wayne Morgan of Nebraska was declared the second place winner. Following him was SSG Seamus Clarkin of Arizona and in fourth, SSG Leslie Dolan of Arkansas. (Mike R. Lau)

CHAPTER 15: The Sniper Match: Sustainment Training Through Practical Precision Shooting

Day 4 saw no sniper matches with the M24 SWS. All of the 900+ competitors in the Wilson Matches were on either the combat pistol or combat rifle ranges. The Scout-Snipers participated in both. The Combat Pistol Match is fired with the standard 9mm or .45 caliber service weapons. In Stage 1, a total of 12 rounds are fired at four Fig. 11 targets. One target is engaged with three rounds on each 4 second exposure beginning with the left most target. Two shots must be placed in the torso scoring area first and then one round in the face scoring area of each target. All firing for this event was done from the standing position at 15 yards. (Mike R. Lau)

The second stage of fire consists of 24 rounds on four each Fig. 11 targets. On the first exposure, all four targets must be engaged with one round each from left to right and then re-engage the first target with 2 more rounds. All of this in 7 seconds. The next exposure lasts 6 seconds and the targets are engaged in the same manner as the first. On the third exposure of 5 seconds, only targets 2, 3, and 4 are engaged with two rounds each. On the last exposure engage the last three targets in the same manner in 4 seconds. (Mike R. Lau)

Like the pistol matches, there are numerous rifle matches that can be selected for the Wilson Matches. The specific Combat Rifle Match chosen has 9 stages requiring 60 rounds for a total of 300 points. Only service rifles such as the M16A1 or A2 are allowed. Stage 1 and 2 are slow and rapid fire prone from 300 yards requiring 6 shots in 6 minutes and 8 shots in 30 seconds on the Fig. 11 target. Competitors then walk to the 200 yard line and assume a kneeling position. (Mike R. Lau)

Four rounds are fired in 25 seconds from kneeling followed by 8 rounds in eight 3 second exposures from the sitting. Five each 2 round exposures of 5 seconds each from the prone position uses up 10 more rounds. The shooters then move to the 100 yard line for stages 6 and 7. From the 100 yard line, six rounds are fired at three targets in 30 seconds from the kneeling. (Mike R. Lau)

CHAPTER 15: The Sniper Match: Sustainment Training Through Practical Precision Shooting

Next 4 rounds are fired at a single target exposed four times from the standing. In Stage 8, shooters advance on line between target exposures. Shooters go from standing to kneeling and fire 8 rounds at two walking and two running targets in a 10 meter wide lane. (Mike R. Lau)

Many foreign countries were represented in the many Combat Matches held throughout the week and these had their own full agenda of interesting events. British soldier is armed with the 5.56mm L85A1. (Mike R. Lau)

Beside British and Canadians, teams from South Africa, Belgium, Mexico and others were present. Many of the foreign countries have telescopes on their assault rifles. The U.S. is one of the few that doesn't. (Mike R. Lau)

Canadian soldier waiting command to load and advance. Rifle is C-7, full auto version of Colt M16A2 with the Elcan optical sight. (Mike R. Lau)

CHAPTER 15: The Sniper Match: Sustainment Training Through Practical Precision Shooting

Stage 9 finds the shooter advancing from the 50 yard line. Three targets of 2 second exposure each are engaged with two rounds. Each exposure is separated by a 5-10 second interval. This completes the Combat Rifle Match. (Mike R. Lau)

Nike™ athletic wear is found in almost all activities (under Brit's helmet). (Mike R. Lau)

Day 5. The Scout-Snipers are tired but their spirits are up as they prepare for the last day's events. (Mike R. Lau)

First comes the cold bore unknown distance match. It is like the first day's match except distance is farther. SFC Gooch uses Unertl 100mm Team Scope w/32X eyepiece to spot hit on iron maiden. (Mike R. Lau)

CHAPTER 15: The Sniper Match: Sustainment Training Through Practical Precision Shooting

Just like on Day 1, SFC Tim Weber on M24 is assisted by his spotter, SGT James Kringlie. Together they plan for a perfect first round hit. SFC Weber mils 40" target as 1.4 mils high. Target distance is estimated to be 726 meters. Tim puts 7+2 on the M3A's BDC. They estimate the wind as 3 mph from 9 o'clock. Windage is 1/2 moa left. SFC Weber fires. The target is spotted as a hit. (Mike R. Lau)

SSG John Hamel estimates target height as 1.45 mils and also estimates target at 726 meters. He puts on almost the same elevation and windage as SFC Weber and gets a first round hit with a "V". The Scout-Snipers are excellent at the cold bore shot and a number of them made first round hits. A shoot off will take place later to determine who will place 1st and who will place 9th. When this is over, SSG John Hamel will be the winner of the cold barrel engagement match for Day 1 and 5. (Mike R. Lau)

The last event of the 1997 Wilson Sniper Match is the Rapid Multiple Target. In this event, each Scout-Sniper team will engage numerous unknown distance targets with the M24, the M16A2, and the 9mm Beretta or SIG. (Mike R. Lau)

An exact number of rounds for each weapon is issued and rules specify which team member can engage which target with what weapon. Before the M16A2 and pistol can be used to engage targets, two long distance iron maidens must be engaged and hit by the team. The targets have been identified to the team. This is a timed event, the prep time has elapsed, and the teams may commence firing when ready. (Mike R. Lau)

CHAPTER 15: The Sniper Match: Sustainment Training Through Practical Precision Shooting 307

Observers will spot on iron maidens and team is to let observer know when they are ready to fire. (Mike R. Lau)

Spectators wait anxiously as the Scout-Snipers work up their data. Civilians are allowed to observe the Wilson Matches. (Mike R. Lau)

Team Texas, SFC Pete Carpentier engages iron maidens at long range as SPC Alan Donaldson corrects windage and elevation. Unknown distance for iron maidens is around 800 meters. (Mike R. Lau)

Another team signals they are ready to fire and engages the far targets with the M24. (Mike R. Lau)

CHAPTER 15: The Sniper Match: Sustainment Training Through Practical Precision Shooting 309

Once the iron maidens are confirmed hit, the spotter moves to his firing position and loads his M16A2 with a magazine of 5.56mm M855. The sniper now shifts his M24 to engage a bank of four falling plate targets at a closer unknown distance. (Mike R. Lau)

With 40 rounds of M855, spotter begins to engage his first bank of 6 falling plate targets at an unknown distance several hundred meters out with the M16A2. A second bank of 6 targets at another unknown distance are engaged with his remaining rounds. (Mike R. Lau)

When the sniper finishes off his bank of targets or fires all twenty 7.62 M118 SB he will engage a near bank of 4 targets with the service pistol and 15 rounds. When all targets are hit, or the team runs out of ammunition, time will be called. (Mike R. Lau)

SFC Tim Sowa checks the time with the observer as SGT Chris Larson unloads his pistol after downing all of his metal targets. The Montana Team placed second in this event. (Mike R. Lau)

CHAPTER 15: The Sniper Match: Sustainment Training Through Practical Precision Shooting 311

All pistol targets are down, but the spotter is still engaging his targets. (Mike R. Lau)

All far targets except one is down and the spotter is out of ammunition. Meanwhile the sniper is still engaging the close targets and has 2 more to go. (Mike R. Lau)

Team Minnesota, SFC Tim Weber and SGT James T. Kringlie accumulated 1472 points for all team matches. SFC Weber also had the second highest individual score accumulated with 515 points. (Mike R. Lau)

SSG Norman Plaat's team, 3/126 Infantry from Michigan placed third in the Multiple Rapid Target Match. His team spotter is PVT Mike Maxfield. SSG Plaat ranked 6th in individual scoring for the Wilson Sniper Match with 465 points. (Mike R. Lau)

CHAPTER 15: The Sniper Match: Sustainment Training Through Practical Precision Shooting

What makes a winner? During each firing event, while others await their relay, MSG Mike Strasburger and SGT Wayne Morgan concentrate on the task at hand. You cannot see the targets from this position, but you can watch the relay on the firing line. Looking for wind indicators, studying the firing positions. How rapidly are they firing? Which shooters finish first and why? How did they score by finishing first? Work out problems with the spotter on wind and range estimation previously encountered. (Mike R. Lau)

Sniper matches are a team effort. To advance the team, the experienced shooter must encourage and assist the new shooter in every way he or she can, even if it is at the expense of his own individual score and the new shooter firing higher scores. Bringing up your partner's score will pay off in the end. SGT Morgan's individual score of 415 gave the team 1st Place and the Logan Trophy. The Nebraska team is a result of one of the best marksmanship programs in the Guard under the leadership of Maj. Tom Brewer. Maj. Brewer and MSG Strasburger won the team trophy at the 1997 Proskopathlon Sniper Match. (Mike R. Lau)

MSG Mike Strasburger does not use his bipod if given a choice. Note the removal of the bipod and the extra low firing position taken with the sandbags. SGT Morgan gets directly behind the rifle to spot the bullet trace. (Mike R. Lau)

Teamwork, concentration, skill, and determination separate the top winners from the rest. Congratulations, MSG Mike Strasburger, winner with the highest individual score of 521 points. Next time we enter a match, we might check who we are shooting next to. He might be Mike Strasburger, the hard core, dedicated, long range professional shooter. (Mike R. Lau)

BIBLIOGRAPHY

Books and Manuals

Ackley, Parker O., **Handbook for Shooters and Reloaders**, Volume I and II, Plaza Publishing, Salt Lake City, Utah, P.O. Ackley, 1962.

Adams, Ronald J.; McTernan, Thomas M.; Remsberg, Charles, **Street Survival, Tactics for Armed Encounters**, Calibre Press Inc., Northbrook, IL, 1980, 13th printing 1989.

Barnes, Frank C., **Cartridges of the World**, 7th Edition, DBI Books, Inc., Northbrook, IL, 1993.

Chandler, Norman, **White Feather**, Norman A, Chandler, Jacksonville, NC, 1997.

Chandler, Norman A. and Chandler, Rocky F., **Death From Afar, A History of USMC Sniping, Vol. 1, 2, 3, and 4**, Norman A, Chandler, Jacksonville, NC, various dates.

Crossman, Edward C., **Military and Sporting Rifle Shooting**, Small Arms Technical Publishing Co., 1932, Wolf Publishing Company, Prescott, AZ, 1986.

Frost, George E., **Ammunition Making**, National Rifle Association of America, Washington, DC, 1990.

Gallagher, CSM James J., USA (Ret.), **Low Intensity Conflict, A Guide For Tactics, Techniques, and Procedures**, Stackpole Books, Harrisburg, PA, 1992.

Gunston, Bill and Spick, Mike; **Modern Air Combat**, Crescent Books, NY 1983.
de Haas, Frank, **Bolt Action Rifles**, 3rd Edn, DBI Books, Inc., Northbrook, IL, 1995.

Hatcher, Julian S., Maj. Gen., USA (Ret.), **Hatcher's Notebook**, The Stackpole Company, Harrisburg, PA., 1962.

Headquarters US Marine Corps, **Operation and Maintenance, Rifle, Sniper, M40A1 and Related Optical Equipment**, TM 00539-13&P/1, US Marine Corps, 30 Nov 1981.

Headquarters US Marine Corps, **M40A1 Sniper Rifle 7.62mm**, FMFRP 0-11A, US Marine Corps, 13 April 1989.

Headquarters, US Marine Corps and Department of the Army, **Organizational and Intermediate Maintenance, Rifle, M16A2 W/E**, TM 05538C-23&P/2A and TM 9-1005-319-23&P, US Govt. Printing Office, August 1987.

Headquarters Department of the Army and Air Force, **Small Arms Ammunition**, TM 9-1305-200 and TO 11A13-1-101, US Government Printing Office, 1961.

Headquarters Department of the Army, **Small Arms Targets and Target Equipment**, TM 9-6920-210-24P, US Government Printing Office, January 1962.

Headquarters Department of the Army and Air Force, **Ammunition For Aircraft Guns**, TM 9-1901-1 and TO 11A-1-39, US Government Printing Office, 1957.

Headquarters Department of the Army, **Machine Gun, Caliber .50, Browning, M2, Heavy Barrel, Flexible, W/E**, TM 9-1005-213-10, US Government Printing Office, 1968.

Headquarters Department of the Army, **Browning Machine Gun, Caliber .50 HB, M2**, FM 23-65, US Government Printing Office, 1991.

Headquarters Department of the Army, **Mission Training Plan, Infantry Platoon/Squad and Scout-Sniper**, ARTEP 7-92-MTP, March 1989, US Government Printing Office, 1992.

Headquarters Department of the Army, **M14 and M14A1 Rifles and Rifle Marksmanship**, FM 23-8, US Government Printing Office, 1974.

Headquarters Department of the Army, **Sniper Training**, FM 23-10, US Government Printing Office, 1989.

Headquarters Department of the Army, **Sniper Training and Employment**, TC 23-14, US Government Printing Office, 1969.

Headquarters Department of the Army, **Operator's Manual 7.62mm M24 Sniper Weapons System SWS**, TM 9-1005-306-10, US Government Printing Office, 1989.

Headquarters Department of the Navy, **Sniping**, FMFM 1-3B, U.S. Marine Corps, US Government Printing Office, 1969.

Headquarters Department of the Navy, **Sniping**, FMFM 1-3B, U.S. Marine Corps, 1984.

Henderson, Charles, **Marine Sniper**, Stein and Day, Briarcliff Manor, New York, 1986.

Hogg, Ivan, edit., **Jane's Security and Counter-Insurgency Equipment, 1994-1995**, Jane's Information Group, Alexandria, VA, 1994.

Hughes, David R., **The History and Development of the M16 Rifle and Its Cartridge**, Armory Publications, Oceanside, California, 1990.

Kronenwetter, Michael, **The War on Terrorism**, The Westport Publishing Group, Edgewood Cliffs, NJ, 1989.

Lacy, John F., **The Remington 700, A History and Users Manual, 25 Years 1962-1987**, John F. Lacy, Elk City, Oklahoma, 1989.

Lapua Ltd., **Sniping and the .338 Lapua Magnum**, Lapua Ltd., Lapua, Finland, no date.

Lonsdale, Mark, **Sniper Counter Sniper**, S.T.T.U., Los Angeles, CA, 1987.

Lonsdale, Mark, **Sniper II**, S.T.T.U., Los Angeles, CA, 1992.

Long, Duncan, **Modern Sniper Rifles**, Paladin Press, Boulder. CO, 1988.

National Rifle Association, **NRA Firearms and Ammunition Fact Book**, National Rifle Association of America, Washington, D.C., 1964.

Owens, James R., M/SGT USMC (Ret), **Leather Sling and Shooting Positions**, JAFEICA Publishing, Milwaukee, Wisconsin, 1994.

Owens, James R., M/SGT USMC (Ret), **Reading the Wind and Coaching Techniques**, JAFEICA Publishing, Milwaukee, Wisconsin.

Owens, James R., M/SGT, USMC (Ret), **Sight Alignment, Trigger Control and "The Big Lie"**, JAFEICA Publishing, Milwaukee, Wisconsin.

Perrett, Bryan, **Weapons of the Falklands War**, Blanford Press, U.K., 1982.

Plaster, John L., Major, USAR (Ret.), **The Ultimate Sniper**, Paladin Press, Boulder, CO, 1993.

Rapp, Burt, **The Police Sniper**, Loompanics Unlimited, Port Townsend, WA, 1988.

Realist Riflescopes, **Telescope/Auto-Ranging ART II Operation Manual**, Menomonee Falls, WI, date unknown.

Remsberg, Charles, **The Tactical Edge, Surviving High Risk Patrol**, Calibre Press Inc., Northbrook, IL, 1986, 10th printing, 1994.

Stevens, R.Blake and Ezell, Edward C., **The Black Rifle**, Collector Grade Publications, Toronto, Canada, 1987.

Sears, Francis W.; Zemansky, Mark W., Young, Hugh D., **University Physics, 6th Edn.**, Addison-Wesley Publishing Co., Reading, Mass., 1982.

Senich, Peter, **One Round War**, Copyright Peter R. Senich, 1996, Paladin Publishing Co., Boulder, CO.

Senich, Peter, **The Complete Book of US Sniping**, Paladin Publishing Co., Boulder, CO., 1988.

Sierra Bullets L.L.C. and John C. Clarke, **Exterior Ballistics Software Version III**, Sierra Bullets and John C. Clarke, 1996.

Sierra Bullets, **Sierra Rifle Reloading Manual**, 4th Edn, 50th Anniversary, Sierra Bullets, L.P., Sedalia, MO, 1995.

Skennerton, Ian, **The British Sniper, British and Commonwealth Sniping and Equipments**, 1915-1983, Ian Skennerton, Australia, 1983.

Stoeger Publishing Company, **The Shooter's Bible, No.88, 1997 Edition**, Wayne, NJ, 1996.

Tobias, Ronald, **They Shoot To Kill, A Psycho-Survey of Criminal Sniping**, Paladin Press, Boulder Co., 1981.

Truby, David J., **Silencers, Snipers, and Assassins, An Overview of Whispering Death**, Paladin Press, Boulder, Colorado, 1972.

U.S. Army Limited War Laboratory, **Installation and Operation for the Adjustable Ranging Telescope (ART) Mounted on the Match Conditioned 7.62MM M-14 Rifle**, Aberdeen Proving Ground, Maryland, September 1968.

US Army National Guard Marksmanship Training Unit, **Match Program, 1997 Winston P. Wilson Rifle-Pistol-LMG-Sniper Championships**.

US Army National Guard Marksmanship Training Unit, **Sniper Field Data Book**, US Army National Guard Scout-Sniper School, Aug 1997.

US Army National Guard Marksmanship Training Unit, **Marksmanship Data Book**, US Army National Guard Scout-Sniper School, Jun 1997.

US Army National Guard Marksmanship Training Unit, **Scout-Sniper School Student Handout**, Feb 1997.

US Army Marksmanship Training Unit, **International Rifle Marksmanship Guide 1970**, US Govt Printing Office, 1972

US Army Marksmanship Training Unit, **Rifle Marksmanship Guide 1970**, US Govt Printing Office, 1972

U.S. Marine Corps, **Scout/Sniper Data Book**, NAVMC, 1979.

U.S. Marine Corps, **First Marine Division Scout Sniper School, Student Outline**, Camp Pendleton, California, unknown date.

U.S. Marine Corps, **Law Enforcement Marksmanship**, S/S Instructor School, Quantico, VA. 1996

War Department, **Browning Machine Gun, Caliber .50, AN-M2, Aircraft, Basic**, TM 9-225, US Government Printing Office, 1951.

Zisa, Thomas, **5.56mm Match Ammunition**, U.S. Army Armament Research, Development and Engineering Center, Close Combat Armaments Center, Picatinny Arsenal, NJ., April 1995.

Magazine Articles and Letters

Ash, Wayne, "Bullet Drift", *Precision Shooting*, April 1991.

Chesnut, Sam, "Rarely Used, But Always There", *The Tactical Edge Magazine*, Winter 1997.

Editors, "Austin's Overlooked Riflemen", *The American Rifleman*, pp.20+, November 1966.

Fairburn, Dick, "Getting the Green Light", *Police,* pp.36+, June 1993.

Fairburn, Dick, "Rural Response", *Police,* pp. 42+, June 1993.

Hansen, Denny, "Savage's 110 FP Tactical Rifle", *S.W.A.T.*, pp. 26+, Aug 1997,

Hantke, Paul, "Proskopathlon 1997", *American Survival Guide*, pp.60+, Dec 1997.

Harris, C.E., "Does NATO's New 5.56mm Round Measure Up, *The American Rifleman*, pp. 42+, May 1982.

McCombs, William E.; Vester, Paul J. Jr.; Smith, J.D.; Yarbor, John W. , "Test Report, Phase One, U.S. Marine Corps 7.62x51mm M118 Special Ball, Product Improvement Program", (NAVSURFWARCENDIV Crane Testing), 22 Aug 1995.

Bray, Ed, Memorandum: "Test of 7.62mm with special 175 gr Bullet", Lake City Army Ammunition Plant, Lake City, MO, 19 October 1994.

Hillen, C.A., VP and General Manager, Winchester Ammunition Div, Olin Corp., "Request For Cost Estimate For Statement Of Work (SOW) 7.62mm M118 Special Ball Improvement Program, Sub Task 308, Cost Estimate Number 21-95, To: Commander, LCAAP", LCAAP, 24 Feb 95.

Hillen, C.A., VP and General Manager, Winchester Ammunition Div, Olin Corp., "Request For Cost Estimate For Statement Of Work (SOW) 7.62mm M118 Special Ball Improvement Program, Sub Task 308, Cost Estimate Number 21-95, To: Commander, LCAAP", LCAAP, 14 Mar 95.

Hillen, C.A., VP and General Manager, Winchester Ammunition Div, Olin Corp., "Request For Cost Estimate For Statement Of Work (SOW) 7.62mm M118 Special Ball Improvement Program, Sub Task 308, Cost Estimate Number 21-95, To: Commander, LCAAP", LCAAP, 21 Mar 95

Mardo, John, ARDEC, Project Leader, "Memorandum For Record, 23 Oct 95; To: Paul Riggs, ARDEC; Subject: M118 Special Ball Improvement Program - First User Accuracy Test".

Chung, Sung K., "Memorandum For Record: To: AMSTA-AR-CCL-B, Paul Riggs, Aeroballistics Engineer, Aeroballistics Branch, MATCD, 2 Jan 96; Subject: Stability of 7.62mm M118 (LR) Ammunition at Extreme Ambient Conditions and Range of Mach Numbers".

Hettel, James W., U.S. Army Research and Development Command, Fire Control and Small Caliber Weapons Systems Laboratory, Dover, NJ., " Technical Report ARSCD-TR-81018, 7.62-mm Match Cartridge Accuracy Improvement Program", July 1981.

Holmes, Maj. Donald C., "NATO's New Rifle Cartridge", **The American Rifleman**, pp. 28+, October 1981.

James, Frank W., "The .300 Whissssper" (sic), **Guns Magazine**, pp. 64+, October 1993.

Karwan, Chuck, "Sniper SNAFU, U.S. Army's M24 SWS Defeated by Design", **Soldier of Fortune**, pp.64+, March 1989.

Kokalis, Peter G., "SNIPER, U.S. Army Adopts Remington M24 SWS", **Soldier of Fortune**, pp. 88+, Jul 88.

Kokalis, Peter G., "Terminal Solution (BAR Interdiction Rifle)", **Soldier of Fortune**, pp. 44+, Jan 96.

Leupold and Stevens, Inc., "Ultra Scope, Model 10X-M3 Rifle Scope, Specifications and Instructions, Publication Part Number: 42620", November 12, 1985.

Lewis, Jack, "A Matter of Accuracy", **Leatherneck,** pp. 19+, May 1996.

Lightbody, Andy, "The International Terrorist", **International Combat Arms,** pp.77+, April 1984.

Nilberg, H.G., "Air Density and Terminal Velocity", **The American Rifleman**, pp.69, September 1965.

Owens, Franklin, "Army Lab Adds ART to Sniping", ***The American Rifleman***, pp.47+, May 1969.

Pagel, Keith, "Some Notes on The Development of the Hi Explosive Fifty Caliber Cartridge," pp.15+, ***Very High Power, 1993 #4,*** .50 Caliber Shooters Assn., 1993.

Pate, James L., "Black Suits, Badges, and Bradleys", ***Soldier of Fortune***, pp.56+, Aug 96.

Pate, James L., "Amateurs and Assassins II", ***Soldier of Fortune***, pp.56+, Dec 96.

Poyer, Joe, "Countering the Modern Terrorist", ***International Combat Arms***, pp.40+, July 1985.

Sterett, Larry S., "A New Sniper Round", ***The Gun Digest 1986***, DBI Books, Inc., Northbrook, Ill, 1985.

Sundra, Jon, "Two New Cartridges (.338 Lapua)", ***Guns Magazine***, pp. 20+, June 1990.

Thomas, Foster, "Mirage", ***The American Rifleman***, pp. 26+, May 1966.

Department of the Army, Small Caliber Ammunition Test Procedure (SCATP-7.62), 7.62mm M118 Special Ball, Mil-C-46934B.

Department of the Army, Small Caliber Ammunition Test Procedure (SCATP-7.62), 7.62mm M852 Match, Mil-C-63450 (AR).

Department of the Army, Small Caliber Ammunition Test Procedure, Caliber. .50 Ball, M33, Mil-C-10190D.

Department of the Army, Small Caliber Ammunition Test Procedure (SCATP-5.56mm Heavy Ball), 5.56mm Ball, M855, Mil-C-63989C.

U.S. Marine Corps, (Title unknown, has early test fire results of Unertl 10X with internal construction drawing), USMC, date unknown.

OIC Scout/Sniper School, U.S. Marine Corps, "Scope Test for M40A1 Sniper Rifle", Quantico, VA, 16 November 1978.

OIC Scout/Sniper School, U.S. Marine Corps, "Bullet Deflection/Glass Fragmentation Test", Quantico, VA, 9 June 1978 and 5 Feb 1985.

Oskin, Lieutenant Richard J., Counter Sniper Support, U.S. Secret Service, "Ballistic Test, Bullet Stability through Window Glass", Administrative and Program Support Branch, U.S. Secret Service, September 23, 1984.

Various authors and articles, ***Guns and Weapons For Law Enforcement***, Various Dates.

Various authors and articles, ***The Police Marksman***, Various Dates.

Walker, Greg, "U.S. Army's New M24 Sniper Rifle", ***International Combat Arms***, March 1989.

Warner, Ken, edit., ***The Gun Digest 1997***, DBI Books, Inc., Northbrook, Ill, 1985

Yarbor, John, "The Development of the Navy .300 Winchester Magnum Cartridge Configuration", ***Precision Shooting Special Edition No. 3, Volume 1***, November 1995

Creighton, Audette, "Testing Rifles and Ammunition", ***Precision Shooting***, Oct, Nov, Dec 1993, Jan, Feb 1994.

BIBLIOGRAPHY

A thought painted on the wall of the Record Journal Building in Oklahoma City by an anonymous rescue "Team 5". (Sam Chesnut)

APPENDIX I:
Ballistics Tables

BALLISTICS TABLE IN **YARDS** FOR
7.62mm M80 Ball, 148 gr FMJBT for 24" Barrel
100 yd. zero, 59 deg F., Elevation ASL, Press 29.53 in, Hum 78%
BC's: .404 (H), .404 (M), .404 (L)

Range YARDS	Velocity (fps)	Maximum Ordinate (inches)	Bullet Path (inches)	Time of Flight (sec)	Come Up to range (MOA)	Come Up cumulative (MOA)
0	2854					
100	2627	+.1	zero	0.1	0	0.00
200	2411	+1.8	-3.5	0.2	2	2
300	2205	+5.5	-13.1	0.4	2.5	4.5
400	2010	+11.4	-29.7	0.5	3	7.5
500	1825	+20.2	-55.0	0.7	3.5	11
600	1653	+32.7	-90.7	0.8	4	15
700	1494	+49.8	-139.2	1.0	5	20
800	1353	+73.2	-203.3	1.2	5.5	25.5
900	1232	+104.2	-286.4	1.5	6.5	32
1000	1134	+145.0	-392.5	1.7	8	40

Come ups rounded to 1/2 MOA.

BALLISTICS TABLE IN **METERS** FOR
7.62mm M80 Ball, 148 gr FMJBT for 24" Barrel
100 yd. zero, 59 deg F., Elevation ASL, Press 29.53 in, Hum 78%
BC's: .404 (H), .404 (M), .404 (L)

Range METERS	Velocity (fps)	Maximum Ordinate (inches)	Bullet Path (inches)	Time of Flight (sec)	Come Up to range (MOA)	Come Up cumulative (MOA)
0	2854					
100	2606	+.2	zero	0.1	0	0.00
200	2372	+2.4	-4.6	0.3	2.5	2.5
300	2149	+6.9	-16.6	0.4	3	5.5
400	1939	+19.3	-37.7	0.6	4	9.5
500	1742	+25.5	-69.8	0.7	4.5	14
600	1562	+41.6	-115.6	0.9	6.5	20.5
700	1400	+64.3	-178.5	1.0	5	25.5
800	1260	+95.6	-262.7	1.5	7.5	33
900	1147	+137.9	-373.2	1.7	8.5	41.5
1000	1063	+193.6	-515.6	2.0	10	51.5

Come ups rounded to 1/2 MOA.

BALLISTICS TABLE IN **YARDS** FOR
.308 Win Match 168 gr HPBT for 24" Barrel
7.62mm M852 Match 168 gr HPBT for 24" Barrel
100 yd. zero, 59 deg F., Elevation ASL, Press 29.53 in, Hum 78%
BC's: .462 (H), .447 (M), .424 (L)

Range YARDS	Velocity (fps)	Maximum Ordinate (inches)	Bullet Path (inches)	Time of Flight (sec)	Come Up to range (MOA)	Come Up cumulative (MOA)
0	2600					
100	2405	+.1	zero	0.1	0	0
200	2219	+2.3	-4.5	0.2	2.25	2.25
300	2039	+6.5	-16.1	0.4	3	5.25
400	1861	+13.6	-36.0	0.5	3.75	9
500	1694	+23.9	-65.9	0.7	4.25	13.25
600	1539	+38.4	-108.0	0.9	4.75	18
700	1399	+58.3	-164.8	1.0	5.5	23.5
800	1276	+85.2	-239.0	1.3	6.25	29.75
900	1173	+120.6	-335.5	1.6	7.5	37.25
1000	1092	+166.5	-456.7	1.8	8.25	45.5

Come ups rounded to 1/4 MOA.

BALLISTICS TABLE IN **METERS** FOR
.308 Win Match 168 gr HPBT for 24" Barrel
7.62mm M852 Match 168 gr HPBT for 24" Barrel
100 yd. zero, 59 deg F., Elevation ASL, Press 29.53 in, Hum 78%
BC's: .462 (H), .447 (M), .424 (L)

Range METERS	Velocity (fps)	Maximum Ordinate (inches)	Bullet Path (inches)	Time of Flight (sec)	Come Up to range (MOA)	Come Up cumulative (MOA)
0	2600					
100	2387	+.3	zero	0.1	0	0.00
200	2185	+3.0	-5.8	0.3	3	3
300	1988	+8.3	-20.3	0.5	3.75	6.75
400	1798	+17.0	-45.4	0.6	4.5	11.25
500	1620	+30.0	-83.3	0.8	5.5	16.75
600	1459	+48.8	-137.1	1.0	6	22.75
700	1316	+75.0	-210.6	1.2	7.25	30
800	1197	+110.7	-308.1	1.5	8.5	38.5
900	1104	+158.4	-434.6	1.8	9.75	48.25
1000	1035	+220.3	-595.1	2.1	11.25	59.5

Come ups rounded to 1/4 MOA.

APPENDIX I: Ballistics Tables

BALLISTICS TABLE IN **YARDS** FOR
.308 Win Match 168 gr HPBT for 26" Barrel
7.62mm M852 Match 168 gr HPBT for 26" Barrel
.30-06 Sprg Match 168 gr HPBT for 24" Barrel
100 yd. zero, 59 deg F., Elevation ASL, Press 29.53 in, Hum 78%
BC's: .462 (H), .447 (M), .424 (L)

Range YARDS	Velocity (fps)	Maximum Ordinate (inches)	Bullet Path (inches)	Time of Flight (sec)	Come Up to range (MOA)	Come Up cumulative (MOA)
0	2700					
100	2504	+.1	zero	0.1	0	0
200	2314	+2.1	-4.0	0.2	2	2
300	2132	+6.1	-14.6	0.4	2.75	4.75
400	1952	+12.4	-32.7	0.5	3.5	8.25
500	1779	+21.8	-60.0	0.7	3.75	12
600	1618	+35.0	-98.3	0.9	4.25	16.25
700	1470	+53.0	-149.9	1.0	5.25	21.5
800	1337	+77.4	-217.7	1.3	5.75	27.25
900	1224	+109.4	-304.9	1.5	6.75	34
1000	1131	+151.4	-415.4	1.8	7.5	41.5

Come ups rounded to 1/4 MOA.

BALLISTICS TABLE IN **METERS** FOR
.308 Win Match 168 gr HPBT for 26" Barrel
7.62mm M852 Match 168 gr HPBT for 26" Barrel
30-06 Sprg Match 168 gr HPBT for 24" Barrel
100 yd. zero, 59 deg F., Elevation ASL, Press 29.53 in, Hum 78%
BC's: .462 (H), .447 (M), .424 (L)

Range METERS	Velocity (fps)	Maximum Ordinate (inches)	Bullet Path (inches)	Time of Flight (sec)	Come Up to range (MOA)	Come Up cumulative (MOA)
0	2700					
100	2487	+.2	zero	0.1	0	0.00
200	2280	+2.7	-5.2	0.3	2.5	2.5
300	2083	+7.6	-18.5	0.4	3.75	6.25
400	1886	+15.5	-41.3	0.6	4	10.25
500	1702	+27.4	-75.9	0.8	5	15.25
600	1534	+44.4	-124.7	1.0	5.5	20.75
700	1382	+68.1	-191.4	1.2	6.5	27.25
800	1251	+100.5	-279.9	1.4	7.75	35
900	1145	+143.9	-395.1	1.7	9	44
1000	1065	+200.6	-541.9	2.0	10.25	54.25

Come ups rounded to 1/4 MOA.

BALLISTICS TABLE IN **YARDS** FOR
.308 Win Match 175 gr HPBT for 24" Barrel
7.62mm M118 LONG RANGE (LR) 175 gr HPBT for 24" Barrel
BC's: .502 (H), .496 (M), .485 (L)
7.62mm M118 SB 173 gr FMJBT for 24" Barrel
100 yd. zero, 59 deg F., Elevation ASL, Press 29.53 in, Hum 78%
BC's: .494 (H), .485 (M), .463 (L)

Range	Velocity	Maximum Ordinate	Bullet Path	Time of Flight	Come Up to range	Come Up cumulative
YARDS	(fps)	(inches)	(inches)	(sec)	(MOA)	(MOA)
0	2600					
100	2424	+.1	zero	0.1	0	0.00
200	2255	+2.3	-4.4	0.2	2.25	2.25
300	2093	+6.5	-15.6	0.4	3	5.25
400	1938	+13.1	-34.8	0.5	3.5	8.75
500	1790	+22.6	-63.2	0.7	3.75	12.5
600	1648	+35.8	-102.4	0.9	4.5	17
700	1515	+53.6	-154.4	1.0	5	22
800	1394	+77.0	-221.7	1.3	5.75	27.75
900	1286	+107.3	-306.8	1.5	6.25	34
1000	1193	+146.1	-413.1	1.7	7.25	41.25

Come ups rounded to 1/4 MOA.

BALLISTICS TABLE IN **METERS** FOR
.308 Win Match 175 gr HPBT for 24" Barrel
7.62mm M118 LONG RANGE (LR) 175 gr HPBT for 24" Barrel
BC's: .502 (H), .496 (M), .485 (L)
7.62mm M118 SB 173 gr FMJBT for 24" Barrel
100 yd. zero, 59 deg F., Elevation ASL, Press 29.53 in, Hum 78%
BC's: .494 (H), .485 (M), .463 (L)

Range	Velocity	Maximum Ordinate	Bullet Path	Time of Flight	Come Up to range	Come Up cumulative
METERS	(fps)	(inches)	(inches)	(sec)	(MOA)	(MOA)
0	2600					
100	2408	+.3	zero	0.1	0	0
200	2225	+2.9	-5.7	0.3	2.75	2.75
300	2049	+8.0	-19.7	0.4	3.75	6.5
400	1882	+16.3	-43.7	0.6	4.5	11
500	1723	+28.3	-79.4	0.8	5	16
600	1572	+45.2	-129.1	1.0	5.5	21.5
700	1435	+68.2	-195.6	1.2	6.5	28
800	1312	+99.0	-282.5	1.5	7.25	35.25
900	1207	+139.4	-393.6	1.7	8.5	43.75
1000	1122	+191.4	-533.2	2.0	9.5	53.25

Come ups rounded to 1/4 MOA.

APPENDIX I: Ballistics Tables

BALLISTICS TABLE IN **YARDS** FOR
.308 Win Match 175 gr HPBT for 26" Barrel
7.62mm M118 LONG RANGE (LR) 175 gr HPBT for 26" Barrel
BC's: .502 (H), .496 (M), .485 (L)
7.62mm M118 SB 173 gr FMJBT for 26" Barrel
00 yd. zero, 59 deg F., Elevation ASL, Press 29.53 in, Hum 78%
BC's: .494 (H), .485 (M), .463 (L)

Range YARDS	Velocity (fps)	Maximum Ordinate (inches)	Bullet Path (inches)	Time of Flight (sec)	Come Up to range (MOA)	Come Up cumulative (MOA)
0	2700					
100	2520	+.1	zero	0.1	0	0
200	2348	+2.0	-4.0	0.2	2	2
300	2182	+5.9	-14.2	0.4	2.75	4.75
400	2023	+12.0	-31.8	0.5	3.25	8
500	1871	+20.8	-57.8	0.7	3.5	11.5
600	1725	+32.9	-93.8	0.8	4	15.5
700	1587	+49.1	-141.4	1.0	4.75	20.25
800	1460	+70.4	-202.8	1.2	5	25.25
900	1344	+98.1	-280.6	1.4	5.75	31
1000	1243	+133.5	-377.7	1.7	6.75	37.75

Come ups rounded to 1/4 MOA.

BALLISTICS TABLE IN **METERS** FOR
.308 Win Match 175 gr HPBT for 26" Barrel
7.62mm M118 LONG RANGE (LR) 175 gr HPBT for 26" Barrel
BC's: .502 (H), .496 (M), .485 (L)
7.62mm M118 SB 173 gr FMJBT for 24" Barrel
100 yd. zero, 59 deg F., Elevation ASL, Press 29.53 in, Hum 78%
BC's: .494 (H), .485 (M), .463 (L)

Range METERS	Velocity (fps)	Maximum Ordinate (inches)	Bullet Path (inches)	Time of Flight (sec)	Come Up to range (MOA)	Come Up cumulative (MOA)
0	2700					
100	2504	+.2	zero	0.1	0	0.00
200	2316	+2.6	-5.1	0.3	2.5	2.5
300	2137	+7.4	-18.0	0.4	3.5	6
400	1965	+14.9	-39.9	0.6	4	10
500	1802	+26.0	-72.7	0.7	4.5	14.5
600	1647	+41.5	-118.2	0.9	5.25	19.75
700	1502	+62.5	-179.1	1.1	5.75	25.5
800	1372	+90.6	-258.5	1.4	6.75	32.25
900	1258	+126.8	-360.0	1.6	7.75	40
1000	1162	+175.0	-487.9	1.9	8.75	48.75

Come ups rounded to 1/4 MOA.

BALLISTICS TABLE IN **YARDS** FOR
5.56mm 55 gr FMJBT M193 Ball for 20" Bbl, M16A2
100 yd. zero, 59 deg F., Elevation ASL, Press 29.53 in, Hum 78%
BC's: .272 (H), .245 (M), .235 (L)

Range YARDS	Velocity (fps)	Maximum Ordinate (inches)	Bullet Path (inches)	Time of Flight (sec)	Come Up to range (MOA)	Come Up cumulative (MOA)
0	3100					
100	2724	+0.0	zero	0.1	0	0
200	2365	+1.7	-3.2	0.2	1.5	1.5
300	2021	+5.5	-12.7	0.4	2.5	4
400	1710	+12.4	-30.7	0.5	3.5	7.5
500	1429	+23.8	-60.7	0.7	4.5	12
600	1221	+42.4	-107.7	0.9	6	18

-17.95 Come ups rounded to 1/2 MOA.

BALLISTICS TABLE IN **YARDS** FOR
5.56mm 62 gr FMJBT M855 Ball for 20" Bbl, M16A2
100 yd. zero, 59 deg F., Elevation ASL, Press 29.53 in, Hum 78%
BC's: .324 (H), .321 (M), .311 (L)

Range YARDS	Velocity (fps)	Maximum Ordinate (inches)	Bullet Path (inches)	Time of Flight (sec)	Come Up to range (MOA)	Come Up cumulative (MOA)
0	3050					
100	2756	+0.0	zero	0.1	0	0
200	2481	+1.6	-3.1	0.2	1.5	1.5
300	2220	+5.1	-11.8	0.3	2.5	4
400	1974	+10.9	-27.6	0.5	3.5	7
500	1740	+19.8	-52.4	0.7	4.5	10.5
600	1527	+33.1	-88.6	0.8	6	14.75
700	1342	+52.4	-139.9	1.0	5.5	20
800	1191	+79.9	-210.6	1.3	11.5	26.25

Come ups rounded to 1/4 MOA.

APPENDIX I: Ballistics Tables

BALLISTICS TABLE IN **METERS** FOR
5.56mm M855 Ball 62 gr FMJBT for 20" Bbl, M16A2

100 yd. zero, 59 deg F., Elevation ASL, Press 29.53 in, Hum 78%
BC's: .324 (H), .321 (M), .311 (L)

Range METERS	Velocity (fps)	Maximum Ordinate (inches)	Bullet Path (inches)	Time of Flight (sec)	Come Up to range (MOA)	Come Up cumulative (MOA)
0	3050					
100	2729	+0.0	zero	0.1	0	0
200	2431	+2.1	-4.1	0.2	2	2
300	2149	+6.4	-15.2	0.4	3	5
400	1884	+13.8	-35.4	0.5	3.75	8.75
500	1637	+25.4	-67.4	0.7	4.75	13.5
600	1419	+43.0	-114.9	0.9	5.75	19.25
700	1239	+69.3	-183.2	1.2	7	26.25
800	1104	+107.4	-278.7	1.5	8.5	34.75

Come ups rounded to 1/4 MOA.

BALLISTICS TABLE IN **YARDS** FOR
.223 Rem 69 gr HPBT Match for 24" Bbl

100 yd. zero, 59 deg F., Elevation ASL, Press 29.53 in, Hum 78%"
BC's: .301 (H), .305 (M), .317 (L)

Range YARDS	Velocity (fps)	Maximum Ordinate (inches)	Bullet Path (inches)	Time of Flight (sec)	Come Up to range (MOA)	Come Up cumulative (MOA)
0	3000					
100	2689	+.1	zero	0.1	0	0
200	2402	+1.7	-3.3	0.2	1.75	1.75
300	2134	+5.4	-12.7	0.4	2.5	4.25
400	1892	+11.7	-29.6	0.5	3.25	7.5
500	1668	+21.3	-56.4	0.7	4	11.25
600	1467	+35.8	-95.6	0.9	4.75	16
700	1295	+56.7	-151.1	1.0	9.25	25.25
800	1158	+86.4	-227.4	1.3	3.25	28.5
900	1059	+127.4	-329.6	1.6	8	36.5
1000	988	+182.0	-462.6	1.9	9.75	46.25

Come ups rounded to 1/4 MOA.

BALLISTICS TABLE IN **YARDS** FOR

.300 Win Magnum 190 gr HPBT Federal Match for 26" Barrel

100 yd. zero, 59 deg F., Elevation ASL, Press 29.53 in, Hum 78%
BC's: .533 (H), .525 (M), .515 (L)

Range YARDS	Velocity (fps)	Maximum Ordinate (inches)	Bullet Path (inches)	Time of Flight (sec)	Come Up to range (MOA)	Come Up cumulative (MOA)
0	3000					
100	2821	+0.0	zero	0.1	0	0
200	2649	+1.5	-2.9	0.2	1.5	1.5
300	2484	+4.5	-10.7	0.3	2	3.5
400	2324	+9.2	-24.1	0.5	2.5	6
500	2171	+16.0	-43.9	0.6	2.75	8.75
600	2022	+25.1	-71.2	0.7	3	11.75
700	1878	+37.1	-106.9	0.9	3.5	15.25
800	1742	+52.6	-152.4	1.0	3.75	19
900	1613	+72.4	-209.4	1.2	4.25	23.25
1000	1490	+97.3	-279.7	1.4	4.75	28
1100	1378	+128.6	-365.8	1.6	5.25	33.25
1200	1278	+167.5	-470.2	1.9	6	39.25

Come ups rounded to 1/4 MOA.

BALLISTICS TABLE IN **METERS** FOR

.300 Win Magnum 190 gr HPBT Federal Match for 26" Barrel

100 yd. zero, 59 deg F., Elevation ASL, Press 29.53 in, Hum 78%
BC's: .533 (H), .525 (M), .515 (L)

Range METERS	Velocity (fps)	Maximum Ordinate (inches)	Bullet Path (inches)	Time of Flight (sec)	Come Up to range (MOA)	Come Up cumulative (MOA)
0	3000					
100	2804	+0.0	zero	0.1	0	0
200	2618	+2.0	-3.8	0.2	2	2
300	2438	+5.7	-13.6	0.4	2.5	4.5
400	2266	+11.5	-30.4	0.5	3	7.5
500	2100	+19.9	-55.3	0.7	3.5	11
600	1940	+31.4	-89.6	0.8	4	15
700	1787	+46.8	-134.9	1.0	4.25	19.25
800	1644	+67.0	-193.3	1.2	5	24.25
900	1509	+93.0	-267.0	1.4	5.75	30
1000	1385	+126.5	-359.1	1.6	6	36
1100	1275	+168.9	-472.9	1.9	7	43
1200	1182	+222.3	-612.4	2.1	8	51

Come ups rounded to 1/4 MOA.

APPENDIX I: Ballistics Tables

BALLISTICS TABLE IN **METERS** FOR
338 (8.6 x 70mm) Lapua Magnum 250gr Lock-Base, 27" Barrel

100 yd. zero, 59 deg F., Elevation ASL, Press 29.53 in, Hum 78%
BC's: .662 (H), .662 (M), .662 (L)

Range METERS	Velocity (fps)	Maximum Ordinate (inches)	Bullet Path (inches)	Time of Flight (sec)	Come Up to range (MOA)	Come Up cumulative (MOA)
0	2952					
100	2795	+0.1	zero	0.1	0	0
200	2645	+2.0	-3.8	0.2	2	2
300	2499	+5.6	-13.5	0.4	2.5	4.5
400	2358	+11.2	-29.9	0.5	3	7.5
500	2221	+19.1	-53.8	0.6	3.25	10.75
600	2089	+29.6	-86.0	0.8	3.5	14.25
700	1961	+43.3	-127.8	1.0	4	18.25
800	1839	+60.7	-180.3	1.1	4.25	22.5
900	1721	+82.4	-245.2	1.3	4.75	27.25
1000	1610	+109.4	-324.0	1.5	5.25	32.5
1100	1505	+142.4	-418.8	1.7	5.5	38
1200	1408	+182.9	-532.2	1.9	6.25	44.25
1300	1319	+232.0	-666.5	2.2	11.25	55.5
1400	1239	+290.9	-824.7	2.4	13.25	68.75
1500	1170	+361.6	-1010.2	2.7	15.25	84

Come ups rounded to 1/4 MOA.

BALLISTICS TABLE IN **YARDS** FOR
Cal..50 Multi-Purpose HEIAP 665 gr. 33" Barrel

100 yd. zero, 59 deg F., Elevation ASL, Press 29.53 in, Hum 78%
BC's: .600 (H), .600 (M), .600 (L)

Range YARDS	Velocity (fps)	Maximum Ordinate (inches)	Bullet Path (inches)	Time of Flight (sec)	Come Up to range (MOA)	Come Up cumulative (MOA)
0	2898					
100	2742	+0.1	zero	0.1	0	0
200	2592	+1.6	-3.1	0.2	1.5	1.5
300	2446	+4.8	-11.4	0.3	2.25	3.75
400	2306	+9.7	-25.5	0.5	2.5	6.25
500	2170	+16.5	-46.1	0.6	3	9.25
600	2038	+25.8	-74.0	0.7	3	12.25
700	1911	+37.8	-110.3	0.9	3.5	15.75
800	1790	+53.0	-156.2	1.0	3.75	19.5
900	1674	+72.2	-212.8	1.2	4	23.5
1000	1565	+95.9	-281.8	1.4	4.5	28
1100	1462	+125.2	-365.0	1.6	5.25	33.25
1200	1367	+161.0	-464.6	1.8	5.5	38.75
1300	1282	+204.4	-582.8	2.0	6	44.75
1400	1206	+256.7	-722.2	2.3	6.75	51.5
1500	1142	+319.3	-885.6	2.5	7.5	59

Come ups rounded to 1/4 MOA.

U.S. Army recruiting poster dated 1994. (Mike R. Lau)

KIM's game. USMC Scout/Sniper School, Quantico, Virginia, 1984. (Kent Gooch)

APPENDIX II:
Leads for Moving Targets

Note: Mil leads were calculated with TOF of .001 accuracy and afterwards they were rounded to nearest .1 for the charts.

MIL LEADS for MOVING TARGETS
YARDS
For **M118 SB, LR, and 175 gr HPBT Match, 24"** Barrel, STP

Range Yards	TOF Sec	Target Speed in FPS				
		1	2	4	6	10
100	0.10	0.00	0.00	0.00	0.25	1.50
200	0.20	0.00	0.00	0.50	1.25	2.50
300	0.40	0.00	0.25	1.25	2.00	3.75
400	0.50	0.00	0.50	1.25	2.00	3.75
500	0.70	0.25	0.50	1.50	2.50	4.25
600	0.90	0.25	0.75	1.75	2.75	4.75
700	1.10	0.25	0.75	1.75	3.00	5.00
800	1.30	0.25	1.00	2.00	3.00	5.25
900	1.50	0.50	1.00	2.00	3.25	5.50
1000	1.80	0.50	1.00	2.25	3.50	5.75

MIL LEADS for MOVING TARGETS
METERS
For **M118 SB, LR, and 175 gr HPBT Match, 24"** Barrel, STP

Range Meters	TOF Sec	Target Speed in fps				
		1	2	4	6	10
100	0.10	0.00	0.00	0.00	0.25	1.50
200	0.30	0.00	0.25	1.00	2.00	3.25
300	0.40	0.00	0.25	1.25	2.00	3.75
400	0.60	0.00	0.50	1.50	2.25	4.25
500	0.80	0.25	0.75	1.50	2.50	4.50
600	1.00	0.25	0.75	1.75	2.75	4.75
700	1.20	0.25	0.75	1.75	3.00	5.00
800	1.40	0.50	1.00	2.00	3.00	5.25
900	1.70	0.50	1.00	2.25	3.25	5.50
1000	2.00	0.50	1.00	2.25	3.50	6.00

MIL LEADS for MOVING TARGETS
YARDS
For **M118 SB, LR, and 175 gr HPBT Match, 26" Barrel, STP**

Range Yards	TOF Sec	Target Speed in FPS				
		1	2	4	6	10
100	0.10	0.00	0.00	0.00	0.25	1.50
200	0.30	0.00	0.00	0.50	1.50	3.00
300	0.40	0.00	0.25	1.00	2.00	3.50
400	0.50	0.00	0.50	1.25	2.00	3.75
500	0.70	0.00	0.50	1.50	2.25	4.00
600	0.80	0.25	0.75	1.50	2.50	4.50
700	1.00	0.25	0.75	1.75	2.50	4.50
800	1.20	0.25	0.75	1.75	2.75	4.75
900	1.40	0.50	0.75	2.00	3.00	5.00
1000	1.70	0.50	1.00	2.00	3.00	5.25

MIL LEADS for MOVING TARGETS
METERS
For **M118 SB, LR, and 175 gr HPBT Match, 26" Barrel, STP**

Range Meters	TOF Sec	Target Speed in fps				
		1	2	4	6	10
100	0.10	0.00	0.00	0.00	0.25	1.50
200	0.30	0.00	0.25	1.00	1.50	3.25
300	0.40	0.00	0.25	1.25	2.00	3.75
400	0.60	0.00	0.50	1.50	2.25	4.00
500	0.70	0.25	0.50	1.50	2.50	4.25
600	0.90	0.25	0.75	1.75	2.50	4.50
700	1.10	0.25	0.75	1.75	2.75	4.75
800	1.40	0.25	0.75	2.00	3.00	5.00
900	1.60	0.50	1.00	2.00	3.25	5.25

APPENDIX II: Leads for Moving Targets

MIL LEADS for MOVING TARGETS
YARDS
For **M852 and 168 gr HPBT MATCH, 26" Barrel**, STP

Range Yards	TOF Sec	Target Speed in FPS				
		1	2	4	6	10
100	0.10	0.00	0.00	0.00	0.25	1.75
200	0.30	0.00	0.00	0.50	1.50	3.25
300	0.40	0.00	0.25	1.25	2.00	3.75
400	0.50	0.00	0.50	1.50	2.25	4.00
500	0.70	0.00	0.50	1.50	2.50	4.50
600	0.90	0.25	0.75	1.75	2.75	4.75
700	1.10	0.25	0.75	1.75	3.00	5.00
800	1.30	0.50	0.75	2.00	3.00	5.25
900	1.60	0.50	1.00	2.00	3.25	5.50
1000	1.90	0.50	1.00	2.25	3.50	6.00

MIL LEADS for MOVING TARGETS
METERS
For **M852 and 168 gr HPBT MATCH, 26" Barrel**, STP

Range Meters	TOF Sec	Target Speed in fps				
		1	2	4	6	10
100	0.10	0.00	0.00	0.00	1.00	2.50
200	0.30	0.00	0.00	1.00	1.75	3.50
300	0.40	0.00	0.25	1.25	2.25	4.00
400	0.60	0.00	0.50	1.50	2.50	4.25
500	0.80	0.25	0.50	1.50	2.75	4.50
600	1.00	0.25	0.75	1.75	2.75	4.75
700	1.30	0.25	0.75	2.00	3.00	5.25
800	1.50	0.50	1.00	2.25	3.25	5.50
900	1.80	0.50	1.00	2.25	3.50	6.00
1000	2.10	0.50	1.25	2.50	3.75	6.25

MIL LEADS for MOVING TARGETS
YARDS
For .300 WIN MAGNUM, 190 gr HPBT, 26" Barrel, STP

Range Yards	TOF Sec	\multicolumn{5}{c}{Target Speed in FPS}
		1	2	4	6	10
100	0.10	0.00	0.00	0.00	0.25	1.75
200	0.21	0.00	0.00	0.50	1.25	2.75
400	0.45	0.00	0.50	1.00	1.75	3.25
500	0.59	0.00	0.50	1.25	2.00	3.50
600	0.73	0.25	0.50	1.25	1.25	3.75
700	0.88	0.25	0.50	1.50	2.25	4.00
800	1.00	0.25	0.50	1.50	2.25	4.00
900	1.23	0.25	0.75	1.75	2.50	4.25
1000	1.42	0.50	0.75	1.75	2.50	4.50
1100	1.63	0.50	0.75	1.75	2.75	4.75
1200	1.86	0.50	1.00	2.00	3.00	5.00

MIL LEADS for MOVING TARGETS
METERS
For .300 WIN MAGNUM, 190 gr HPBT, 26" Barrel, STP

Range Meters	TOF Sec	\multicolumn{5}{c}{Target Speed in fps}
		1	2	4	6	10
100	0.11	0.00	0.00	0.00	0.50	1.75
200	0.23	0.00	0.00	0.50	1.25	2.75
300	0.36	0.00	0.25	1.00	1.75	3.25
400	0.50	0.00	0.50	1.25	2.00	3.50
500	0.65	0.00	0.50	1.25	2.00	3.50
600	0.81	0.25	0.50	1.50	2.25	3.75
700	1.00	0.25	0.50	1.50	2.50	4.25
800	1.18	0.25	0.75	1.50	2.50	4.55
900	1.40	0.25	0.75	1.75	2.75	4.50
1000	1.62	0.50	0.75	1.75	2.75	4.75
1100	1.90	0.50	1.00	2.00	3.00	5.25
1200	2.13	0.50	1.00	2.00	3.25	5.25

APPENDIX II: Leads for Moving Targets

MIL LEADS for MOVING TARGETS
YARDS
For **M855 62 gr SS109, 20" Barrel, STP**

Range Yards	TOF Sec	Target Speed in FPS				
		1	2	4	6	10
100	0.10	0.00	0.00	0.00	0.50	1.50
200	0.20	0.00	0.00	0.50	1.25	2.50
300	0.40	0.00	0.25	1.00	1.75	3.25
400	0.50	0.00	0.50	1.25	2.00	3.50
500	0.70	0.00	0.50	1.50	2.25	4.00
600	0.80	0.25	0.75	1.50	2.50	4.25
700	1.00	0.25	0.75	1.75	2.50	4.50
800	1.30	0.50	0.75	2.00	3.00	5.25

MIL LEADS for MOVING TARGETS
METERS
For **M855 62 gr SS109, 20" Barrel, STP**

Range Meters	TOF Sec	Target Speed in fps				
		1	2	4	6	10
100	0.10	0.00	0.00	0.00	0.50	1.75
200	0.20	0.00	0.00	0.75	0.50	3.00
300	0.40	0.00	0.25	1.00	1.75	3.50
400	0.60	0.00	0.50	1.50	2.25	3.75
500	0.70	0.25	0.50	1.50	2.50	4.25
600	1.00	0.25	0.75	1.75	2.75	4.50
700	1.20	0.25	0.75	2.00	3.00	5.00
800	1.50	0.50	1.00	2.00	3.25	5.50

KIM's game. USMC Scout/Sniper School, Quantico, Virginia, 1984. (Kent Gooch)

APPENDIX III: Wind Drift Tables

Wind Deflection is given in inches so that you can extrapolate values between yardages and wind speeds. Use the Sierra Ballistics Program on your computer to get correct windage adjustments for your altitude and temperature.

Wind Drift Table in INCHES
Range in YARDS
For **M118 SB, LR, and 175 gr HPBT Match, 26" Barrel, STP**

Range Yards	\multicolumn{6}{c}{3 o'clock or 9 o'clock Crosswind Speed}					
	2	4	6	8	10	12
100	0.1	0.3	0.4	0.5	0.7	0.8
200	0.6	1.1	1.7	2.3	2.8	3.4
300	1.3	2.6	4.0	5.3	6.6	7.9
400	2.4	4.9	7.3	9.7	12.2	14.6
500	4.0	7.9	11.9	15.8	19.8	23.7
600	5.9	11.8	17.7	23.7	29.6	35.5
700	8.4	16.8	25.2	33.5	41.9	50.3
800	11.4	22.8	34.2	45.7	57.1	68.5
900	15.0	30.1	45.1	60.2	75.2	90.3
1000	19.3	38.6	57.9	77.2	96.6	115.9

Wind Drift Table in INCHES
Range in METERS
For **M118 SB, LR, and 175 gr HPBT Match, 26" Barrel, STP**

Range Meters	3 o'clock or 9 o'clock Crosswind Speed					
	2	4	6	8	10	12
100	0.10	0.30	0.50	0.70	0.80	1.00
200	0.70	1.40	2.00	2.70	3.40	4.10
300	1.60	3.20	4.80	6.40	8.00	9.60
400	3.00	5.90	8.90	11.80	14.80	17.70
500	4.80	9.60	14.40	19.20	24.10	28.90
600	7.20	14.50	21.70	28.90	36.20	43.40
700	10.30	20.60	30.90	41.20	51.50	61.80
800	14.10	28.10	42.20	56.30	70.40	84.40
900	18.60	37.20	55.80	74.40	93.00	111.60
1000	23.90	47.80	71.60	95.50	119.40	143.30

Wind Drift Table in INCHES

Range in YARDS

For .300 Win Mag 190 gr. Federal Match, 26" Barrel, STP

Range Yards	3 o'clock or 9 o'clock Crosswind Speed					
	2	4	6	8	10	12
100	0.10	0.20	0.30	0.40	0.60	0.70
200	0.50	0.90	1.40	1.80	2.30	2.70
300	1.00	2.10	3.10	4.20	5.20	6.30
400	1.90	3.80	5.80	7.70	9.60	11.50
500	3.10	6.20	9.30	12.40	15.50	18.60
600	4.60	9.20	13.90	18.50	23.10	27.70
700	6.50	13.00	19.60	26.10	32.60	39.10
900	11.60	23.20	34.90	46.50	58.10	69.70
1000	14.90	29.80	44.70	59.60	74.60	89.50
1100	18.80	37.50	56.30	75.10	93.80	112.60

Wind Drift Table in INCHES

Range in METERS

For .300 Win Mag 190 gr. Federal Match, 26" Barrel, STP

Range Meters	3 o'clock or 9 o'clock Crosswind Speed					
	2	4	6	8	10	12
100	0.1	0.3	0.40	0.50	0.70	0.80
200	0.5	1.1	1.60	2.20	2.70	3.30
300	1.3	2.5	3.80	5.10	6.30	7.60
400	2.3	4.7	7.00	9.30	11.60	14.00
500	3.8	7.5	11.30	15.10	18.90	22.60
600	5.6	11.3	16.90	22.60	28.20	33.80
700	8	16	24.00	32.00	39.90	47.90
800	10.9	21.8	32.60	43.00	54.40	65.30
900	14.4	28.7	43.10	57.40	71.80	86.10
1000	18.5	37	55.50	74.00	92.50	111.00
1100	23.3	46.7	70.00	93.40	116.7	140.10
1200	28.9	57.8	86.70	115.70	144.6	173.50

APPENDIX III: Wind Drift Tables

Wind Drift Table in INCHES
Range in METERS
For **5.56mm M855, 62 gr. SS109, 20" Barrel, STP**

Range Meters	3 o'clock or 9 o'clock Crosswind Speed					
	2	4	6	8	10	12
100	0.20	0.40	0.60	0.90	1.10	1.30
200	0.90	1.80	2.70	3.70	4.60	5.50
300	2.20	4.40	6.50	8.70	10.90	13.10
400	4.10	8.30	12.40	16.50	20.60	24.80
500	6.90	13.80	20.80	27.70	34.60	41.50
600	10.70	21.40	32.10	42.90	53.60	64.30
700	15.70	31.30	47.00	62.60	78.30	93.90
800	21.80	43.50	65.30	87.10	108.80	130.60

APPENDIX IV: Manufactures Mentioned in Book

Accuracy International-North America, P.O. Box 5257, Oak Ridge, TN 37831
Arms Tech, Ltd., 5133 N. Cantral Ave., Phoenix, AZ 85012
Barrett Cal..50 distributor: CFI 2557 E. Loop 820 N., Fort Worth, TX 76118
Bausch and Lomb, Sports Optics Division, 9200 Cody, Overland Park, KS 66214
Black Hills Ammunition 1997, PO Box 3090, Rapid City, SD 57709-3090
Burris Co. Inc., 331 East 8th Street, Greeley, CO 80631
Corbon, 1311 Industry Rd., Sturgis, SD 57785
Dakota Arms, Inc., HC 55, Box 326, Sturgis, SD 57785
Eagle (drag bags) Distributor: Dub Ball, AWG Tactical, 405-751-6427
Federal Cartridge Company, 900 Ehlen Drive, Anoka, MN 55303-7503
Iron Brigade Armory, 100 Radcliffe Circle, Jacksonville, NC 28546
Harris Engineering Inc., Barlow KY 42024
Hart Rifle Barrels, Inc., PO Box 182, Lafayette, NY 13084
H-S Precision, Inc., 1301 Turbine Drive., Rapid City, SD 57701
K & P Gun Company, 1024 Central Ave., New Rockford, ND 58356
Lapua Ltd., PO Box 5, FIN-62101, Lapua, Finland
Laser Technology, Inc., 7070 S. Tuscon Way, Englewood, CO 80112
Leupold Law Enforcement Products, PO Box 688, Beaverton, OR 97075-0688
Litton Electro-Optical Systems, 3414 Herrmann Drive, Garland, TX 75041-6188
L.O.D. Training Associates, Inc., 3545 Omeara Drive, Houston, TX 77025
McMillan Fiberglass Stocks, Inc., 21421 North 14th Ave. AZ 85027
MWG Co., PO Box 97-1202, Miami, FL 33197
National Target Company, PO Box 2152, Rockville, MD 20847-2152
North American Integrated Technologies, 590 Menlo Drive, Ste 8, Rocklin, CA 95765
Obermeyer Rifled Barrels, 23122 60th St., Bristol, WI 53104
Premier Reticles, Ltd., 920 Breckinridge Lane, Winchester, VA 22601
Redfield, 5800 E. Jewell Ave, Denver CO 80224
Remington Arms Company Inc., Law Enforcement and Government Sales, PO Box 700, Madison, NC 27025-0700
The Robar Companies, Inc., 21438 North Seventh Ave., Suite B, Phoenix, AZ 85027
Savage Arms, Inc., 1997, 100 Springdale Rd., Westfield, MA 01085
Schmidt & Bender, P.O. Box 134, 438 Willow Brook Rd., Meriden NH 03770
Springfield Inc., 420 West Main St., Geneso, IL 61254
Sturm, Ruger and Co., Inc., Lacey Place, Southport Connecticut 06490
S.W.F.A., Inc., P.O. Box 69, Desoto, TX 75123.
Swarovski Optik North America, One Wholesale Way, Cranston, RI 02920

Tasco, 7600 N.W. 26th St., Miami, FL 33122

Texas Brigade Armory, 906 Middle Run, Duncanville TX 75137

Winchester Ranger Law Enforcement Ammunition, Winchester/Olin Corporation, East Alton, IL 62024

Winchester Rifles and Shotguns, U.S. Repeating Arms Company, 275 Winchester Ave., New Haven CT 06511-1970

INDEX

.22 Long Rifle, 66
.300 Whisper, 143-45
.300 Winchester Magnum, 21, 24, 37, 40, 73, 90, 131-37
.338 Lapua Magnum, 137-39
.50 Caliber Sniper Rifle, 139-42
1997 Wilson Sniper Matches, 22, 40, 277, 278
40X, 40
5.56/.223 Cartridge, 67, 128
5R Groove Barrels, 23
7.62x51mm Sniper Cartridge, 105-20
700 Police Sniper, 40
7mm Remington Magnum, 61, 184

A-Square Hannibal, 138
Accu-Range, 2, 165
Accu-Trac, 2, 165
Accuracy International, 137, 138
Adjustable Butt Plate, 76
Adjustable Cheek Pieces, 78
Air Density, 151
Air Force, 55
AK-47, 55
AN/PVS-10 Sniper Sight, 35
Anemometers, 191
Angle Firing Chart, 181-85
Angle of Jump, 151
Angle of Movement, 209
AR-15, 52
AR-15/M16, 79, 90
Arlington, TX Police Department, 69, 233, 239
Armament Technology, 49
Army Scout-Sniper, 19-50
ART-MPC Scope, 20, 165
Atkinson Gun Company, 2
Audette, Creighton, 108
Austin, TX Police Department, 179
AWC, 84

Ballistic Coefficient (BC), 154
Ballistics Tables, 159
Barrel Length v. Velocity, 156
Barrel Life, 37
Barrett, 139, 142, 144, 271, 273
Bausch & Lomb, 83
BDL Floorplate Assembly, 2
Bedding Block, 26
Benning, GA, 117
Berger Bullets, 25
Bisonite, 26
Black Hills, 121, 122, 126
Black September, 54
Branch Davidians, 55
Brevex, 138
Brewer, Maj. Tom, 313
British System of Wind Reading, 202
Brno 138
Brookfield Precision Company, 30
Brownell, 89
Bullet Drift, 158
Bullet Path, 150
Bullet Path Change with Head or Tail Wind, 190
Burke, LCPL Jim, 205-08
Burris, 51, 52, 83, 165
Bushnell Laser Range Finder, 61, 174, 175

CamelBak™, 283
Camp Lejuene, NC, 13
Camp Pendleton, CA, 13, 14, 15, 247
Camp Perry, OH, 205
Camp Robinson, AR, 22, 39, 40, 45, 47, 58, 277
Chandler, Norm, 72, 240
Cloward, 29, 31
Clymer Reamers, 85, 86
Colt AR-15, 67
Come Ups, 150, 171-74
Cooper, Malcolm, 72

Corbon, 126, 130, 143
Crane, IN, 117, 134
Custom Sniper Rifle, 71-94
Cyalume Lightsticks, 252

Dakota, 24, 138
Dallas, TX Police Department, 67, 72
Dallas-Fort Worth Airport, 60
Danzac, 119
Davis, Dick, 4
Delta Force, 54
Department of Energy, 55
Desert Storm, 65, 271
Detachable Magazines, 79
Determining Amount of Lead, 209
Devcon, 6, 26, 88
Douglas, 3, 90
Drag, 151
Dragunov System, 166
Dupont, 89

Eagle Drag Bags, 262
"Elephant Valley", 205
ERMA, 138
Exocet Missile, 271
External Ballistics, 149-62
External Mount Assembly (EMA), 30, 31

Fabrique Nationale, 141, 144
Fackler, Dr. Martin L., 100
Falkland Islands, 271
Falklands War, 219
FBI Academy, 55
FBI Hostage Rescue Team, 55
Federal Cartridge Company, 61, 121, 122, 126, 128, 134, 135, 136
Federal Gold Medal Match, 59
Federal Match Round, 168 grade, 107, 117, 267
Federal Match Round, 175 grade, 171
Fort Benning, GA, 39, 46
Fort Bragg, NC, 20, 54, 55
Fort Wolters, TX, 8, 60
Fort Worth Rifle and Pistol Club, 69
Frankford Arsenal, 110, 128

Garmin, 59
Geltmacher, SSGT Dan "Genghis", 13, 15

Ghillie Suits, 261
Gravity, 151
Grid Positioning System (GPS), 59

H&K G3 Rifle, 20
H-S Precision, 2, 26, 39, 89
Hall Custom Action, 81
Hardigg Industries, 41
Harrington & Richardson Ultra Rifle, 2
Harris Bipod, 77, 93, 258
Harris Engineering, 80
Hart Barrels, 3, 85, 92, 117
Hatcher's Notebook, 158
Hathcock, Gysgt. C. N. Carlos, 17, 48, 72, 203, 205-208, 213, 251
Hawkins Position (Prone), 247
Heym, 138
High Efficiency Low Reflect (HELR), 9
Hornady, 122, 130
Hunting Shack Manufacturing, 135

IMI, 122
Iraqi Armored Vehicles, 271
Iron Brigade Armory, 84
Israel, 21
Italian 6.5 Mannlicher-Carcano, 64

JLK Bullets, 25
Johnson, Ken, 85
Jones, J.D., 143

K&P Barrels, 85
"Keep in Memory" (see KIM's game), 289
Kennedy, President John F., 53
Keppler, 138
Kevlar, 5, 26
KIM's game (see "Keep in Memory"), 289
Kitton Electro0Optical Systems, 35
Knight SR-25, 79
Kreiger Barrels, 85, 90
Kwik Klip, 80

L.O.D., 55, 84, 89
Lake City Army Ammunition Plant, 17, 40, 112, 116
Lake City Arsenal, 110
Lake City LR, 121
Land, Capt. Jim, 203
Lapua, 135

INDEX

Laser Filter, 30
Laser Range Finders, 174-76
Laser Technology, 175
Leica, 175
Leupold, 23, 27, 28, 29, 31, 59, 81, 82, 91, 93, 94, 167, 267
Light Conditions, 158
Line of Departure (LOD), 150
Line of Sight (LOS), 150
Little, CPL Sean, 5, 6, 7, 8, 9, 84, 111, 261
Litton Electro-Optical Systems, 35
Log/Data Book, 213-42
Los Angeles, CA Police Department, 56
Lubalox, 126
Lyman, 88

M-16, 55
M118, 8, 9, 16, 17
M118 Ammunition, 111, 115
M118LR, 39, 40, 117, 122
M118 Match, 20, 39, 107
M118 Special Ball, 21, 101, 117, 118, 119, 181
M14/M1A, 79
M16 A1/A2, 67
M16A2, 15, 32
M193, 128, 130
M21, 20
M21 Sniper Weapons System, 19
M24, 6
M24 Sniper Weapons System, 19-50
M3A Scope, 28, 29
M40, 6, 8
M40A1, 1-18, 22, 79, 92, 94
M49 20X Telescope, 11
M70/Unertl Rifle, 2, 205, 206
M852 Match, 39, 112, 115, 117, 118, 119
M8555 Ball, 130
Magellan, 59
Marine Corps Scout/Sniper, 1-18
Mark 4 M3, 30
Mauser, 75, 138
Maximum Ordinate, 149
McMillan, 74, 84
McMillan's Fiberglass Stock, 3-4, 5, 10, 59, 78, 88, 92, 93
Medulla Oblongata, 97

Meopta, 166
Mid-Range Trajectory, 149
MIL-DOT, 8, 10
MIL-DOT Reticle, 166-67, 203
Minute of Angle (MOA), 151
Mirage, 192-99
Morgan, Sgt. Wayne, 313
Moving Targets, 205-12
Munich, Germany, 54
Muzzle Brakes, 80
Muzzle Velocity, 151

National Guard Scout-Sniper School, 58
Naval Marine Corps Reserve Center, Houston, TX, 7
Navy SEALs, 55, 93, 134, 139, 141, 142
NECO, 119
Night Vision Optics, 35
NightForce, 83, 87, 166
NM140A1, 141
Norma, 122
North American Integrated Technologies, 175
North Hollywood Shoot-out, 55, 67
Nosler Ballistic Tip, 126
Nylon Coated Rods, 15

Obermeyer Barrels, 85
Offsetting Factor, 152
Oklahoma City, OK Police Department, 51-70
Oswald, Lee Harvey, 53, 64
Owens, M/SGT Jim, 190, 199

Pachmayer, 5, 88
Parallax, 9
Paramount Custom Action, 81
Parker Hale, 77, 80
Pelican Travel Vault, 37
Picatinney Arsenal, 116
Plaster, Maj. John, 164, 165, 181
Police Sniper, 51-70
Police Sniper (PSS) Rifles, 23
Precision Shooting Magazine, 82, 108, 158
Precision Shooting Magazine's Special Edition No. 3, Vol. 1, 134
Premier Reticles, 28, 83, 220
Proskopathlon™ 267, 277, 313

Quantico, VA, 2, 12, 13, 16, 22, 46, 66, 117

Raine, Inc., 215
Range Estimation, 163-78
Raufoss, 140, 141, 144
Redfield, 31, 32, 83, 88, 165
Redfield 1", 10
Redfield 3x9 Telescope, 2
Relative Humidity, 152
Rem-Tuf, 24
Remington's Custom Shop, 2, 3, 121, 122
Remington 600, 2
Remington 700 ADL/BDL, 2, 6, 7, 10, 24, 37, 40, 59
Remington 700 Barrel, 92
Remington 700 Receivers, 87, 91, 92
Remington 700 Short Action, 86, 92
Remington 700 Trigger, 3
Remington 700/40X, 2, 3
Remington 700P, 72, 79
Remington Extractor, 8, 37, 75
Research Armament Industries, 142
Retained or Remaining Velocity, 151
Robar, 79, 84, 89
Rock, Mike, 23
Rogers, John, 21, 23, 24, 26, 37, 39
Ruger, 75
Ruger .44 Magnum Carbine, 53
Ruger Mini-14, 51, 52, 59, 67

Sako Extractors, 7-8, 75
Savage, 147
Schmidt and Bender, 83
Schofield Barracks, HI, 5
Scope Mounts, 82
Secret Service, 55, 60, 61, 184
Sendero, 40
Shepherd, 166
Sherwin Williams, 89
Shooters Choice, 15
Shooting Through Glass, 101-04, 126
Sierra, 17, 40, 110, 113, 114, 115, 116, 119, 134, 135, 136, 143, 173
Sierra 168 grade Match, 183
Sight Adjustment, 163-78
Silencer Ammunition, 142-43
Simmons, 175

Simrad, 29, 35
Sniper Bolt Operation, 249
Sniper Match, The: Sustainment Training, 277-314
Sniper Tip: Engaging Very Close Targets, 257
Somalia, 272
Soviet AK-47, 23
Speed of Sound, 154
Speed of Target, 208
Springfield Armory, 83, 166
Steyr, 21, 79, 81
Stolle Custom Action, 81
Strasburger, MSG Mike, 313
Swarovski, 83, 166, 175

Tactical Scenarios, 243-76
Talon Manufacturing, 126
Tasco, 83, 175
Temperature Affects Air Density, 152
Temperature Affects Ammunition, 152
Terminal Ballistics, 95-104
Texas Brigade Armory, 59, 78, 84, 85, 90, 91, 243, 276
Thomas, Chris, 28, 30
Time of Flight, 150
Trans-Sonic Region, 154, 155
Trigger Pull Rate, 24
Trigon Technologies, 216

U.S. Cavalry, 59
U.S. Optics, 83
Ultimate Sniper, The, 164, 181
Unertl, 14, 20, 55, 87, 92, 173, 252, 273
Unertl 10-X Scope, 8, 10
Unertl 100mm Team Scope, 304
University of Texas Sniper Incident, 179
Unknown Distance Targets, 164
Uphill and Downhill Shooting, 179-86

Versa Pod, 80
Vietnam, 48, 76, 128, 205

Waco, TX, 7
Weatherby, 138
Weaver, 30, 83
Weber, Andy, 49
Whitman, Charles, 179, 251

INDEX

Wichita Engineering Company, 5, 88
Wimbledon Cup Match, 205
Winchester, 121, 122
Winchester 70, 89, 92, 142
Winchester M70, 2, 6, 7, 8, 10, 59, 75, 205
Winchester Ranger 168 grade HPBT match, 124
Winchester Ranger Frangible, 131
Wind Gauges, 191
Wind Observations, 192

Wind Reading and Sight Adjustment, 187-204
Wind Speed, 209
Wound Ballistics Laboratory, 100

XM777 Steel Penetrator, 128
XM852 Match Ammunition, 111

Yarbor, John, 134, 136

Zero Distance, 150

This Marine Corps recruiting poster was sent to the author by a local Marine recruiter around 1984 after I signed up for a course at a local community college. The poster is very difficult to locate and was short lived. Surrounding the poster is author's Marine Corps patch collection. The small shoulder tab, second from the bottom on the left, reads "USMC SCOUT SNIPER". It was given to author by Jim Furgeson. Jim said only 11 were made in Okinawa for the instructors of the Scout/Sniper school there. (Mike R. Lau)